Geometric Product Specification and Verification:
Integration of Functionality

Geometric Product Specification and Verification: Integration of Functionality

Selected Conference Papers of the 7th CIRP International Seminar on Computer-Aided Tolerancing, held at the École Normale Supérieure de Cachan, France, 24–25 April 2001

Edited by

Pierre Bourdet and Luc Mathieu

Laboratoire Universitaire de Recherche en Production Automatisée,
École Normale Supérieure de Cachan,
Cachan, France

SPRINGER-SCIENCE+BUSINESS MEDIA, B.V.

A C.I.P. Catalogue record for this book is available from the Library of Congress.

ISBN 978-90-481-6342-7 ISBN 978-94-017-1691-8 (eBook)
DOI 10.1007/978-94-017-1691-8

Printed on acid-free paper

The 7th CIRP International Seminar on Computer-Aided Tolerancing

Ecole Normale Supérieure de Cachan, 93235 Cachan, France
April 24-25, 2001

Sponsored by
CIRP (International Institution for Production Engineering Research)
PSA-Peugeot-Citroën
Renault S.A.
Dassault Systèmes
UNM (Union de la Normalisation de la Mécanique)
ENS de Cachan

PREFACE

The Computer Aided Tolerancing (CAT) is an important topic in the field of mechanical designing and production. Even if the means of production have been greatly improved and if the variability of production systems is better and better controlled, the dispersions of the characteristics manufactured remain. They directly influence the assembly and the quality of the way mechanisms work. So as to better take them into account within design phases, it is of interest to be able to describe and measure them with no ambiguity of potential error. The subject of the present document "Geometric Product Specification and Verification : Integration of Functionality" focuses on the importance of the description model of faulty parts. Both in the fields of research and of industrial development, efforts have been made in recent years to provide designers with the computer tools of description and design based upon the sole ideal geometry of products. To take tolerancing of manufacturing defects and clearance into consideration right from the stage of design, one has to define geometric models that are capable of specifying the functional geometry of the faulty and non-faulty parts with accuracy, and to adapt them to the expectations of the people involved in the process of the design and the manufacturing of the products.

ISO experts round the world have realized how important this is. Knowing that the only industrial tool to describe tolerances actually is the standardized graphic language, which has become legally compulsory for all the exchanges between the actors of the company, it has become necessary to match this graphic language with a consistent description of the concepts used.

In 1996, ISO created a new technical committee, the ISO/TC213 in charge of the standards on geometrical product specification. This committee followed JHG ISO/TC 3-10-57 which coordinated the technical committees ISO/TC3 "adjustment", ISO/TC57 "Metrology and properties of the surfaces" and the subcommittee ISO/TC10/SC5 "Specification and tolerance". To provide industrialists with complete and coherent standards in the field of specification and geometrical inspection of products (GPS) is the aim of this new committee. In 1996, ISO published a technical report FD CR ISO/TR 14638 fixing the diagram of standards to create or to revise.

In order to answer this ambitious aim, it is no more possible to create or develop standards based only on engineers and experts' know-how. One now has to have a global and theoretical approach of the geometrical specification and verification problem. On the basis of French research results, a model to describe the micro and macro geometry has been worked out. The basic concepts are described in the document ISO/TR 17450-1. The model is original because of its declarative method describing the process of tolerance specification and the process of tolerance verification in inspection.

The contents of this book originate from a collection of selected papers presented at the 7th CIRP International Seminar on Computer Aided Tolerancing (CAT), organized by the LURPA (The laboratory for automated manufacturing) of the Ecole Normale

Supérieure de Cachan, France in April 24-25, 2001. The CIRP (Collège International pour la Recherche en Production or International Institution for Production Engineering Research) plans this seminar every two years.

This book focuses in particular on Geometrical Product Specification and Verification which is an integrated tolerancing view and metrology proposed for ISO/TC213. Common geometrical bases for a language allowing to describe both functional specification and inspection procedures are provided. An extended view of the uncertainty concept is also given.

Geometric Product Specification and Verification: Functionality Integration gives an excellent resource to anyone interested in computer aided tolerancing, it is intended for a wide audience including:

- Researchers in the fields of product design, Computer Aided Process Planning (CAPP), precision engineering, quality, inspection and dimensional and geometrical tolerancing (Professors and graduate student).
- Technicians of Standardization interested in the evolving ISO standards for tolerancing in mechanical design, manufacturing and inspection.
- Practitioners; designers, design engineers, manufacturing engineers, staff in R&D and production departments (e.g. automotive, aerospace, machines, –).
- Software developers for CAD/CAM/CAX and computer aided tolerancing (CAT) application packages.
- Instructors and students of graduate (masters and Ph.D.), professional development and undergraduate courses in design.
- Individuals interested in design, assembly, manufacturing, precision engineering, inspection, and CAD/CAM/CAQ.

The book is organized into nine parts following the Editor's Preface. The first one Geometrical Product Specification: ISO, research and industrial views presents a synthesis of the existing and ongoing works on the subject of international standardization, as well as in research and industrial applications in the field of geometric specification. Part 2 deals with the Models for tolerance representation required for specification and simulation. Part 3, Tolerance analysis covers a more traditional activity of tolerancing, i.e., the simulation of the values of geometric functional conditions on the basis of the tolerance values imposed on the parts of the mechanism, while still considering the three-dimensional aspects and the statistics of tolerancing and the flexible parts. Part 4, Tolerance synthesis consists in determining the part specifications required to comply with the geometric functional conditions of the mechanism. Part 5, Tolerance in manufacturing more specifically deals with the quality of part handling within the process of manufacturing and the three-dimensional simulation of manufacturing processes. Part 6, Tolerance management shows a new aspect of the interest of tolerancing within the design process. Part 7, Tolerance verification covers the problems of identification of geometric elements and of the calculation of uncertainty. In part 8, Industrial applications and CAT systems, the problems and the solutions encountered essentially in the fields of car and plane industry are presented. To finish with, the last part Tolerancing standard focuses upon the exchanged of tolerancing data.

We want to express our sincere thanks to the authors for this contribution, to the members of the international program committee and the organizing committee, in particular Mrs M. Dos Santos and K. Frattini for their effort to make this book published.

Pierre BOURDET

Luc MATHIEU

Geometrical Product Specification: ISO, research and industrial views

Model for tolerance representation

Tolerance analysis

Tolerance synthesis

Tolerance in manufacturing

Tolerance management

Tolerance verification

Industrial applications and CAT systems

Tolerancing standard

An Integrated View of Geometrical Product Specification and Verification

Vijay Srinivasan
IBM Corporation and Columbia University
New York, U.S.A.
vasan@us.ibm.com

Abstract: ISO/TC 213 is a technical committee that standardizes tolerancing and related metrological practices. During the past five years it has focused its attention on harmonizing and integrating these two areas. In that process it recognized the need for linking these two areas to product functionality and has published a vision for directing its future efforts towards it. This paper presents an integrated view of tolerancing and metrology, and describes the new vision of ISO/TC 213.
Keywords: ISO, Standards, GPS, Specification, Verification

1. INTRODUCTION

Since the summer of 1996, ISO/TC 213 on "Dimensional and Geometrical Product Specifications and Verification" [ISOTC213] has been working towards harmonizing previously standardized practices in tolerancing (specification) and related metrology (verification). Initial impetus for this work came from perceived gaps and contradictions in the "chain" of standards that dealt with dimensional and geometric tolerance specifications and their verifications using metrological instruments, systems and procedures [Bennich, 1994]. It was also reinforced by the needs of rapidly expanding CAD/CAM/CAQ marketplace that placed a high premium on mathematical formalism so that reliable and compact software can be developed to support computerized applications in these areas.

In the course of its work ISO/TC 213 recognized that a GPS (Geometrical Product Specification) language can be developed using an elegant classification of symmetry groups and that this could impact a functional feature taxonomy, datum definition and, as an added benefit, parameterization [Srinivasan, 1999b]. It also uncovered sets of dual operations in the specification and verification procedures. These dual operations include partitioning, sampling, filtering and fitting. It is thus possible to harmonize the (previously unspoken) set of operations involved in tolerance specifications with those employed during inspection of a workpiece. Finally, ISO/TC 213 began to espouse and expand the notion of uncertainty beyond just measurements.

Armed with the experience gained thus far, ISO/TC 213 recently published its vision for the next generation of the GPS language (see Annex at the end of the paper for a reproduction of this "vision document"). The objective of the improved GPS system is to provide engineering tools for economic management of variability in products and processes. ISO/TC 213 envisions an enriched, math-based GPS language that will allow expression of requirements relating to a wide range of workpiece function. It will use "uncertainty" as an economic tool to enable optimum allocation of resources amongst

P. Bourdet and L. Mathieu (eds.),
Geometric Product Specification and Verification: Integration of Functionality, 1-11.

specification, manufacturing and verification. To ease the burden of inevitable verbosity of the resulting GPS language, it will strive towards a simplification using consistent logic and a "default" concept. This paper presents an integrated view of geometrical product specification and verification, and describes the new vision of ISO/TC 213.

Using a scientific structure borrowed from physics, the rest of the paper is organized as follows. In section 2 we start with a definition of tolerancing semantics as a *gedanken experiment*. This is followed in section 3 by a description of a *duality principle* that provides a one-to-one mapping between specification and verification. Then in section 4 product functionality is linked to specification and verification thorough a *generalized uncertainty principle*. Section 5 summarizes and concludes the paper.

2. A GEDANKEN EXPERIMENT

The works of ISO committees responsible for tolerancing (specification) and related metrology (verification) were combined in mid-1990s to form the mission of ISO/TC 213. One may wonder why these were united and why others, such as those dealing with manufacturing, were left out of this unification. The decision was not arbitrary, and it has to do with the fact that specification and verification are strongly related to each other both in theory and in practice. This section will be devoted to explaining one aspect of this strong relationship.

The GPS language has a syntax whose graphical representation is indicated on a view of a nominal model of a part; this nominal model consists of ideal surfaces. The semantics of the GPS language is, however, based on an abstract geometric model of a workpiece; this model consists of non-ideal surfaces. In his original work Requicha [Requicha, 1983] defined the semantics of the GPS language as a "theoretical inspection procedure". In his scheme

> *"A (model of a) part P satisfies a tolerance specification T if and only if there exists a decomposition of ∂P (boundary of P) into subsets G_i, called actual surface features, such that*
>
> - $\cup G_i = \partial P$.
> - *There exists a one-to-one correspondence between G_i and the nominal surface features F_i of S.*
> - *Each G_i satisfies the assertion A_{ij} associated with its corresponding F_i."*

Requicha's work provided the first formalism for defining the semantics of tolerancing - that is, the GPS language - as a gedanken experiment. It was an important intellectual milestone in the theoretical evolution of the GPS language in the pre-TC 213 era. What is impressive is that even today we depend on this formalism to define tolerancing semantics, with some additional operations defined to elaborate what Requicha termed "actual surface features".

Figure 1 is a reproduction of an illustration from a recent document ISO/TS 17450-1 [ISO17450-1]. Under the heading "NOMINAL DESIGN" it describes how the syntax of the GPS language is presented. Using Requicha's terminology, we would say that the nominal model *S* is decomposed into nominal features F_i (these are the ideal features in

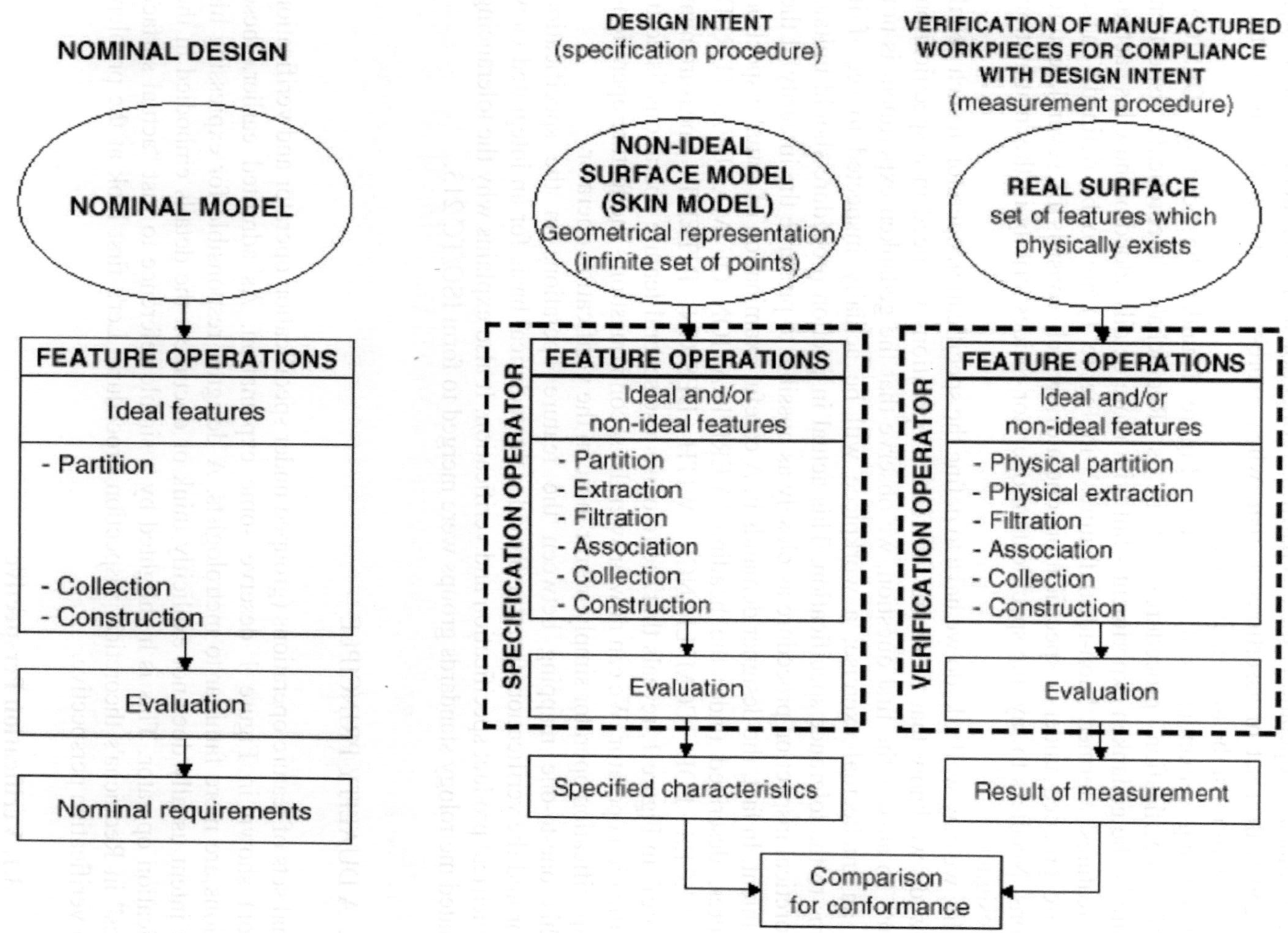

Figure 1. ISO/TC 213's comparison between tolerancing and metrology.

Figure 1 obtained by operations such as partition, collection and construction). Then each F_i is tagged with a set of assertions A_{ij} that, when expressed using graphical or other symbology, will form the tolerancing syntax.

Tolerancing semantics is illustrated in Figure 1 under the heading "DESIGN INTENT (specification procedure)". This is Requicha's "theoretical inspection procedure" - the gedanken experiment - but with more details added on how his "actual surface features" (termed "non-ideal features" in Figure 1) are to be defined. These details are grouped under a specification operator, which consists of several feature operations. Needless to say, the specification operator resides strictly in the imagination of the designer.

Why, we may well ask, do we have to define the specification operator in such detail and how do we know that these are the feature operations we need in a specification operator? To answer the first question, we observe that the gedanken experiment is not merely an intellectual exercise; a workpiece will be actually inspected to see if it conforms to the tolerance specification. This actual inspection procedure should imitate the theoretical inspection procedure as closely as possible to preserve the integrity of the design intent. In fact, the designer demands it. A careful examination of actual inspection procedures, described under the heading "VERIFICATION OF MANUFACTURED WORKPIECES FOR COMPLIANCE WITH DESIGN INTENT (measurement procedure)" in Figure 1, reveals that metrologists use several feature operations listed in the verification operator. We can now answer the second question - the feature operations in the specification operator simply mirror those in the verification operator.

This one-to-one mapping between the feature operations in the specification operator and the verification operator provides the technical basis for an integrated view of geometrical product specification and verification. It also explains why the tolerancing and related metrology standards groups were merged to form ISO/TC 213.

3. A DUALITY PRINCIPLE

The dual sets of feature operations (grouped under specification operator and verification operator) shown in Figure 1 deserve some explanation. As admitted earlier, these operations are more familiar to metrologists. A designer responsible for expressing his design intent usually does not explicitly think or document the details embodied in the specification operator. This is highlighted by a simple reference to just "actual surface features" in Requicha's theoretical inspection procedure. Let's first look at the problem from a verification perspective.

3.1. Verification Perspective

We start by considering how a metrologist perceives a manufactured workpiece, that is, a physical object. This is addressed under "REAL SURFACE" in Figure 1. Intuitively, a *real* surface is the set of infinite number of points that separate a workpiece from its surrounding. Given any physical workpiece, it is, of course, impossible to come up with a mathematical representation of this set; we need some additional information about the resolution at which the real surface is perceived. ISO/TC 213 defines at least two ways to characterize this resolution - one is a *mechanical* real surface and the other is an *electro-magnetic* real surface.

Roughly speaking, a mechanical real surface is the set of all points on a real surface that a spherical probe of finite radius r can touch; an electro-magnetic real surface is the set of all reflection points on a real surface by electro-magnetic radiation with a specified wavelength λ. In both cases, we have a *nesting parameter* - radius r in the case of mechanical real surface and wavelength λ in the case of electro-magnetic real surface. The nesting arises from the fact that a real surface obtained with a smaller value for the nesting parameter contains more information than the one obtained with a larger value for the parameter. Theoretically, a mathematical model that approximates the real surface can be obtained within any measure of closeness by choosing the nesting parameter very close to zero.

We can now describe various feature operations listed under the verification operator in Figure 1. No ordering of operations is implied in this list. In Figure 2 the results of various feature operations are illustrated. A real surface corresponding to a specified nesting parameter is *partitioned* into *real integral features*. These features still contain infinite number of points. During actual inspection, however, we sample only a finite number of points on these features. These are called *extracted* integral features. It turns out that sampling alone is insufficient to extract a feature; it should be accompanied by some smoothing to remove noise and unwanted details from the measured data. Therefore, techniques for extracting information on real integral features involve both sampling and some *filtration*. Ideal surfaces are then fitted to the extracted and filtered data points - this fitting process is called *association*. Results of most of these feature operations are illustrated in Figure 2.

Sometimes the geometric objects are combined together using a *collection* operation; for example, two cylindrical features may be collected together to form a pattern of two cylindrical features. Features may also be subjected to a *construction* operation; for example, two planar faces may be intersected to obtain a straight edge feature.

3.2. Specification Perspective

The rationale for spelling out the details of feature operations in the verification operator is clear - the metrologist needs them to do his job. In the course of implementing these operations several decisions - the nesting parameter, sampling density, filtering scheme, etc - are made by the metrologist. Can the designer remain an uninterested bystander while these decisions are being made? The answer seems to be clearly no, because these decisions may impact the ultimate functionality of the workpiece.

A solution to this problem proposed by ISO/TC 213 is to postulate a principle of duality. Under this principle the specification operator and verification operator are duals of each other - one imagined by the designer and the other implemented by the metrologist. These operators contain identical sets of feature operations as shown in Figure 1. By explicitly defining these feature operations many of the previously unspoken details have been brought out into the open. ISO/TC 213 believes that this openness greatly reduces the possibility for unpleasant surprises for the designer when the metrologist implements the feature operations.

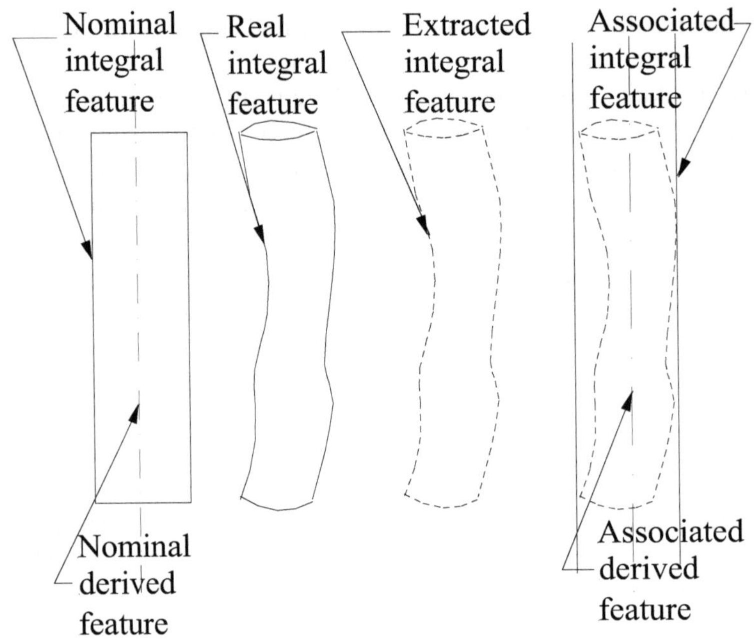

Figure 2. Results of various feature operations. This example is for a cylindrical feature.

But does this *principle of duality* violate the time-honored *principle of verification independence*, that is, the designer specifies only what is required of a part and not how to check it? To answer this question we need to look at the specification operator in some detail. Partition, collection and construction operations are not controversial. They are clearly needed in the specification operator. That leaves extraction, filtration and association for further scrutiny.

Association is a geometrical fitting operation. It can be defined mathematically as a constrained optimization problem. We need it, and this a crucial argument, to define datums and position tolerances. Datum specification implies a substitution of ideal geometrical elements in place of non-ideal features. Position tolerance, in particular RFS (Regardless of Feature Size) position tolerance, specification defines allowable variations in the relative position of axis, center, or median plane of a feature. This implies a fitting of an ideal surface to a non-ideal surface feature even at the gedanken experiment stage.

Having established the need for an association operation in a specification operator we then observe that we will also need extraction and filtration operations to prepare the input for the optimization algorithms. One may argue that all these three operations can be dispensed with if a physical operation is substituted - for example, a surface plate to establish a planar datum - in the verification operator in the place of a "datum" operation in the specification operator. A flaw in this argument is that it makes the physical apparatus as the primary definition of a verification process whereas ISO/TC 213 strives to give primacy to mathematical definitions. As a fundamental rule ISO/TC 213 has decided to base all its definitions on mathematics - as opposed to definitions by examples - so that product functions can be simulated using mathematical models.

It therefore seems possible to observe the principle of duality without compromising the principle of verification independence. Note that we allow the verification operator to be implemented in multiple ways - for example, a datum by a fitting algorithm or by a surface plate - and these are distinguished, as we will see in the next section, by the amount of uncertainty. In fact, the guiding philosophy of ISO/TC 213 is to standardize only the specification operator. The verification operator will implement the standardized feature operations in the specification operator with careful accounting of the amount of uncertainty in each step of the implementation.

4. A GENERALIZED UNCERTAINTY PRINCIPLE

The notion of uncertainty is by now well entrenched in metrology. It is based on a basic principle in metrology that all measurements have inherent uncertainty and therefore this uncertainty should be quantified and reported along with the result of any measurement. As ISO/TC 213 looked at the integrated picture of specification and verification it realized that the notion of uncertainty should be generalized beyond that of just measurement.

In successful product development processes one starts with customer requirements and breaks this down to product functional specifications and then to detailed design specifications. Broadly speaking the function of a product is dependent on its geometry, material properties and operating conditions. The supplier of the product may not have much control over its operating conditions but he does have control over its geometry and material properties. The scope of ISO/TC 213 is restricted to geometry. It is charged by ISO with developing a language for detailed specification of product geometry and methods for verifying whether manufactured workpieces meet these specifications. Therefore a major, though not all, responsibility for satisfying the intended function of the product rests with the geometric specification language and verification methodology standardized by ISO/TC 213.

When one looks at the integrated picture the natural question to ask is whether the finished workpieces when assembled will perform the intended function. Here ISO/TC 213 postulates a generalized uncertainty principle: there will always be some uncertainty as to whether the finished product will perform the intended function. This uncertainty has several components. At the geometric level the uncertainty has been divided into (1) correlation uncertainty (2) specification uncertainty and (3) measurement uncertainty [ISO17450-2]. Before we proceed further it should be pointed out that these terms and concepts are still evolving and are subject to modification and refinement as the work of ISO/TC 213 progresses. Never the less, some basic ideas have emerged as front runners.

4.1. Correlation Uncertainty

Correlation uncertainty arises from the fact that the intended functionality and the controlled geometric characteristics may not be perfectly correlated. A good designer tries to achieve a very high level of correlation but it may never reach 100%. There are several reasons for this. First, the chemical and physical phenomena governing a product's functionality may be quite complex and may only be partially understood.

Second, mathematical modeling of the understood phenomena and solutions to resulting mathematical equations may be approximate. Third, all these may be confounded by interactions of geometry with material properties and operating conditions.

Thus far we have spoken only about correlation. But to control function we implicitly assume *causality*, a condition stronger than correlation. Although we speak of correlating a product's functionality to its geometry, we intend to control the function through geometry (and, by other standards, material properties and operating conditions). Function is deemed to be the effect - geometry, material properties and operating conditions are the causes. Engineers rely upon considerable prior knowledge - whether tacit or explicit - and some controlled experiments to establish causality. Statistically speaking all these are wrought with some uncertainties, however small they may be.

4.2. Specification Uncertainty

Specification uncertainty results from incorrect or incomplete application of geometric specifications. A diligent designer can reduce this uncertainty to an insignificant amount. This assumes that the GPS language defined by ISO/TC 213 is complete and correct. This assumption can be called into question. Our standards development process is in a continuous improvement mode and we may never claim perfection in these standards. Some allowance has to be made for this type of uncertainty as well.

4.3. Measurement Uncertainty

Measurement uncertainty [GUM, 1995] is the best known of the three types of uncertainties mentioned earlier. It is a statistical parameter associated with the result of a measurement and it characterizes the dispersion of the values that could be attributed to what is being measured. This parameter may be, for example, a standard deviation (or a given multiple thereof) or the half-width of an interval having a stated level of confidence. It has several components, too numerous to discuss here.

In a verification process the measurement uncertainty may be further compounded by poor or intentionally different interpretation of the specifications. Because of this, ISO/TC 213 has proposed a further classification of measurement uncertainty into method uncertainty and implementation uncertainty. Method uncertainty arises from the difference between the actual application of the specification operator and its mapping into a verification operator. Implementation uncertainty arises from physical deviations in applying a verification operator.

4.4. Relative Importance of Uncertainties

What is the benefit of the generalized uncertainty principle? We argue that for optimal allocation of resources in an engineering enterprise one should take an integrated view of various processes employed and should ascertain their relative importance from the overall business perspective. This is enabled in part by applying the generalized uncertainty principle. The overall uncertainty can be broken down into its component factors and their relative magnitudes can be compared. For example, if we know that correlation uncertainty is the dominant uncertainty affecting a product's quality then more resources should be applied to correct this as opposed to buying more precise coordinate measuring machines.

Before leaving this section, let's make a few general observations. Uncertainty may be viewed as an annoyance but it is a reality we have to live with. The vision of ISO/TC 213 is to turn this unpleasant fact into an economic tool. It believes that by openly acknowledging the prevalence of uncertainty better decisions can be made in allocating resources in an engineering enterprise.

5. SUMMARY AND CONCLUDING REMARKS

This paper described recent efforts of ISO/TC 213 in integrating product functionality, its geometric specification and verification. Starting with a definition of tolerancing semantics as a gedanken experiment, a tentative link between a theoretical inspection procedure and an actual inspection procedure was established. This link was strengthened by postulating a duality principle; it provided a one-to-one mapping between operations in a specification operator and a verification operator. It was then asserted that only the operations in the specification operator need to be standardized; the verification operator would then merely implement their dual operations in the specification operator - albeit with some inevitable uncertainty. This then led to a postulation of a generalized uncertainty principle that connected product functionality with its geometric specification and verification. It was a natural consequence of taking an integrated view of functionality, specification and verification.

It should be emphasized that we have just begun the task of integrating product functionality with its geometric specification and verification. As ISO/TC 213's vision statement reproduced at the end of the paper indicates, it is a challenge that would take considerable effort. Researchers in computer-aided tolerancing have several important roles to play in this effort.

First, they can strengthen the link between tolerancing and metrology. Geometric sampling, fitting and filtering, which form the bulk of computational metrology, have become even more important now and they are a source of several interesting problems.

Second, a language is beginning to emerge to convey the concept of generalized uncertainty. This opens up a wide area for research. Correlation uncertainty, in particular, is an unchartered territory. Standards don't tell us how to find this. They merely provide a language to describe it. This should be an area for some intense research.

Finally, pervasive use of uncertainty in geometric specification and verification will necessitate consistent use of statistics all through specification, production and verification, and in relating back to functionality. It is indeed a most welcome step [Srinivasan, 1999a].

6. ACKNOWLEDGMENT AND A DISCLAIMER

The author is deeply indebted to his colleagues in ISO/TC 213 for the technical material presented in this paper. However, opinions expressed here are his own and not that of ISO or any of its member bodies.

REFERENCES

[Bennich, 1994] P. Bennich, "Chains of Standards - A New Concept in GPS Standards", *Manufacturing Review,* The American Society of Mechanical Engineers, Vol. 7, No. 1, pp.29-38, 1994.

[GUM, 1995] *Guide to the Expression of Uncertainty in Measurement*, ISO, 1995.

[ISOTC213] http://www.ds.dk/isotc213

[ISO17450-1] ISO/TS 17450-1, Geometrical Product Specifications (GPS) - General concepts - Part 1: Model for geometrical specification and verification.

[ISO17450-2] ISO/DTS 17450-2, Geometrical Product Specifications (GPS) - General concepts - Part 2: Operators and uncertainties.

[Requicha, 1983] A.A.G. Requicha, "Toward a Theory of Geometric Toleracing", *International Journal of Robotics and Automation*, Vol. 2, No. 4, pp. 45-60, 1983.

[Srinivasan, 1999a] V. Srinivasan, "Role of Statistics in Achieving Global Consistency of Tolerances", In F.Houten and H.Kals (Eds), *Global Consistency of Tolerances, Proceedings of the 6th CIRP International Seminar on Computer-Aided Tolerancing,* Kluwer Academic Pub., pp. 395-404, 1999.

[Srinivasan, 1999b] V. Srinivasan, "A Geometrical Product Specification Language Based on a Classification of Symmetry Groups", *Computer-Aided Design,* Vol. 31, No. 11, pp.659-668, 1999.

ISO/TC 213 N 355 Annex 1 **ISO/TC 213/AG 1 N 65**

Next generation of the Geometrical Product Specifications (GPS) language

The vision for an improved engineering tool

This integrated GPS system for specification and verification of workpiece geometry is an improved engineering tool for product development and manufacturing. This GPS system is necessary, as companies are rapidly moving ahead with new technologies, new manufacturing processes, new materials and visionary products in an environment of international outsourcing.

Objective

The objective of the improved GPS system is to provide tools for the economic management of variability in products and processes. This will be achieved by the use of a more precise method of expressing workpiece functional requirements, complete and well defined specifications, and integrated verification approaches. This improved GPS system will clarify the current practices and be harmonized with the work of other relevant Technical Committees (TC) of the International Organization for Standardization (ISO). This harmonization will, for example, enable better integration with 3D CAD/CAM/CAQ-systems.

The improved GPS system will be based on the experience from the use of current practices and traditions. The legal and technical contents of existing drawings will be left intact, realizing that there is a vast domain of existing specifications, which cannot be changed without the explicit or implicit consent of those responsible.

Proper implementation of the improved GPS system will *reduce costs* by avoiding the manufacture of inadequate workpieces due to incompletely defined specifications.

Controlling function

The intended function of a product can be ensured by controlling the geometry and material properties of the workpiece(s) making up the product. GPS is the language for controlling geometry only and its evolution will be based on computable mathematics and correct, consistent logic using a generic set of rules, that can be applied to all types of specifications.

The challenge for the future is to enrich the GPS language to allow expression of requirements relating to a wide range of workpiece functions.

Proper implementation of the improved GPS system is a prerequisite for the continuous improvement of *product quality* and *time to market*.

Uncertainty - An economical tool

The improved GPS system will use "uncertainty" as the "currency" for quantifying:

a) how well the specification expresses the functional requirements;

b) what ambiguities exists in the specification itself;

c) the uncertainty of measurement.

Proper implementation of the improved GPS system will enable *optimum economical allocation* of resources amongst specification, manufacturing and verification.

Simplification

The improved GPS language will be richer, more precise and, therefore, more verbose. However, in most cases, the complexity of drawing indications will not increase due to the consistency of the logic and the "default" concept.

There will be a global default for each type of GPS specification, which is based on simplicity and minimization of total cost. In addition, there will be a number of shorthand indications covering commonly occurring workpiece functions, e.g. fits.

Proper implementation of the improved GPS system within a company is important for *surviving in global competition*.

International standardization

This effort is being spearheaded by ISO/TC 213 "Dimensional and geometrical product specifications and verification".

Secretariat of ISO/TC 213
Danish Standards Association
Kollegievej 6
DK-2920 Charlottenlund
Denmark
Tp.: +45 39 96 61 01
Fax.: +45 39 96 61 02
E-mail: hhk@ds.dk
Internet: http://www.ds.dk/isotc213/

Global view of geometrical specifications

Alex BALLU
a.ballu@lmp.u-bordeaux.fr
LMP - UMR 5469 CNRS - Université Bordeaux 1
351 cours de la Libération - 33405 Talence Cedex - France,
Luc MATHIEU – Jean-Yves DANTAN
mathieu@lurpa.ens-cachan.fr - dantan@lurpa.ens-cachan.fr
LURPA - ENS de Cachan
61 Avenue du Président Wilson - 94235 Cachan Cedex - France,

Abstract : This paper deals with geometrical specifications, more particularly, with the geometrical concepts used.

The answers to the following questions are approached: Why do we need specifications? What is a geometrical specification? What is specified? What are the geometrical features used? What are the characteristics?

A review of specification shows their diversity, but also the common geometrical bases. Even if the diversity of specifications is great, a specification could be considered as a characteristic limited in a given range, this characteristic is defined on one or more features and these features are obtained from the "skin model" by some basic operations which are common to the different types of specifications.

The main difference between the specification is based on the level of filtration of the skin model. The errors which are specified depend on the level of filtration. This global view homogenise the specifications from the micro geometry to the macro geometry.

Keywords : dimensional specification, tolerance zone, vectorial tolerancing, parametric specification, specification model

1. WHY DO WE NEED SPECIFICATIONS?

The inherent inaccuracies of manufacturing machines imply differences between the ideal shape of the parts of a mechanism and the produced shape. Moreover, these differences are variable when a series of mechanisms is considered.

This well-known fact has a consequence, the need to control the geometrical variations of the products to manage their quality.

For this purpose, it is necessary to be able to express geometrical specifications and to have a coherent model for the description of the non-ideal geometry.

The companies usually use the standardised graphical specifications on the isolated parts. But these specifications, elaborated since many years from solutions to specific

P. Bourdet and L. Mathieu (eds.),
Geometric Product Specification and Verification: Integration of Functionality, 13-24.

needs for the mechanical engineering industries, find their limits. The encountered difficulties are mainly:

- the integration of these solutions in the CAX systems of design, manufacturing and metrology,
- the adequacy of these solutions to the functional needs.

In the same way, the solutions coming from research are various and generally adapted to specific problems of tolerancing.

The industrial needs, the rise of the CAX software and the development of coordinate metrology justified our research and an evolution of the standards.

2. WHAT IS A GEOMETRICAL SPECIFICATION?

In order to make a state of the art of the geometrical specifications, we start from the following postulate: **a geometrical specification describes a condition, which must be satisfied by a part or a set of parts. This condition is expressed from a geometrical characteristic between geometrical features or on geometrical features**. The whole of geometrical features to consider is the whole of non-ideal surfaces of the part ("skin model") [Ballu et al., 1993], [ISO/TR 17450-1, 2000]). The non-ideal surface of a part is a closed surface modelling the interface material - environment of the part. It is defined by the set of all the points belonging to the interface.

The concerned specifications are those which refer to the limitation of the geometrical variations of the manufactured products to satisfy functions and not to define the nominal geometry [Clément et al., 1999], [Mathieu et al., 1997], [Voelcker, 1993], [Walker et al., 1993].

3. WHAT IS SPECIFIED?

According to the functional requirements, the functional conditions and the geometrical specifications are of various types. It is possible to classify these types of specifications according to the known geometrical errors [Ballu et al., 1993], [Voelcker, 1993].

At first, there are the surface texture and the size of the features. The objective is to ensure a certain level of quality of contact and an adjustment of the joints. Then, it should be noted that not only the surface texture and the size have an influence on the contact in a joint, but also the form. These first three types of specifications are intrinsic specifications on features. They allow the specification for the good functioning of joints in an isolated way.

The orientation and the location ensure the relative positioning of the features, and thus a specification between surfaces of joints.

These five types of specification are more or less independent, indeed a standardised specification of location specifies generally at the same time an orientation, a form and even a surface texture according to the applied filtration. [Mullins et al., 1995]

The last type of specification is a type of specification with a significant dependence between the various errors. It consists in considering overall various errors. That often makes it possible to better express the functional requirements. These specifications are specifications by gauge.

3.1. Surface texture of a feature

The surface texture of a feature is defined by a function of the deviations between two profiles. The two profiles are defined from a same feature which is more or less filtered. A specification of surface texture consists in defining the tolerance value of the function of the deviations.

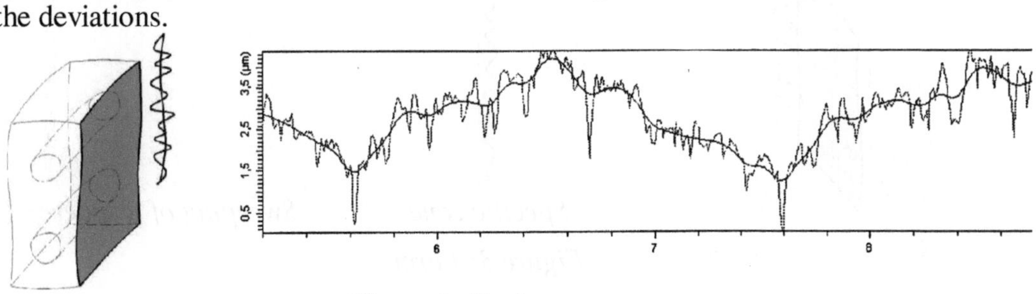

Figure 1; Surface texture

3.2. Size of a feature

The size of a feature is often seen wrongly as the size of an ideal feature. Whereas the basic feature considered is a non-ideal feature. Then, the size can be defined:

- locally, by distances between two opposed points, diameters of circles or diameters of spheres, [Hillyard et al., 1998]
- globally, by the association of ideal features, for example of a least squares cylinder for a nominally cylindrical surface [Henzold, 1993].

The specification consists in defining the limits of the local or global size.

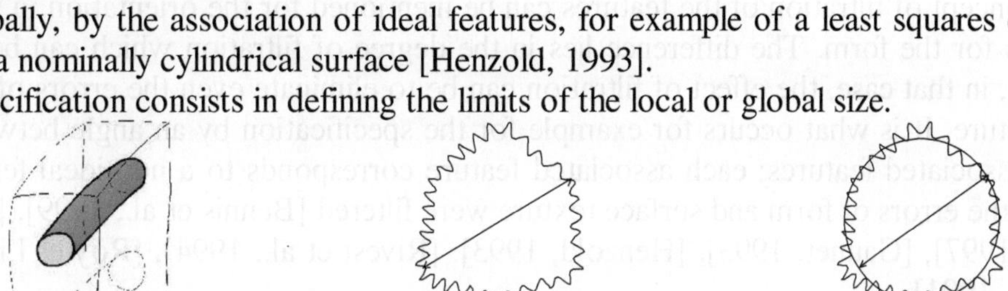

Figure 2 : Size

3.3. Form of a feature

The ideal form of a feature is given by the nominal geometry. The specification of the form of a feature is generally defined by a minimal space in which this feature must be included. The feature is often a filtered feature, i.e. some errors of this feature are not taken into account, this filtration relates to the micro geometry. Various types of filters are used with various bandwidths. A majority of people filters roughness and considers the errors of waviness when it is possible to distinguish them. The standards do not indicate anything about this subject.

The space can be generated according to various methods:

- specific definition for each type of feature in the standards [Ballu et al., 1993],
- generic definition by offset [Requicha, 1983], [Requicha, 1993], [Robinson, 1998], [Rossignac et al., 1986],

  generic definition by sweeping of a sphere (with or without evolution of the diameter), [Etesami, 1993], [Fleming, 1987].

According to the type of feature and according to these various methods, the spaces created can be more or less identical.

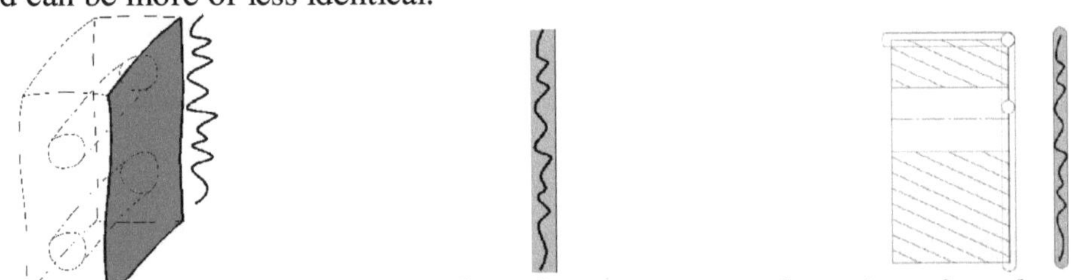

Specific zone *Sweeping of a sphere*

Figure 3; Form

3.4. Orientation of a feature

The ideal orientation of a feature is given by the nominal geometry. Two solutions can be considered to specify the orientation of a feature, in defining:

- a minimal space constrained in orientation, in which the feature shall be contained,
- an angle between two features associated to the non-ideal features whose value shall be contained in a range .

The concept of filtration of the features can be mentioned for the orientation in the same way as for the form. The difference lies in the degree of filtration which can be higher. Indeed, in that case, the effect of filtration can be to eliminate even the errors of form of the feature. It is what occurs for example for the specification by an angle between two ideal associated features; each associated feature corresponds to a non-ideal feature for which the errors of form and surface texture were filtered [Bennis et al., 1999], [Clement et al., 1997], [Gaunet, 1993], [Henzold, 1993], [Rivest et al., 1994], [Roy et Li., 1999], [Wirtz, 1991].

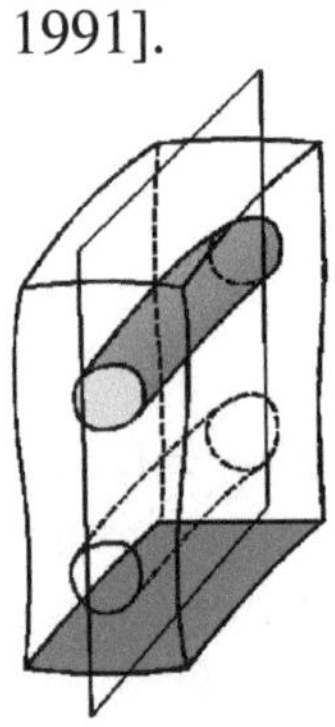

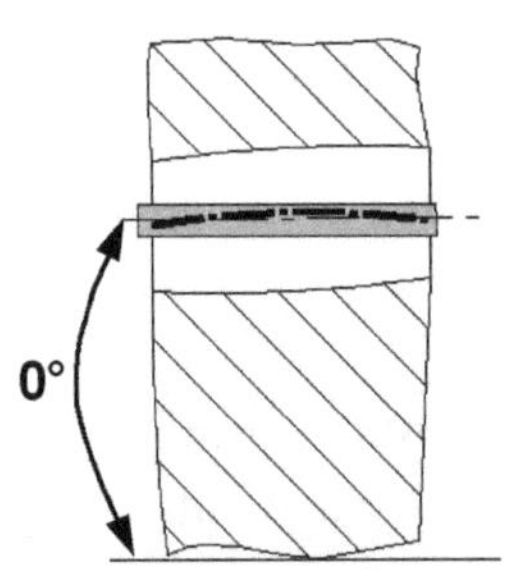

Orientation by zone
Standard specification

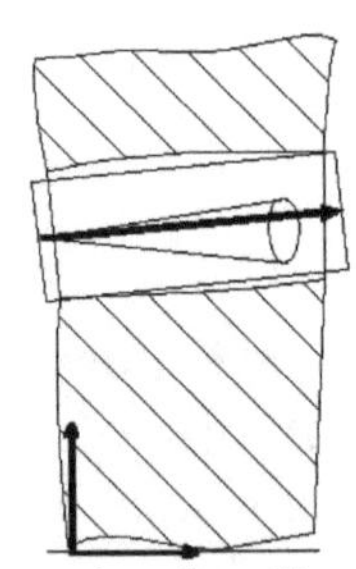

Orientation by dimension
Vectorial tolerancing

Figure 4; Orientation

3.5. Location of a feature

The ideal location of a feature is given by the nominal geometry. Two solutions can be considered to specify the location of a feature, in defining:

- a minimal space constrained in orientation and location, in which the feature shall be contained [Ballu et al., 1993], [Farmer, 1985],
- a distance between two ideal features associated to the non-ideal features with location constraints, whose value shall be contained in a range [Bennis et al., 1999], [Clement et al., 1997], [Gaunet, 1993], [Henzold, 1993], [Rivest et al., 1994], [Roy et al., 1999], [Wirtz, 1991].

One more time, the concept of filtration can be evoked. In that case, the filtration could eliminate also the errors of orientation of the features.

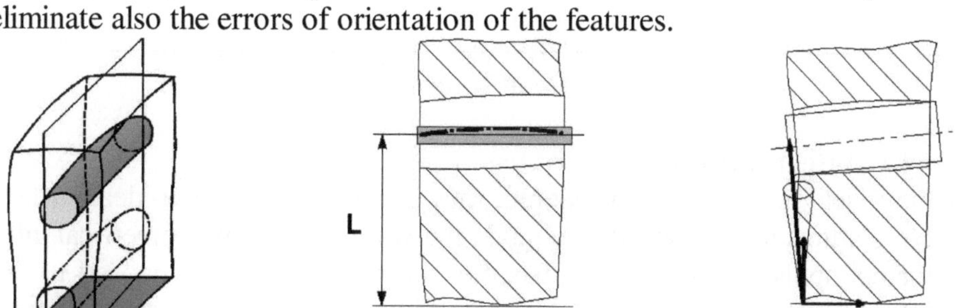

By zone *By dimension*

Standard specification *Vectorial tolerancing*

Figure 5; Location

3.6. Whole of errors

To answer to some functions and to directly satisfy geometrical conditions, the concept of gauge is used. It consists in defining a virtual state simulating the tolerated limits of the feature or the features of the part which have an influence on the function. For example, the simplest gauge considered in the standards is the specification with an envelope requirement applied to a cylindrical shaft. This specification translates a condition of assembly between a shaft and a hole. It includes: the errors of surface texture, form and size of the toleranced feature. The specification means that the specified feature must remain inside a cylinder of a given diameter. In a more general way, the gauge defines by an ideal geometrical shape, the volume in which the toleranced feature must be. The gauge takes into account various errors according to the degree of filtration applied [Etesami, 1991], [Jayaraman et al., 1989], [Srinivassan et al., 1989].

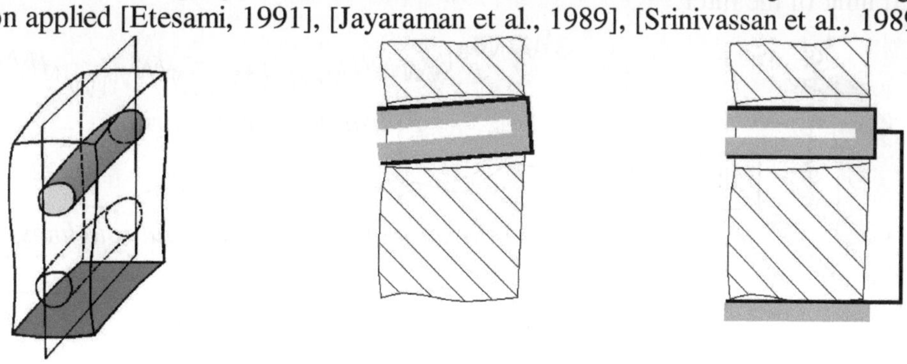

Figure 6; Gauge

4. WHAT ARE THE GEOMETRICAL FEATURES USED?

Many types of features are specified, but the main difference to be pointed out is based on the concept of filtration applied to the non-ideal features. The filtration has a direct consequence on the type of specified error. By filtration, it is understood, from the "skin model" reduction of the level of knowledge of a feature. Filtration makes it possible to eliminate successively errors, of surface texture (with roughness and waviness), of form, of size, of orientation and location.

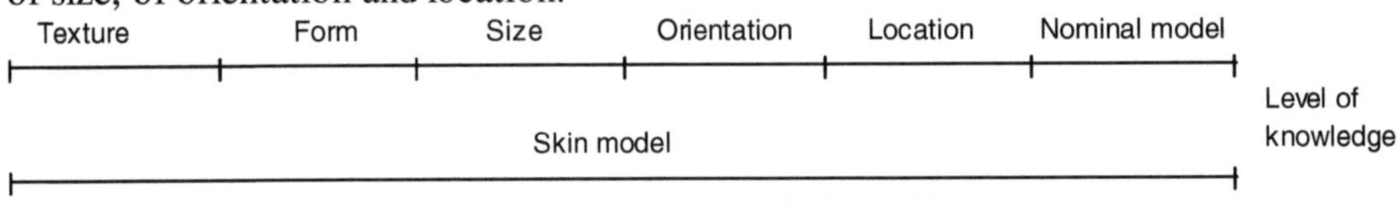

Figure 7; Scale of knowledge of features

4.1. Integral feature

At first, non-ideal features are considered, they are parts of the non-ideal surface of the part. These features are supposed to contain the whole of useful geometrical information for the current applications of mechanics, i.e. these features are supposed to be known with a sufficient degree of accuracy. These parts of the non-ideal surface of the part are called integral features.

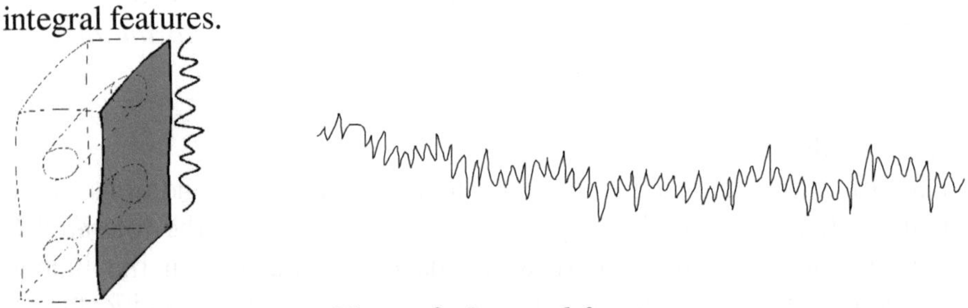

Figure 8; Integral feature

4.2. Smoothed feature

The first level of filtration consists in smoothing the feature, i.e. eliminating whole or part of the errors of surface texture. At the extreme, only the errors of form are preserved. It should be seen that this filtration is more or less high according to the selected limit of the filter.

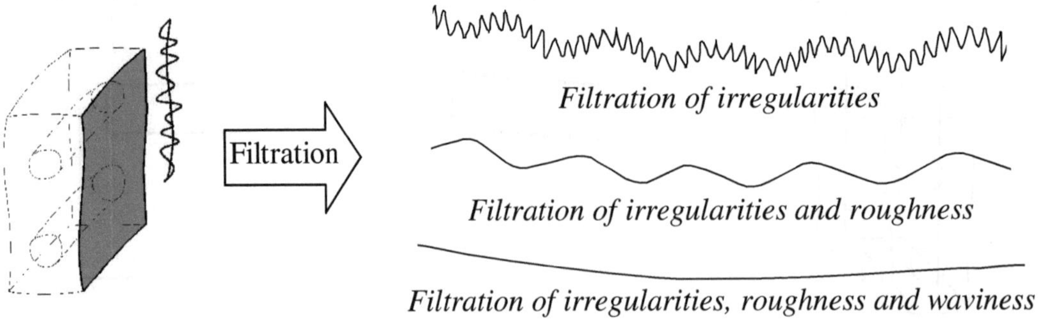

Figure 9; Smoothed feature

Its various degrees of filtration thus make it possible to eliminate what is conventionally called errors of the 4th order, 3rd order, 2nd order and 1st order for the profiles. One often speaks about irregularities, roughness and waviness.

The limit between these four orders is not always obvious. The limits of the bandwidths are difficult to define precisely, the separation is often subjective.

4.3. Substitute and limited feature

A second and a third level of filtration consist in eliminating the errors of form, so as to obtain ideal features whose only orientation and/or location are preserved. For this purpose an ideal feature is associated with the non-ideal feature according to a particular criterion (least squares, Tchebychev, ...) [ISO 17450-1, 2000]. The second level of filtration corresponds to the case of an associated feature, which is not completely constrained in orientation, and it thus preserves errors of orientation. The third level of filtration corresponds to the case of an associated feature, which is completely constrained in orientation, it then preserves only its errors of location.

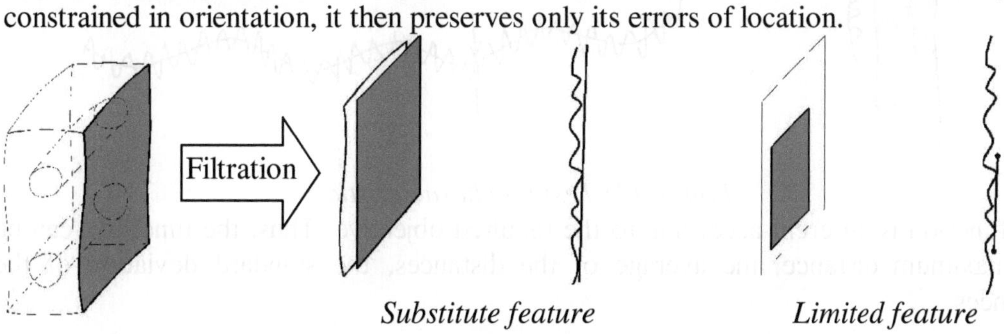

Figure 10; Substitute and limited features

During the second or third level of filtration of a dimensional feature (as a cylinder or two parallel planes) the error of size could be eliminated adding a constraint on the size. The feature considered in the specification could be:

- An ideal feature of infinite area (for example an illimited plane) called substitute feature,
- An ideal feature limited by the area of the corresponding non-ideal feature (for example a plane limited by four close plane faces) called limited feature,
- An ideal feature limited by a functional zone (for example a segment of the axis of a least squares cylinder) also called limited feature,
- a point allowing to locate the feature (for example the "central " point of a limited plane).

4.4. Nominal feature

At last, a fourth level of filtration can be defined; it consists in eliminating any errors and thus obtaining perfect nominal features from all points of view. This level of filtration does not give practical interest for the study of the specifications but it makes it possible to show the continuity of the concept of filtration from the "skin model" to the nominal model.

5. WHAT ARE THE CHARACTERISTICS?

The concept of characteristic is significant for the analysis of the specifications, their verification and the setting of the manufacturing process. Each specification must be defined as a condition on a characteristic.

5.1. Texture characteristic

The errors of surface texture are characterised by a function of the distances between:

- the non-ideal feature whose some errors of surface texture can be filtered,
- the same feature such as the filtration of the errors of surface texture is higher.

The characteristic is a characteristic between two non-ideal features.

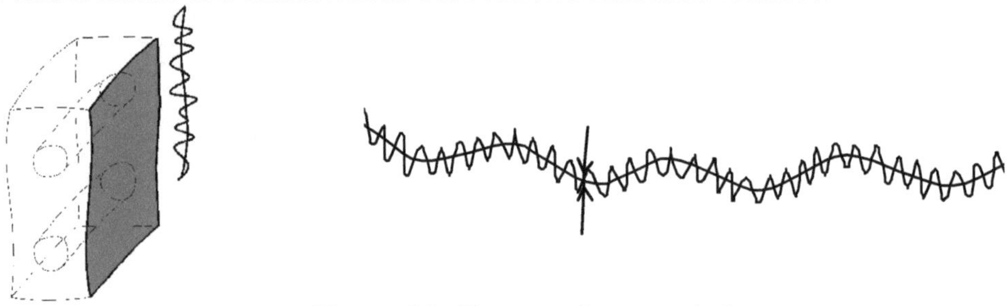

Figure 11; Texture characteristic

The function is different according to the required objective. Thus, the functions can be the maximum distance, the average of the distances, the standard deviation of the distances, ...

5.2. Size characteristic

We distinguished two types of size: local size and global size (Figure 2). The local size is defined by the distance between two opposite points, by the diameter of an inscribed circle or an inscribed sphere.

The global size is defined by the size of a feature whose errors of form and surface texture are filtered. It is the size of a feature of the type cylinder or two parallel planes. These features are substitute features.

In both cases, the characteristic is an intrinsic characteristic of an ideal feature: distance between two points, distance between two planes (the distance between 2 points or two planes can be regarded as intrinsic to the point pair or the plane pair), a diameter of a circle, a sphere or a cylinder.

Different criteria of association of the substitute feature can be used according to the functional objective.

5.3. Form characteristic

The error of form is characterised by the maximum distance between:

- the non-ideal feature whose some errors of surface texture can be filtered,
- the same feature whose errors of form and surface texture are filtered.

The characteristic is a characteristic between a non-ideal feature and an ideal feature. It corresponds to half of the size of the tolerance zone such as defined in the standards.

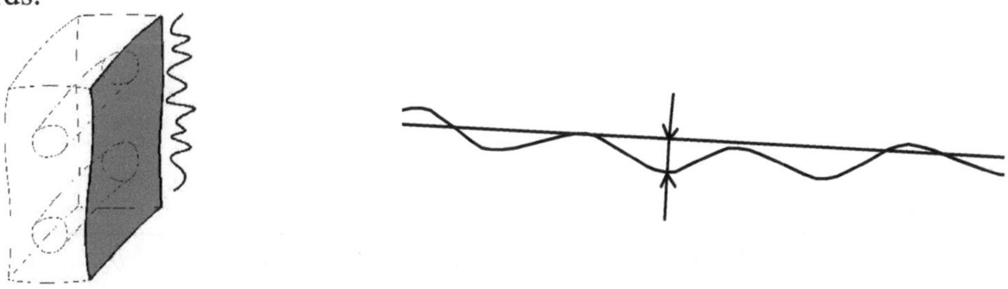

Figure 11; Form characteristic

The different criterion of association of the ideal feature can be used according to the required functional objective. The ISO standards consider the minmax objective, the measuring machines often use the least squares objective.

5.4. Orientation and location characteristic

The error of orientation is characterised by a characteristic between:

- the non-ideal feature whose errors of surface texture and/or of form can be filtered,
- the same feature whose errors of surface texture, form and orientation are filtered.

The error of location is characterised by a characteristic between:

- the non-ideal feature whose errors of surface texture, form and/or orientation can be filtered,
- the same feature whose errors of surface texture, form, orientation and location are filtered.

According to cases, the characteristic is:

- the maximum distance between the two features [Ballu et al., 1993], [Henzold, 1993], [Srinivassan, 1993],
- the angle (the distance for the location) between the two features if the form is filtered for the two features [Bennis et al., 1999], [Clement et al., 1997], [Gaunet, 1993], [Henzold, 1993], [Rivest et al., 1994], [Roy et al., 1999], [Wirtz, 1991].

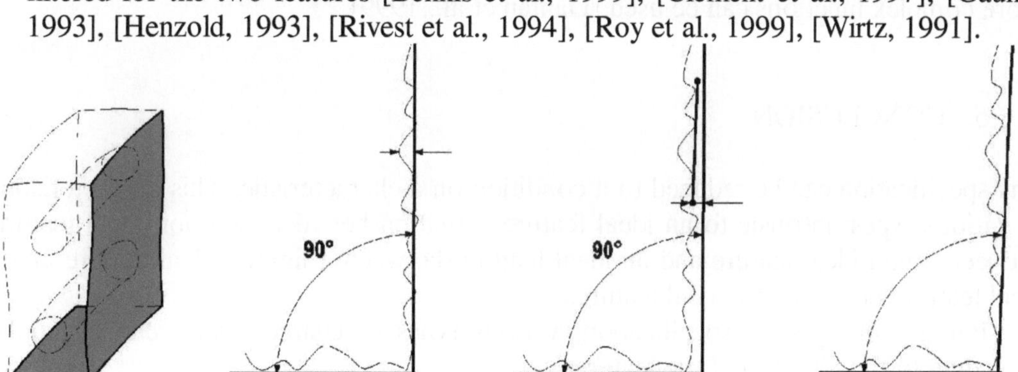

by maximum distance *by maximum distance on limited feature* *by angle*

Figure 12; Orientation characteristics

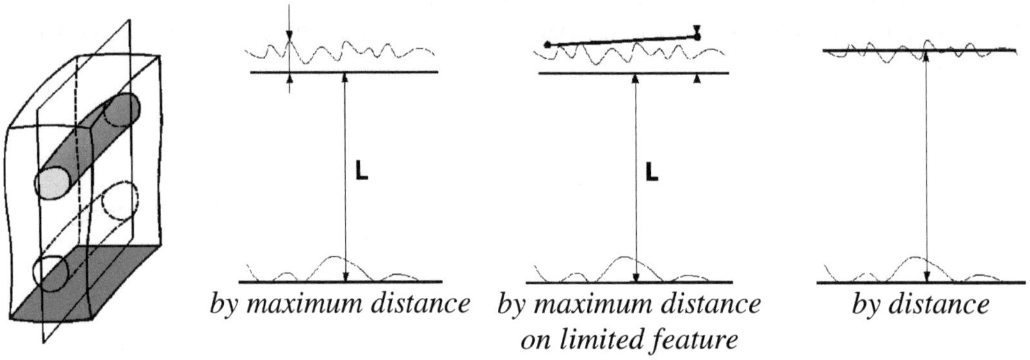

by maximum distance *by maximum distance on limited feature* *by distance*

Figure 13; Location characteristics

In the first case, the characteristic corresponds to half of the size of the tolerance zone such as defined in the standards. It should be noticed that this characteristic is not always defined in the standards, for example for the position of two groups of holes simultaneously according to two different tolerances.

In the second case, it corresponds to the concept of vectorial tolerancing defines by Wirtz, or to parametric tolerancing [Wirtz, 1991], [Henzold, 1993], [Hoffman, 1982].

5.5. Gauge characteristic

Currently, the concept of characteristic on the gauges is not developed. During the verification, only control is carried out, no measurement is done. However, it is necessary to consider a characteristic even for the gauges in order to be able to evaluate how much a part is good or bad, for example at ends of setting of machine. The characteristic on a gauge can be defined as the distance between:

- the non-ideal feature whose errors of surface texture, form and/or orientation can be filtered,
- the same feature whose errors of surface texture, form orientation and location, are filtered (gauge) [Etesami, 1991].

More complex functions can be used [Dantan et al., 1999].

6. CONCLUSION

Any specification can be reduced to a condition on a characteristic. This characteristic is of various types: intrinsic to an ideal feature, situation between two non-ideal features, between a non-ideal feature and an ideal feature, between a limited ideal feature and an ideal feature, between two ideal features.

For a same type of specification, various types of characteristics can be applied according to the filtration of the features.

According to the degree of filtration of the features, there is independence or not between the specifications.

REFERENCES

[ISO/TS 17450-1, 2000] Geometric Product Specification (GPS) – General concepts – Part 1: Model for geometrical specification and verification.

[Ballu et al., 1993] A. Ballu and L. Mathieu, "Analysis of dimensionnal and geometric specifications : Standard and model"; In: Proc. of 3 rd CIRP Seminar on Computer Aided Tolerancing; Cachan; France; April 27-28; 1993.

[Bennis et al., 1999] F. Bennis, L. Pinos, C. Fortin, « Analysis of positional tolerance based on the assembly virtual state », In: Proc. of CIRP Seminar on Computer Aided Tolerancing, University of Twente, Netherlands, March 1999.

[Clément et al., 1997] A. Clement, A. Riviere, P. Serre, C. Valade, "The TTRS: 13 constraints for dimensioning and tolerancing"; In: Proc. of 5 rd CIRP Seminar on Computer Aided Tolerancing; Toronto; Canada; April 27-29; 1997.

[Clément et al., 1999] A. Clement, A. Riviere, P. Serre, C. Valade, "Global consistency of Dimensioning and Tolerancing"; In: Proc. of CIRP Seminar on Computer Aided Tolerancing, University of Twente, Netherlands, March 1999.

[Dantan et al., 1999] J.Y. Dantan, A. Ballu, "Functional and product specification by Gauge with Internal Mobilities", In: Proc. of CIRP Seminar on Computer Aided Tolerancing, University of Twente, Netherlands, March 1999.

[Etesami, 1991] F. Etesami, « Position Tolerance Verification Using Simulated Gaging », The International Journal of Robotics Research, Vol. 10, n°4, pp. 358-370, 1991.

[Etesami, 1993] F. Etesami, « A mathematical model for geometric tolerances », Journal of Mechanical Design, Vol. 115, n°1, pp. 81-86, 1993.

[Farmer, 1985] L. Farmer, « Datums, Tolerances and NC Machine Tools », SME Technical Paper MM 85-670, 1985.

[Fleming, 1987] A. Fleming, "Analysis of uncertainties and geometric tolerances in assemblies of parts", PhD Thesis, University of Edimburgh, 1987.

[Gaunet, 1993] D. Gaunet, "Vectorial tolerancing model", In: Proc. Of 3 rd CIRP Seminar on Computer Aided Tolerancing, Cachan, France, April 27-28, 1993.

[Henzold, 1993] G. Henzold, « Comparison of Vectorial Tolerancing and Conventional Tolerancing », In: Proc. of International Forum on Dimensional Tolerancing and Metrology, CRTD-Vol. 27, pp. 147-160, 1993.

[Hillyard et al., 1978] R.C. Hillyard, I.C. Braid, « Characterizing non-ideal shapes in term of dimensions and tolerances », Computer Graphics, Vol. 12, n°3, pp. 234-238, 1978.

[Hoffman, 1982] P. Hoffman, «Analysis of tolerances in process inaccuracies in discrete part manufacturing », Computer Aided Design, Vol. 14, n°2, pp. 83-88, 1982.

[Jayaraman et al., 1989] R. Jayaraman, V. Srinivasan, « Virtual boundary requirements », IBM Journal of Research and Development, Vol. 33, n°2, pp. 90-104, 1989.

[Mathieu et al, 1997] L. Mathieu, A. Clément, P. Bourdet, « Modeling, representation and processing of tolerances, tolerances inspection : a survey of a current hypothesis » In: Proc. of CIRP Seminar on Computer Aided Tolerancing, Toronto, Canada, April 1997.

[Mullins et al., 1995] S. H. Mullins, D. C. Anderson, « Algebraic representation of geometric and size tolerances », In Proc. of ASME Design Engineering Technical Conferences, Vol. 1, DE-Vol. 82, pp. 309-316, 1995.

[Requicha, 1983] A.A.G. Requicha, "Toward a theory of geometric tolerancing", The International Journal of Robotics Research, Vol.2, n°4, pp.45-60, 1983.

[Requicha, 1993] A.A.G. Requicha, « Mathematical Meaning and Computational Representation of Tolerance Specifications », In: Proc. of International Forum on Dimensional Tolerancing and Metrology, CRTD-Vol. 27, pp. 61-68, 1993.

[Rivest et al., 1994] L. Rivest, C. Fortin, C. Morel, "Tolerancing a solid with a kinematic formulation", Computer Aided Design, Vol.26, n°6, pp.465-476, 1994.

[Robinson, 1998] D.M. Robinson, "Geometric tolerancing for assembly", PhD thesis, Cornell University, May 1998.

[Rossignac et al., 1986] R. Rossignac, A. A. G. Requicha, "Offsetting operations in solid modelling", Computer Aided Geometric Design, Elsevier Science Publishers, Vol. 3, pp. 129-148, 1986.

[Roy et al., 1999] U. Roy, B. Li, "Representation and interpretation of geometric tolerances for polyhedral objects", Computer Aided Design, Vol.31, n°4, pp.273-285, 1999.

[Srinivassan et al., 1989] V. Srinivassan, R. Jayaraman, "Conditional tolerances"; In: IBM Journal of Research and Development; Vol. 33; n° 2; pp 105-124; 1989.

[Srinivassan, 1993] V. Srinivassan, « Recent Efforts in Mathematization of ASME/ANSI Y14.5M », In: Proc. of 3rd CIRP Seminar on Computer Aided Tolerancing, Eyrolles, pp.223-232, 1993.

[Voelcker, 1993] H. B. Voelcker, « A Current Perspective on Tolerancing and Metrology », In: Proc. of International Forum on Dimensional Tolerancing and Metrology, CRTD-Vol. 27, pp. 49-60, 1993.

[Walker et al., 1993] R. K. Walker, V. Srinivassan, « Creation and Evolution of the ASME Y-14.5.1M Standard », In: Proc. of International Forum on Dimensional Tolerancing and Metrology, CRTD-Vol. 27, pp. 19-30, 1993.

[Wirtz, 1991] A. Wirtz, "Vectorial tolerancing for production quality control and functional analysis in design", In: Proc. of Annals of the CIRP, Pennstate, USA, 1991.

3D Functional Tolerancing & Annotation : CATIA tools for Geometrical Product Specification

Dominique Gaunet.
Dassault Systemes
9, quai Marcel Dassault -BP 310
92156 Suresnes Cedex,France.
dominique_gaunet@ds-fr.com

Abstract: In this paper we present some Dassault Systemes geometric product specification tools based on the use of the Technologically and Topologically Related Surface (TTRS) model. Ensuring the consistency of the specifications to the standards, these 3D tools help the user in the dimensioning & tolerancing definition, enabling to reduce the cost of part and product design and increasing their quality. The meaning of 3D specifications is capture enabling its exploitation by downstream application. Some direct benefits of the use of the TTRS model are shown.
Keywords: CAD/CAM, TTRS, Geometric Product Specifications, CATIA.

1. INTRODUCTION.

To reduce the design costs and to improve the quality of their products, most of the companies are looking for using the Geometric Product Specification standards.
The correct use of Dimensioning & Tolerancing standards relying on 2D drawings is a big step to the quality, but it is not sufficient. Because they are only 2D symbols attached to 2D geometry on the drawings, the dimensions and tolerances have to be redefined or converted several times for each downstream 3D application, such as tolerance analysis, manufacturing or inspection, that is using parts or assemblies geometric variations as input data. These manual entries increase the risk of mistakes and misunderstandings.
Maintaining the consistency between 2D drawings and 3D design of parts and products costs a lot (it can raise 50% of the total price of a part for small production).
The ability to put tolerances and dimensions directly on the 3D geometric features, handled by a robust mathematical model is another step to the total quality.

P. Bourdet and L. Mathieu (eds.),
Geometric Product Specification and Verification: Integration of Functionality, 25-33.

2. TOPOLOGICALLY AND TECHNOLOGICALLY RELATED SURFACES (TTRS) MODEL.

A TTRS is a pair of surfaces (or TTRS) belonging to a unique part or product and associated by functional relations [Charles et al. 1989], [Clement et al. 1991], [Desrochers 1991], [Clement et al. 1994].

2.1. TTRS classes

The TTRS model defines 7 classes of Euclidian surfaces by considering the degrees of freedom that let the surface globally unchanged (See table I).

Surface Classes	Degrees of freedom
 **1-** Spherical surface	**3** 0 translation 3 rotations
 **2-** Planar surface	**3** 2 translations 1 rotation
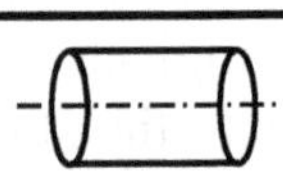 **3-** Cylindrical surface	**2** 1 translation 1 rotation
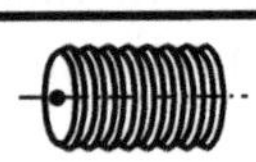 **4-** Helical surface	**1** 1 combined translation and rotation
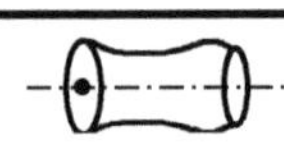 **5-** Surface of revolution	**1** 0 translation 1 rotation
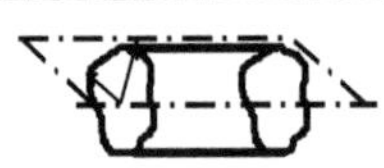 **6-** Prismatic surface	**1** 1 translation 0 rotation
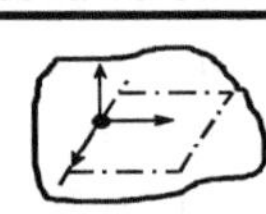 **7-** Complex surface	**0** 0 translation 0 rotation

Table I : Surface classes.

2.2. TTRS combination

The TTRS model by using the association operator holds a commutative group structure and defines 28 configurations corresponding to 44 reclassification cases (See table II).
Lets consider for instance the configuration that corresponds to the association of two cylindrical surfaces :

Cylindrical surface 1 ∪ Cylindrical surface 2

- If the cylindrical surfaces are coaxial, the global surface made of the set of the two surfaces is cylindrical class (there are 2 degrees of freedom, 1 translation along the common axis and 1 rotation around the common axis, that let the set of the 2 surfaces globally unchanged).
- If the cylindrical surfaces are parallel, the global surface is prismatic class (there is 1 degree of freedom in translation along the axis direction that lets the set of the 2 surfaces globally unchanged).

- Otherwise, the global surface is complex class (there is no degree of freedom).

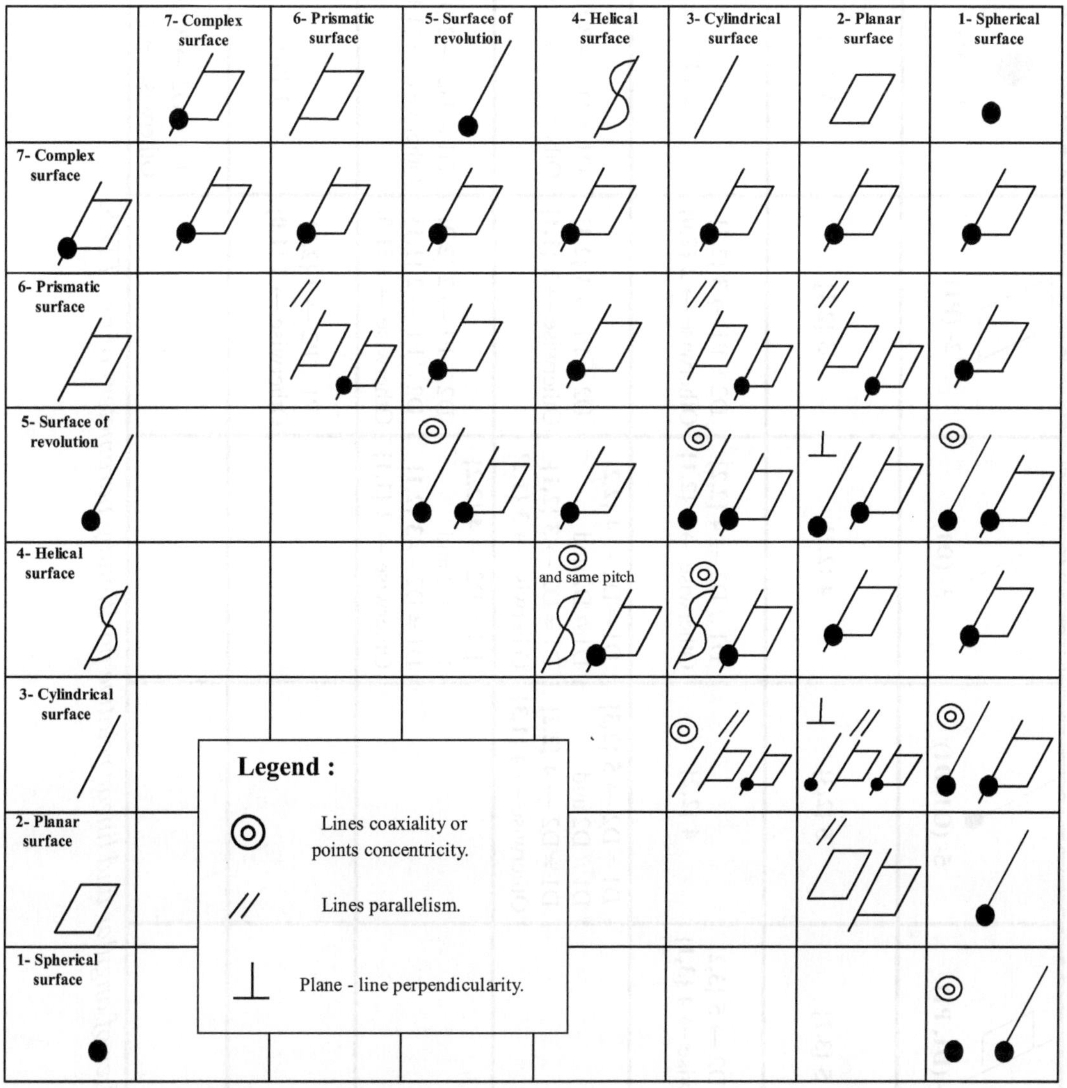

Table II : Association and reclassification cases of TTRS.

2.3. Relative positioning parameters of two TTRS.

The number and the types of relative positioning parameters of two TTRS are totally defined by their own degrees of freedom (See table III) [Gaunet 1994] :

$$np = 6 + ddl_{TTRS1 \cup TTRS2} - ddl_{TTRS1} - ddl_{TTRS2}$$

	7- (O1, D1, P1)	6- (D1, P1)	5- (O1, D1)	3- (D1)	2- (P1)	1- (O1)
7- (O2, D2, P2)	**6 {3,3}**	**5 {3,2}**	**5 {2,3}**	**4 {2,2}**	**3 {2,1}**	**3 {0,3}**
6- (D2, P2)		D1 // D2 → **5 {3,2}** Otherwise → **4 {3,1}**	**4 {2,2}**	D1 // D2 → **4 {2,2}** Otherwise → **3 {2,1}**	D2 // P1 → **3 {2,1}** Otherwise → **2 {2,0}**	**2 {0,2}**
5- (O2, D2)			D1 = D2 → **5 {2,3}** D1 // D2 and D1 ≠ D2 → **4 {2,2}** Otherwise → **4 {1,3}**	D1 = D2 → **4 {2,2}** D1 // D2 and D1 ≠ D2 → **3 {2,1}** Otherwise → **3 {1,2}**	D2 ⊥ P1 → **3 {2,1}** Otherwise → **2 {1,1}**	O1 ∈ D2 → **3 {0,3}** Otherwise → **2 {0,2}**
3- (D2)				D1 = D2 → **4 {2,2}** D1 // D2 and D1 ≠ D2 → **3 {2,1}** Otherwise → **2 {1,1}**	D2 ⊥ P1 → **2 {2,0}** D2 // P1 → **2 {1,1}** Otherwise → **1 {1,0}**	O1 ∈ D2 → **2 {0,2}** Otherwise → **{1 {0,1}**
2- (P2)					P1 // P2 → **3 {2,1}** Otherwise → **1 {1,0}**	**1 {0,1}**
1- (O2)						O1 = O2 → **3 {0,3}** Otherwise → **1 {0,1}**

Table III: Number of angular and linear relative positioning parameters of two TTRS.

2.4. Some direct benefits of TTRS model use.

To define the position or orientation tolerances of a feature in regards to a datum system is equivalent to define the allowed limits of the relative positioning parameters of the two TTRS that correspond to the toleranced feature and the datum system.
So the TTRS model is well adapted for position and orientation tolerancing.

Check of datum system consistency.

Lets consider a datum system made of primary, secondary and tertiary datum features.
If $TTRS_{d1}$, $TTRS_{d2}$ and $TTRS_{d3}$ respectively correspond to the primary, secondary and tertiary datum features, the datum system will be consistent (not over defined) when the following rules are satisfied :

$$\textit{class of } TTRS_{1} < \textit{class of } TTRS_{d1} \cup TTRS_{d2}$$
$$\textit{class of } TTRS_{d1} \cup TTRS_{d2} < \textit{class of } (TTRS_{d1} \cup TTRS_{d2}) \cup TTRS_{d3}$$

Check of useless datum feature.

Lets consider the positioning of a feature in regard to a datum system made of primary, secondary and tertiary datum features.
If $TTRS_t$ corresponds to the toleranced feature and if $TTRS_{d1}$, $TTRS_{d2}$ and $TTRS_{d3}$ respectively correspond to the primary, secondary and tertiary datum features, the datum features will all be useful for the positioning when the following rules are satisfied :

$$\textit{class of } TTRS_t \cup TTRS_{d1} < \textit{class of } TTRS_t \cup (TTRS_{d1} \cup TTRS_{d2})$$
$$\textit{class of } TTRS_t \cup (TTRS_{d1} \cup TTRS_{d2}) < \textit{class of } TTRS_t \cup ((TTRS_{d1} \cup TTRS_{d2}) \cup TTRS_{d3})$$

If we are only looking for the orientation of the toleranced feature, the datum features will all be useful when the following rules are satisfied (where *nap* means number of angular relative positioning parameters) :

$$\textit{nap of } TTRS_t \cup TTRS_{d1} < \textit{nap of } TTRS_t \cup (TTRS_{d1} \cup TTRS_{d2})$$
$$\textit{nap of } TTRS_t \cup (TTRS_{d1} \cup TTRS_{d2}) < \textit{nap of } TTRS_t \cup ((TTRS_{d1} \cup TTRS_{d2}) \cup TTRS_{d3})$$

Cylindrical, circular or spherical tolerance zone availability.

Knowing if a cylindrical tolerance zone can be defined for the positioning of a cylindrical surface is easy. You just have to look at the number of positioning parameters that are involved for the association of the cylindrical surface with the datum system TTRS. If 2 angular or 2 linear positioning parameters are involved, the tolerance zone can be cylindrical.
It's quite the same for a spherical surface. If 3 linear parameters are involved, the tolerance zone can be spherical. If only 2 parameters are involved, the tolerance zone can be circular.

3. CATIA TOOLS

By taking advantage of the TTRS modeler, Dassault Systemes offers, in the CATIA product line, a set of tools that provide assistance to the user along the geometric product specification phases.

3.1. Specification tools

Both the V4 CATIA 3D Functional Dimensioning & Tolerancing product (FD&T) and the V5 CATIA 3D Functional Tolerancing & Annotation product (FT&A) help the user in the correct use of standards by giving the following major capabilities :

- Proposal of applicable tolerance types regarding the selected surfaces.
- Proposal of tolerance options when applicable.
- Dimensioning & Tolerancing rules verification.
- Automatic support of annotation syntax.
- GUARANTEE of semantic & syntactic validity and consistency of the tolerancing, through the part assembly life cycle.

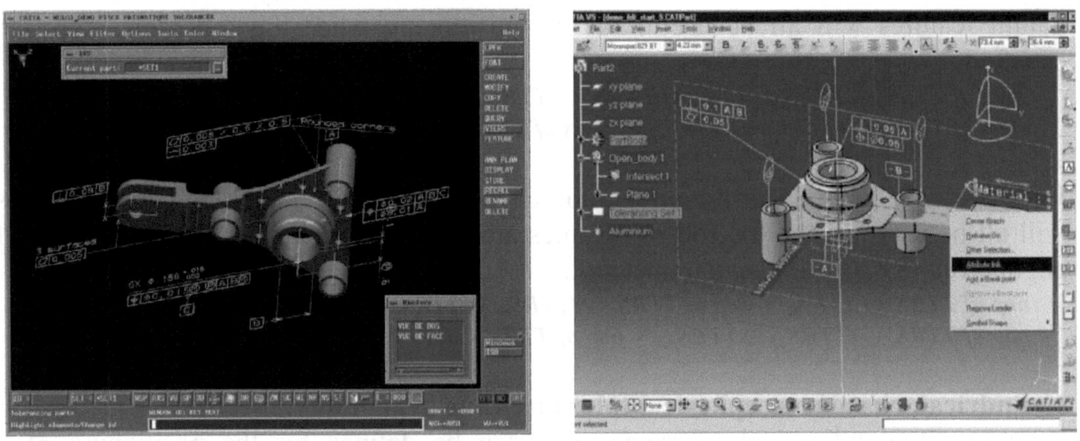

Figure 1 : V4 FD&T and V5 FT&A products illustration.

3.2. Communication tools

Several ways to communicate the Dimensioning & Tolerancing information are provided :

- Through 2D drawings.
  This is the usual way of communication but the difference, here, is that all the geometric product specifications are defined in the 3D and simply extracted on demand with the geometry in the 2D. The master model is the 3D.

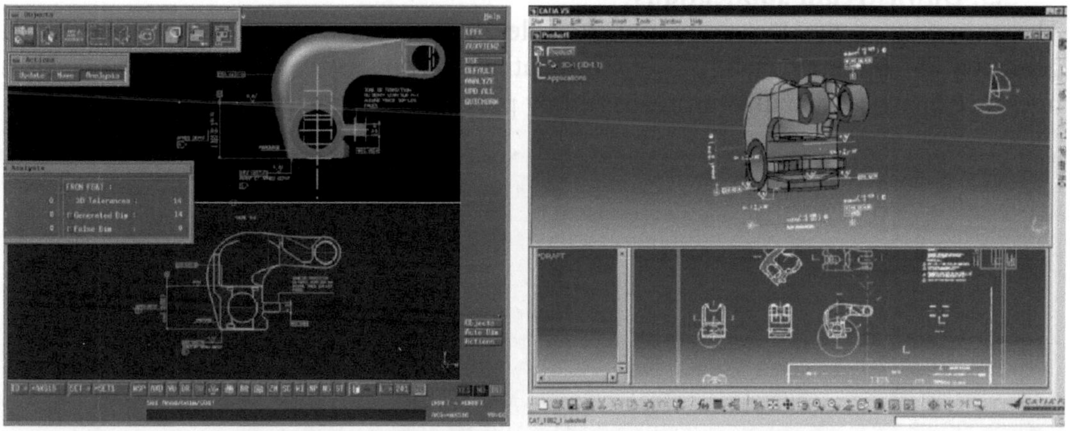

Figure 2 : V4 and V5 2D extraction illustration.

- Through VRML format files.
- Through the use of the DMU Dimensioning & Tolerancing Review product. This product offers advanced display and query capabilities that help the user to access and understand the specifications :
  - ✓ Display of annotation in 3D and/or in a 2D table.
  - ✓ Associative highlight query between annotations and geometry.
  - ✓ Highlight of all the geometry that supports 3D specifications.
  - ✓ Filter by features, views and annotation planes, default tolerances, tolerance types and sub-types, tolerance values.

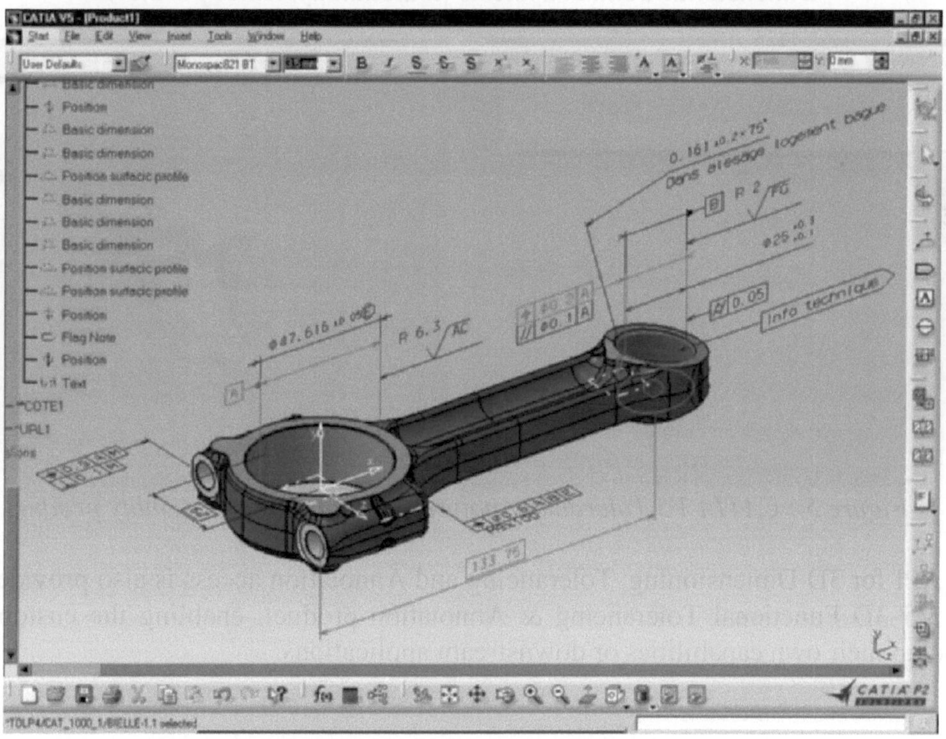

Figure 3 : Associative query highlight illustration.

3.3. Downstream applications

The meaning of the Dimensioning & Tolerancing data and the links with the 3D geometry is captured enabling its exploitation by downstream applications, such as tolerance analysis, manufacturing, assembly process planning and inspection.

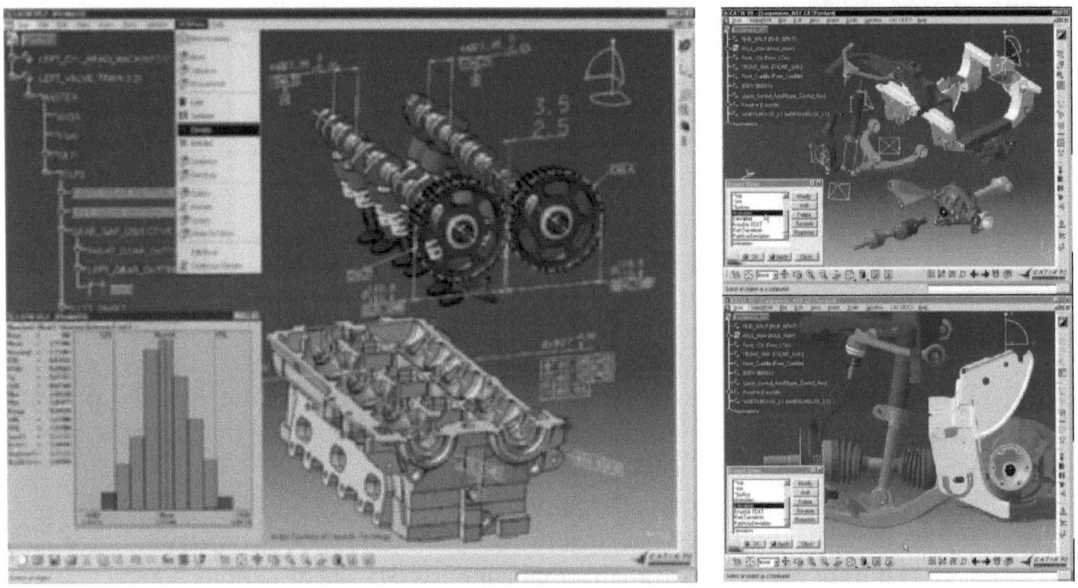

Figure 4 : CATIA V5 Tolerance Analysis CAA application Dimensional Control Systems (Dassault Systemes partner).

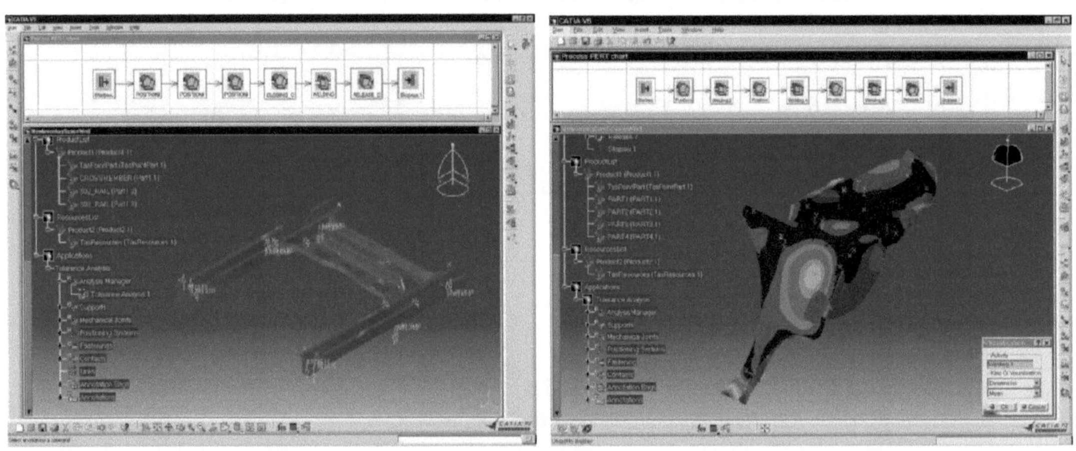

Figure 5 : CATIA V5 Tolerance Analysis of Deformable Assembly product.

An API for 3D Dimensioning, Tolerancing and Annotation access is also provided with the V5 3D Functional Tolerancing & Annotation product, enabling the customers to develop their own capabilities or downstream applications.

4. PERSPECTIVES

Research and development are continuing on several topics such as dimension transfer for manufacturing, functional analysis (3D chain dimensions search) [Dufosse 1990] and tolerance synthesis [Gaunet 1995].

REFERENCES

[Charles et al. 1989] Charles B.; Clement A.; Desrochers A.; Pelissou P.; Riviere A.; "Toward a computer aided functional tolerancing model", International Conference on CAD/CAM and AMT in Israël, C.I.R.P. Sessions on tolerancing for function in a CAD/CAM environment, Proceedings, Vol. 2, Jérusalem ,Israël, 11 - 14 décembre 1989.

[Clement et al. 1991] Clement A.; Desrochers A.; Riviere A.; "Theory and Practice of 3-D Tolerancing for Assembly", C.I.R.P. INTERNATIONAL WORKING SEMINAR on COMPUTER-AIDED TOLERANCING, Mai 1991.

[Desrochers 1991] Desrochers A.; "Modèle conceptuel du dimensionnement et du tolérancement des mécanismes. Représentation dans les systèmes de CFAO", thèse de doctorat, Ecole Centrale Paris, Septembre 1991.

[Clement et al. 1994] Clement A.; Riviere A.; Temmerman M. ; "Cotation tridimensionnelle des systèmes mécaniques - Théorie & pratique", Edition PYC, 1994.

[Gaunet 1994] Gaunet D. ; "Modèle formel de tolérancement de position. Contributions à l'aide au tolérancement des mécanismes en CFAO.", thèse de doctorat, Ecole Normale Supérieure de Cachan, février 1994.

[Dufosse 1993] Dufosse P.; "Automatic Dimensioning and Tolerancing", 3rd CIRP Seminar on Computer Aided Tolerancing, avril 1993.

[Gaunet 1995] Gaunet D.; "FD&T, une fonction interactive d'aide à la cotation fonctionnelle des mécanismes", seminaire « Tolérancement et Chaînes de cotes », Cachan, 8 et 9 février 1995.

Areal Coordinates: The Basis of a Mathematical Model for Geometric Tolerances

S. Bhide, J. K. Davidson, and J. J. Shah
Department of Mechanical & Aerospace Engineering,
Arizona State University, Tempe, AZ 85282-6106, USA
J.Davidson@asu.edu

Abstract: A mathematical model is presented for representing the tolerances of planar surfaces. The model is compatible with the ASME/ISO Standards for geometric tolerances. Central to the new model is a Tolerance-Map® [1], a hypothetical volume of points which corresponds to all possible locations and variations of a segment of a plane which can arise from tolerances on size, form, and orientation. This model is one part of a bi-level model that we are developing for geometric tolerances. The new model makes stackup relations apparent in an assembly, and these can be used both to allocate size and orientational tolerances and to identify sensitivities. All stackup relations can be met for 100% interchangeability or for a specified probability.
Keywords: geometric tolerances, affine geometry, barycentric coordinates, areal coordinates

1. INTRODUCTION

The purpose of this paper is to suggest a mathematical model for geometric tolerances that can incorporate the modern rules governing geometric dimensioning and tolerancing [ASME Standard, 1994] [ISO 1101, 1983]. A number of mathematical models for geometric tolerances have been produced, all with the objective of developing computer tools to help designers to analyze and specify geometric tolerances. Examples are [Gossard, *et al.*, 1988], [Requicha, 1983], [Turner, 1990], [Clément, *et al.*, 1996], [Chase, *et al.*, 1998], and [Shah, *et al.*, 1998]. A more detailed survey of the literature appears in [Davidson, *et al.*, 2000]. The first result of our own work appeared in [Mujezinović, 1999]. A related paper appeared in [Davidson and Shah, 2000].

2. THE TOLERANCE-MAP (T-MAP) FOR FACES

The main feature of the local model is a Tolerance-Map® (T-Map®), a hypothetical Euclidean point-space, the size and shape of which reflect all variational possibilities for

---
[1] Patent pending; provisional patent No. 60/120,961 (1999).

P. Bourdet and L. Mathieu (eds.),
Geometric Product Specification and Verification: Integration of Functionality, 35-44.

a target feature. It is a one-to-one mapping of points from all the variational possibilities of a feature within its tolerance-zone. These variations are determined by the tolerances that are specified for size, position, and orientation. A tolerance on form is represented by an internal component-region of a T-Map.

2.1. The basis-tetrahedron

The T-Map for any combination of tolerances on a feature is constructed from a basis-tetrahedron and described with areal coordinates. Classical descriptions of this subject, a form of affine geometry, are in [Coxeter, 1969]. We choose to position the four basis-points σ_1, σ_2, σ_3, and σ_4 as in Fig. 1. Points $(\sigma_1,\sigma_2,\sigma_3)$ and $(\sigma_1,\sigma_2,\sigma_4)$ form two joined congruent isosceles triangles at right angles and of aspect ratio unity, i.e. their heights and base-length $\sigma_1\sigma_2$ are the same. At the four basis-points we place four masses λ_1, λ_2, λ_3, and λ_4 that may be positive or negative. So long as $\lambda_1+\lambda_2+\lambda_3+\lambda_4=1$, the position of σ, the centroid of these masses, is determined uniquely by the linear combination

$$\sigma = \lambda_1\sigma_1 + \lambda_2\sigma_2 + \lambda_3\sigma_3 + \lambda_4\sigma_4; \tag{1}$$

it can be made to assume any position in three-space by varying λ_1, λ_2, λ_3, and λ_4. For example, when $\lambda_1 \ldots \lambda_4$ are all positive, σ identifies a point *inside* tetrahedron $\sigma_1\sigma_2\sigma_3\sigma_4$. The four masses $\lambda_1 \ldots \lambda_4$ are the areal coordinates of σ. These are the coordinates that we use in describing T-Maps for faces (planar surfaces) on parts.

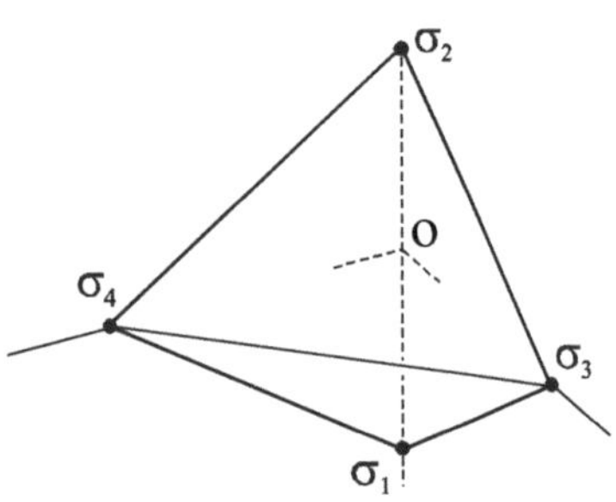

Figure 1. The reference tetrahedron. $\sigma_1\sigma_2=O\sigma_3=O\sigma_4=t$.

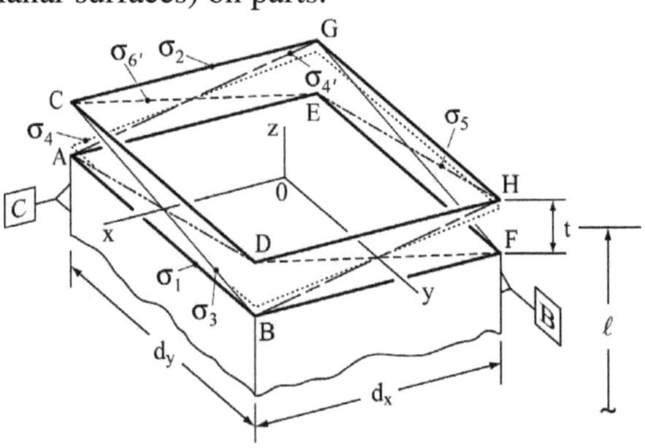

Figure 2. The end of a rectangular bar with a size tolerance t.

2.2. The Tolerance-Map (T-Map) for a Face

In Fig. 2 is shown the tolerance-zone on size at the end of a rectangular bar of cross-sectional dimensions $d_x \times d_y$ $(d_x < d_y)$. To apply the concept of areal coordinates, above, consider that the planes labelled σ_1, σ_2, σ_3, and σ_4 are four *basis-planes* that define the space of our three-dimensional set of planes in the tolerance-zone of Fig. 2. We map these to the four vertices of the tetrahedron in Fig. 1 (shown dotted in Fig. 3). Then the points on the line-segment $\sigma_1\sigma_2$ in Fig. 3 represent the parallel planes that are perpendicular to the *z*-axis in Fig. 2 and lie between σ_1 and σ_2. The points on the line-segment $\sigma_2\sigma_3$ in Fig. 3 correspond to a partial axial pencil of planes in the tolerance-zone, i.e. those planes in Fig. 2 that are parallel to the x-axis, pass through line CG, and also lie in the tolerance-zone between σ_2 and σ_3.

There is a fundamental linear relationship between any partial axial pencil of planes in the tolerance-zone for a face and its corresponding line-segment of points in the Tolerance-Map. This relationship is embodied in eqn (1), and it derives from the extremely small ranges for orientational variations in a tolerance-zone that are imposed by tolerance-values; it can be used to further construct the T-Map in Fig. 3. For example, the planes σ_3 and $\sigma_{4'}$ form limits to an axial pencil of planes about line BG in Fig. 2, every plane of which is at the boundary of the tolerance-zone. As a plane is rotated from σ_3 to $\sigma_{4'}$ about BG in Fig. 2, the consecutive sequence of planes corresponds to the consecutive points in Fig. 3 that lie on the line-segment between points σ_3 and $\sigma_{4'}$. The remaining edges of the T-Map can be found in a similar manner by repeating the construction for the remaining seven combinations of the six vertices. The result is the octahedron in Fig. 3; it is the T-Map for the entire tolerance-zone in Fig. 2.

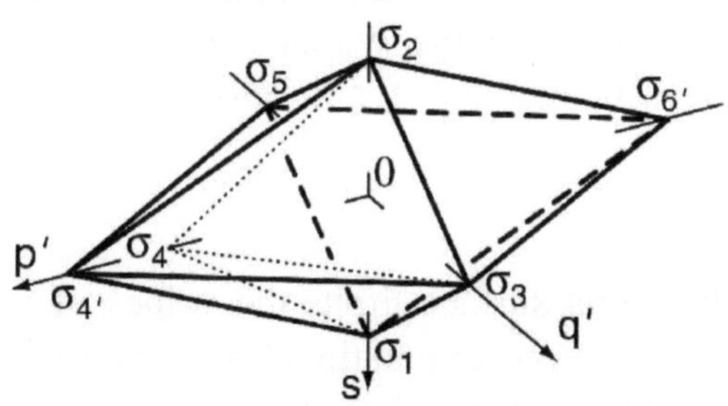

Figure 3. The Tolerance-Map® for the tolerance-zone on the bar in Fig. 2.

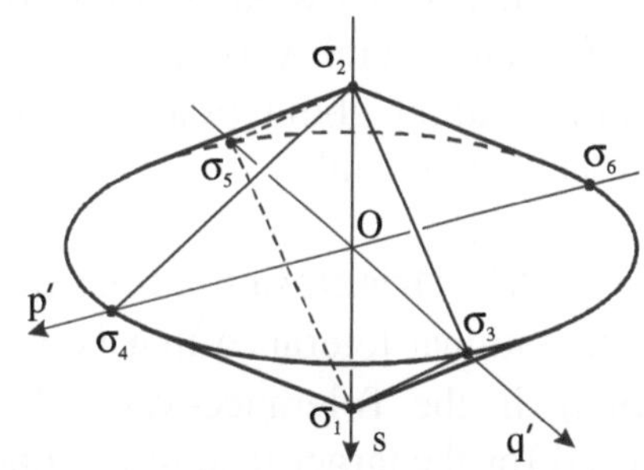

Figure 4. The Tolerance-Map for a round bar.[i]

When the cross-sectional dimensions of the bar in Fig. 2 are $d_x = d_y = d$, point $\sigma_{4'}$ in Fig. 3 moves inward to coincide with σ_4, $O\sigma_{6'} = O\sigma_4$, and rhombus $\sigma_3\sigma_{4'}\sigma_5\sigma_{6'}$ becomes the square $\sigma_3\sigma_4\sigma_5\sigma_6$. The T-Map for a size tolerance t on the length of a round bar is the dicone shown in Fig. 4. It is obtained by rotating $\Delta\sigma_1\sigma_2\sigma_3$, itself representing all the planes in the tolerance-zone of Fig. 2 which are parallel to the x-axis, a full turn about line $\sigma_1\sigma_2$.

Although for affine geometry it is only necessary that the positions for basis-points $\sigma_1 \ldots \sigma_4$ be independent, we have chosen the shape for the basis-tetrahedron shown in Fig. 1 for three principal reasons; additional details about these reasons are described in [Davidson, *et al.*, 2000]. First, the two isosceles triangles permit an orthogonal Cartesian frame of reference (p',q',s) to be overlain on the T-Maps in Figs. 3 and 4; the Cartesian frame is related to a subset of the coordinates (p,q,r,s) that describe the position of a plane in the tolerance-zone (p, q, and r represent direction ratios of a line that is normal to the plane and $-s$ is equal to the distance of the plane from the origin of the xyz-frame in Fig. 2). This permits the summing of T-Maps in assemblies and the computing of volumes with coordinates that have consistent units and consistent directions. Second, the orientational variations are decoupled from parallel variations in the tolerance-zone, thereby simplifying greatly the interpretation of a T-Map. Measures in the directions of the p'- and q'-axes of any T-Map are proportional to rotations about the y- and x-axes in the tolerance-zone, respectively ($p' = pd$, $q' = qd$ for a square or a round bar), yet parallel displacement in the tolerance-zone resides entirely with the s-coordinate in the T-Map. The third reason for the shape of the basis-tetrahedron is that it allows the use of traditional formulae for the computation of metric quantities (length, area, volume, etc.).

Because of our choice in positioning the basis-points σ_1, σ_2, σ_3, and σ_4 as in Fig. 1, the dipyramidal shape in Fig. 3 for a rectangular face in a tolerance-zone conforms with diagrams and spaces presented by others for representing size-tolerances. [Whitney, *et al.*, 1994] obtained the shape using an intuitive argument. [Roy and Li, 1999] used inequalities to establish a *variation zone* of acceptable ranges for the p, q, and s coordinates of any plane in the tolerance-zone relative to limiting positions. [Giordano, *et al.*, 1999] get a dipyramidal *deviation space* using the same method. [Teissandier, *et al.*, 1999] use Minkowski addition to obtain a singly compound dipyramid in deviation space for an assembly of two parts. However, the existing works contain neither recommendations about floating tolerance-zones nor methods for developing stackup relations in an assembly.

2.3. Orientational tolerances

An orientational tolerance that is applied to the target surface limits the values of p' and/or q' in the Tolerance-Map. For example, when an orientational tolerance t'' is specified for the target face of a round bar, the T-Map in Fig. 4 becomes truncated by a circular cylinder of radius t'' with its axis along $\sigma_1\sigma_2$.

3. AN ASSEMBLY OF TWO SQUARE OR TWO ROUND PARTS

In this section we develop the T-Maps for the accumulated tolerances for the end-faces of two square or two cylindrical parts that are stacked coaxially and have different lateral dimensions and size tolerances t_1 and t_2, as shown in Fig. 5. The functional feature for the assembly is the round end-face of Part 2, and the reference is Datum A. Each individual part is gaged with one end as reference and the other end as its target feature; the parts are stacked reference-to-target. In Figs. 6(*a*), (*b*), and (*c*) are shown half-sections (in the $q's$-plane) of three T-Maps: one for each of the target-faces on the individual parts and one accumulation-map for the target face on the assembly. The dimensions of the accumulation-map are in terms of t_1 and t_2. Figure 4 can be used to depict half the $q's$-section $\sigma_{1f}\sigma_{2f}\sigma_{3f}$ of the diconical functional T-Map for the target face; its dimensions are specified with t_f, a value that depends on the functional requirements that are desired for that face in the completed assembly. For comparisons to have validity, all of these T-Maps must be scaled to be conformable with a *single* reference tetrahedron. Since functional requirements for the target face in the assembly control the assignments of tolerances to the individual parts, we select *its* T-Map, and the T-

Map for Part 2, to be constructed on the basis tetrahedron $\sigma_1\sigma_2\sigma_3\sigma_4$ in Figs. 1, 3, and 4 in which the aspect ratio is unity. But the T-Map for Part 1 has an aspect ratio less than unity because the larger lateral dimension reduces the ratio of orientational variations to positional variations in its tolerance-zone relative to the ratio for the tolerance-zone for Part 2. The accumulation T-Map for the assembly is formed as the Minkowski sum of two square dipyramids (or two dicones) represented by Figs. 6(*a*) and 6(*b*); the result is the singly compound dipyramid (or dicone). Half of its *q's*-section is shown in Fig. 6(*c*).

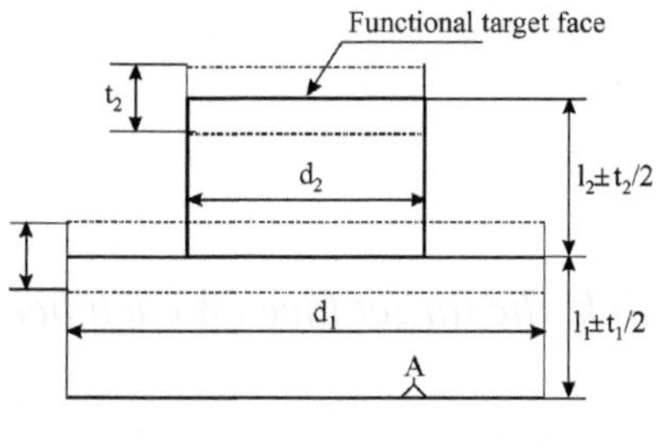

Figure 5.[i] *A central assembly (no offset) of Parts 1 and 2 with their individual tolerance-zones.* $t_1=t_2$; *the functional feature is the round face on Part 2.*

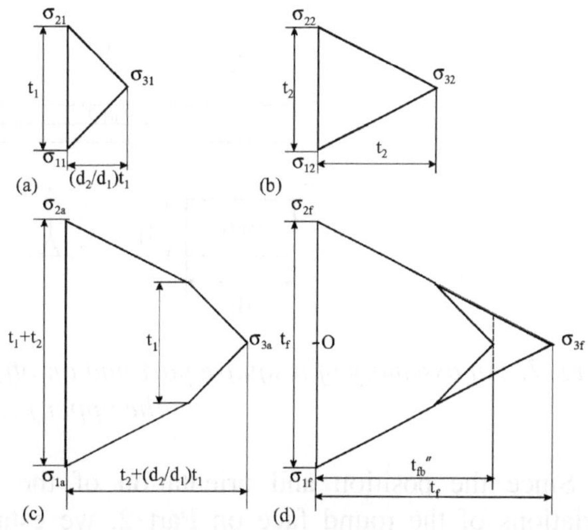

Figure 6.[i] *(a) and (b) Central half-sections of the T-Maps for each of the parts in Fig. 5. (c) A half-section of the accumulation map, point* σ_{3a} *corresponding to the* CW*-most plane. (d) A half-section of the functional T-Map and, inscribed in it, the accumulation map (heavy solid line).*

To avoid specifying excessively tight tolerances on the individual parts, we recommend that the accumulation T-Map be scaled up/down until it can be inscribed in the functional map; the result is shown in Fig. 6(*d*) as a half-section. The stackup condition relating size tolerances then becomes

$$t_f = t_1 + t_2 . \tag{2}$$

4. A NON-COAXIAL ASSEMBLY

Consider the assembly of a square part and a link that contains an offset b between its two end-faces (Fig. 7). Size tolerances t_1 and t_2 are specified for the dimensions that separate the faces on the parts. Our objective is to obtain the accumulation T-Map for the assembly when the functional target surface is the top of the boss of diameter d_2.

Because of the offset b, which acts as a lever-arm, there is an amplification of orientational variations in the tolerance-zone of Part 1. The accumulation of tolerances with such an offset was analysed also by [Laperrière and Lafond, 1999] using closed kinematic chains.

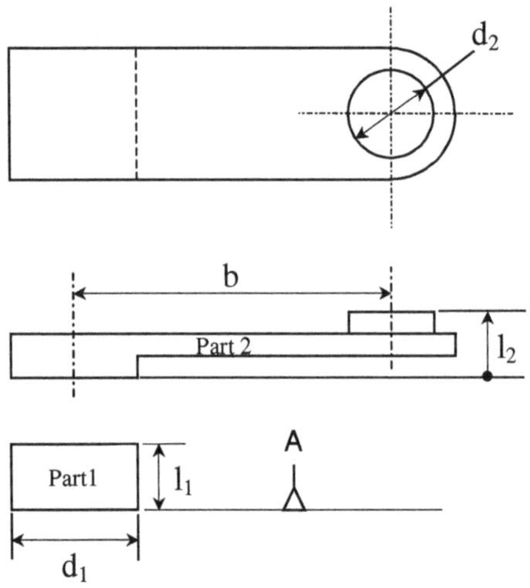

Fig. 7. An assembly of a square part and an offset link, the target face on each being the upper face.

Since the position and orientation of the upper surface on Part 1 affects the variations of the round face on Part 2, we establish three tolerance-zones and three coordinate frames, as shown in Fig. 8. The zone for Part 2 lies within the accumulation zone (labeled t_a), but is drawn with exaggerated geometry for clarity in defining variables. The j-frame is located in the tolerance-zone for Part 1. The k-frame is attached to Part 2, but the i-frame is attached to Part 1. The tolerance-zone for Part 2 and the k-frame are shown rotated an angle, $-q_1$. The nominal planes within the three tolerance-zones are labeled σ_N. The accumulation zone will be described in the i-frame. When the homogeneous transformation $[A_{ik}]$ is used to locate the k-frame in the i-frame, its inverse

$$[A_{ik}]^{-1} = \begin{bmatrix} 1 & 0 & -p_1 & (-bp_1q_1 - p_1s_1) \\ 0 & 1 & -q_1 & (-bq_1^2 - q_1s_1) \\ p_1 & q_1 & 1 & (bq_1 + s_1) \\ 0 & 0 & 0 & 1 \end{bmatrix} \tag{3}$$

is used to transform the coordinates of a plane from the k-frame to the i-frame. When $(p_2, q_2, 1, s_2)$ are the coordinates for any plane in the tolerance-zone for Part 2, as expressed in the k-frame, then its coordinates in the accumulation zone are obtained from $[A_{ik}]^{-1}$ as

$$[p_i\ q_i\ 1\ s_i] = [p_2\ q_2\ 1\ s_2][A_{ik}]^{-1}$$
$$= [p_1 + p_2\quad q_1 + q_2\quad 1\quad s_1 + bq_1 + s_2], \tag{4}$$

in which small quantities of second order and higher are neglected.

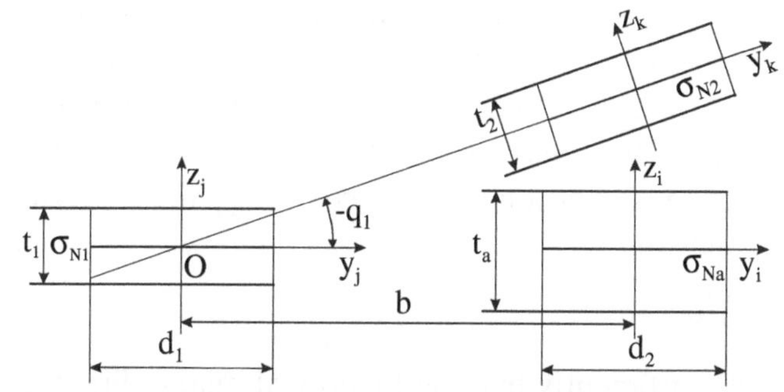

Figure 8.[i] Tolerance-zones and coordinate frames for the assembly in Fig. 7.

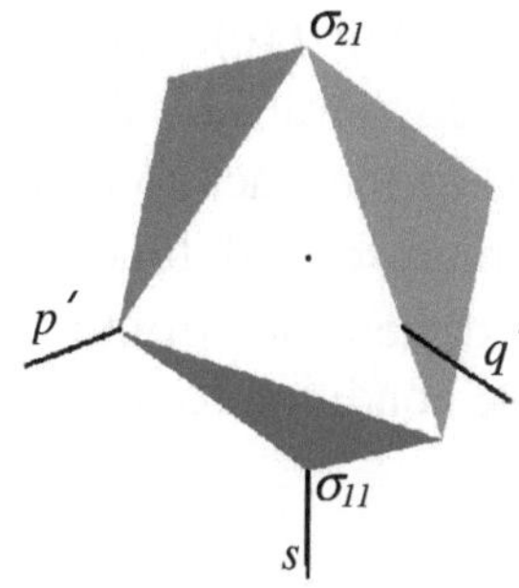

Figure 9. A modified T-Map for Part 1 in Fig. 7 which accounts for the offset.

To convert the coordinates (4) to the Cartesian coordinates for T-Maps, we take b=0 so that the T-Maps reduce to those in Fig. 6. Comparison of these coordinates to the dimensions in Fig. 6 shows that $p'_i=d_2p_i$ and $q'_i=d_2q_i$ (i=1,2). Then, so long as we neglect higher order terms, the second of eqns (4) leads to the Cartesian coordinates p'_i, q'_i, and s_i, in the i-frame, i.e.

$$(p'_i, q'_i, 1, s_i) = (d_2p_1 + d_2p_2\ ,\ d_2q_1 + d_2q_2\ ,\ 1\ ,\ s_1 + bq_1 + s_2) \tag{5}$$

that locate points in the accumulation T-Map. And they are also the elements of the Minkowski sum of T-Maps for Parts 1 and 2. Clearly p'_2, q'_2, and s_2 are the elements (coordinates) for the T-Map of Part 2, and this T-Map is a right-circular dicone with an aspect ratio of unity, as shown in Fig. 4. What remains, coordinates

$$(d_2p_1,\ d_2q_1,\ s_1 + bq_1), \tag{6}$$

then must be the elements for the T-Map of Part 1 that includes the effect of the offset b. When b=0, these coordinates reduce to those for the right-square dipyramid (Fig. 3 in which $d_2 = d_1 = d$) with its two principal vertices on $\sigma_{11}\sigma_{21}$. Consequently, for $b\neq 0$, the modified T-Map for Part 1 can be obtained by displacing vertically every point of the right square dipyramid by an amount proportional to the q'-coordinate of the point. When, additionally, we make this modified T-Map conformable with the target surface by reducing every p'_1- and q'_1-coordinate by the ratio d_2/d_1, the result is an oblique rhombic dipyramid (Fig. 9) that is symmetrical about its geometric center.

The accumulation map for the parts in Fig. 7 is obtained from the Minkowski sum of the right-circular dicone in Fig. 4 and the oblique dipyramid in Fig. 9, both shown with dashed lines in the $q's$ cross-section of Fig. 10. The full $q's$-section of the

Minkowski sum is shown with the heavy line. Half of the *p's* -section is Fig. 6(*c*), but no contact occurs with the functional map. Also shown (narrower line) in Fig. 10 is a section of the circumscribed functional T-Map, truncated on the left, that touches the accumulation map at two generator-lines.

For b=0, half of the *q's*-section of the accumulation map for the two parts is represented by Fig. 6(*c*), but, as offset b is allowed to increase, point σ_{3a} in Figs. 6(*c*) and (*d*) moves downward by the amount bt_1/d_1. A small amount of offset, i.e. $b \leq b_0$, causes no change to the stackup relation $t_f = t_1 + t_2$ in eqn (2) because point σ_{3a} does not contact the circumscribed functional map. But, when $b=b_0$, point σ_{3a} touches line $\sigma_{1f}\sigma_{3f}$. Using dimensions from Fig. 6 and applying proportion to $\Delta O\sigma_{1f}\sigma_{3f}$ in Fig. 6(*d*), we find that the condition for touching corresponds to

$$b_0 = \tfrac{1}{2}(d_1 - d_2). \tag{7}$$

When $b>b_0$, point σ_{3a} moves downward sufficiently that the functional map in Fig. 6(*d*) must enlarge in order to remain circumscribed. The results are the *q's*-sections of the T-Maps that are shown in Fig. 10. Since the downward displacement of point σ_{3a} is controlled by orientational variations at the top face on Part 1, we can generalize the amount bt_1/d_1, which appears in the shorter dimension labeled in Fig. 10, to bt''_1/d_1 (Fig. 11), thereby including coupling between the orientational tolerance t''_1 at Part 1 and the size-limit for Part 2.

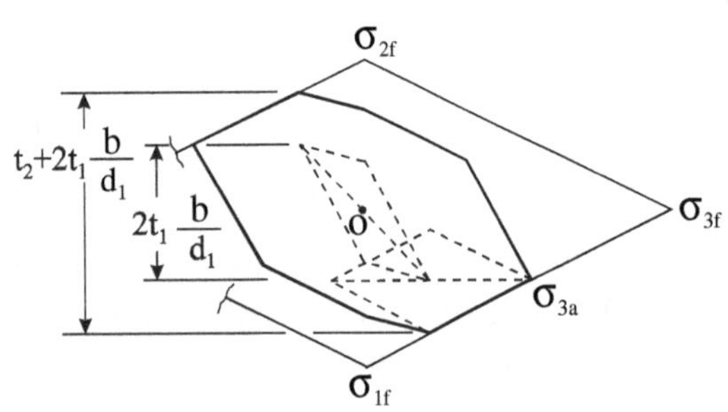

Figure 10. The q's-section of the accumulation map for the parts in Fig. 7 when $b > b_0$. Drawn for $t_1 = t_2$.

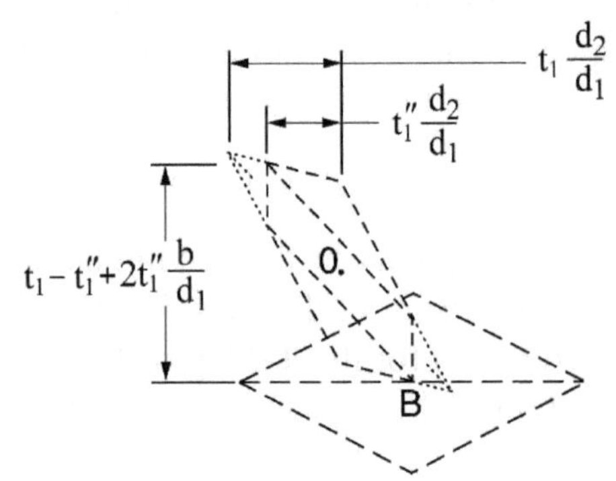

Figure 11. The The q's-section of the two operands from Fig. 10, now including an orientational tolerance t''_1.

When an orientational tolerance t''_1 is applied to the target face on Part 1, there are two modifications to its T-Map. First, the rhombic base of the dipyramid in Fig. 9 is less oblique than when $t''_1=t_1$. Second, the oblique dipyramid becomes truncated by a square prism of side $d_2t''_1/d_1$ with its axis along $\sigma_{11}\sigma_{12}$. Both effects are illustrated in Fig. 11 with the the *q's*-sections of the two operands that are used in the Minkowski sum for the accumulation map when $t''_1<t_1$. The T-Map for Part 1 then is the volume of the dipyramid that remains inside the truncating prism. For the T-Map of Part 1 (Fig.

11), the offset b causes all points (such as B) having a q'-coordinate of $d_2t''_1/d_1$ to move down by the amount bt''_1/d_1. The amount of offset, $b=b_0$, for which point B touches line $\sigma_{1f}\sigma_{3f}$ in Fig. 6(d) remains that in eqn (7). Consequently, the accumulation of variational possibilities in the axial direction can be obtained by increasing the size of the functional dicone in Fig. 6(d) by twice the amount that B moves down less twice the amount it moves to first contact line $\sigma_{1f}\sigma_{3f}$. Therefore, the stackup condition becomes

$$\left.\begin{aligned} t_f &= t_1 + t_2 + 2\left(b\frac{t_1''}{d_1} - b_0\frac{t_1''}{d_1}\right) \\ &= t_1 + t_2 + t_1''\left(2\frac{b}{d_1} + \frac{d_2}{d_1} - 1\right). \end{aligned}\right\} \quad (b > b_0) \qquad (8)$$

We note that, when $t_1'' = t_1$, and offset $b=b_0$, eqn (8) reduces to $t_f = t_1 + t_2$, i.e. eqn (2) for the accumulation map in Fig. 6(c).

5. CONCLUSION

We have presented selected features of a new mathematical model for geometric tolerances, and we have illustrated some of these with an example. The new model is based on a convex volume of hypothetical points which we call a Tolerance-Map (T-Map). T-Maps can be resolved into component-regions and manipulated geometrically both to give T-Maps for assemblies and to illustrate the quantitative effect of interdependencies that are stipulated in the [ASME Standard, 1994].

We regard T-Maps, their use, and their development to be a local model for tolerances. For implementation in software, there is also a need for a global model to treat such issues as sufficiency of datums, constraint and freedom, and an algebra for degrees of freedom. A first stage of this work can be found in [Kandikjian *et al.*, 2001].

ACKNOWLEDGEMENTS

The authors are grateful for funding provided to this project by National Science Foundation Grant #DMI-9821008. They also are grateful to Mr. Michael Marchak, a student at Arizona State University, whose work led to the correct form of eqn (8).

REFERENCES

[ASME Standard, 1994] ASME Y14.5M.; "Dimensioning and Tolerancing"; The American Society of Mechanical Engineers, NY.

[Chase, *et al.*, 1998] Chase K., Gao J., Magelby S., Sorensen C. (1998). "Including geometric feature variations in tolerance analysis of mechanical assemblies"; In: *IIE Transactions*, **28**, pp 795-807.

[Clément, *et al.*, 1996] Clément, A., Rivière, A., and Serré, P.; "The TTRS: a common declarative model for relative positioning, tolerancing and assembly"; In: *MICAD Proc.*, **11**, No. 1-2, pp. 149-264.

[Coxeter, 1969] Coxeter, H. S. M.; *Introduction to Geometry*; 2nd ed. Wiley.

[Davidson, *et al.*, 2000] Davidson, J.K., Mujezinović, A., and Shah, J. J.; "A New Mathematical Model for Geometric Tolerances as Applied to Round Faces", In: *CD Proc., ASME Des. Technical Conf's.*, #DETC00/DAC-14249; Baltimore, MD.

[Davidson and Shah, 2000] Davidson, J.K. and Shah, J.J.; "Geometric Tolerances: A New Application for Line Geometry and Screws"; In: *CD Proc. for the Symp. Commem. the Legacy, Works, and Life of Sir Robert S. Ball*, Cambridge, UK.

[Giordano, *et al.*, 1999] Giordano, M., Pairel, E., and Samper, S.; "Mathematical representation of tolerance zones"; In: *Proc., 6th CIRP Int'l Seminar on Computer-Aided Tolerancing*, pp. 177-86.

[Gossard, *et al.*, 1988] Gossard, D. C., Zuffante, R. P. and Sakurai, H.; "Representing dimensions, tolerances, and features in MCAE systems"; In : *IEEE Comp. Gr. and Appl.* **8**(2), pp. 51–59.

[ISO 1101, 1983] "Geometric tolerancing—Tolerancing of form, orientation, location, and run-out—Generalities, definitions, symbols, and indications on drawings"; International Organization for Standardization.

[Kandikjian *et al.*, 2001] Kandikjian, T., Shah, J.J., and Davidson, J.K.; "A mechanism for validating dimensioning & tol. schemes in CAD systems"; *CAD*, (in press).

[Laperrière and Lafond, 1999] Laperrière, L. and Lafond, P.; "Tolerance analysis and synthesis using virtual joints"; In: *Proc., 6th CIRP Int'l Seminar on Computer-Aided Tolerancing*, pp. 405-414.

[Mujezinović, 1999] Mujezinović, A.; "A new mathematical model for representing geometric tolerances"; MSc thesis, Arizona State Univ., Tempe, AZ, May, 1999.

[Requicha, 1983] Requicha, A. A. G.; Toward a theory of geometric tolerances; *Int. J. of Robotics Research*, **2**(4), pp. 45–60.

[Shah, *et al.*, 1998] Shah J. J., Yan Y., Zhang B-C.; "Dimension and tolerance modeling and transformations in feature based design and manufacturing"; In: *J. Integrated Manufacturing*, **9**, N5, pp. 475-488.

[Teissandier, *et al.*, 1999] Teissandier, D., Delos, V., and Couetard, Y.; "Operations on polytopes: application to tolerance analysis"; In: *Proc., 6th CIRP Int'l Seminar on Computer-Aided Tolerancing*, pp. 425-434.

[Turner, 1990] Turner, J. U.; "Exploiting solid models for tolerance computations"; In: *Geometric Modeling for Product Engineering*, pp. 237–258, North-Holland.

[Whitney, et al., 1994] Whitney, D. E., Gilbert, O. L., and Jastrzebski, M.; "Representation of geometric variations using matrix transforms for statistical tolerance analysis in assemblies"; In: *Research in Engineering Design*, **6**, pp. 191-210.

---

[i] These figures were drawn by A. Mujezinović

Curves for Profile Tolerance zone boundaries

T. M. Kethara Pasupathy *and* ***Robert G. Wilhelm***
Center for Precision Metrology and Intelligent Manufacturing
Department of Mechanical Engineering and Engineering Science
The University of North Carolina at Charlotte
9201 University City Blvd., Charlotte, NC 28223-0001
Email: rgwilhel@uncc.edu

Abstract: This paper presents a representation scheme and tools for the analysis of profile tolerances. Innovations in the design of small mechanical parts and complex function parts have made new requirements for profile tolerance definitions. In particular, design intent sometimes requires that the outermost boundaries of profile tolerance zones vary in shape along the surface of single features. A new scheme is required to empower the designer to geometrically specify design intent unambiguously.

The first part of the paper describes designer requirements and application of parametric curves. B-splines are shown to be a suitable scheme for specifying profile tolerance zones for two-dimensional features. We then present a procedure that can be used to verify the conformance of actual features specified by such a scheme.
Keywords: profile tolerances, tolerance zone, varying boundary, parametric-curves

1. INTRODUCTION

Profile tolerances are used for specifying the tolerance zones for features that can be defined by both implicit and freeform curves and surfaces [Standard, 1994]. The specification imparts size, form and orientation of the tolerance zone, or in other words, the zone is fixed with respect to the datum reference frame [Al Neu, 1995]. Profile tolerance zones are obtained by offsetting every point of the feature as per the specified tolerance value [Standard, 1994]. The zone that is obtained by offsetting differs from the zone generated by other types of tolerances (like form, size etc.,) in the sense that the boundaries can be mathematically obtained by translation and scaling. It should be noted that the translation of a free form curve might not generate a uniform tolerance zone as shown in Figure 1. Offsets seem to be a better choice for defining the zone boundaries as they mimic the nominal feature in many cases. Limiting our discussion to two-dimensional (2-D) parts in this paper, offsets for 2-D parts can be well defined using n-offsets as described in [Wayne, 1983]. These n-offsets are rigid and cannot convey the designer's intent when singularities are present. Problems that arise with n-offsets for cusps and self-intersection are described in [Wayne, 1983]. [Wilhelm, 1996] shows that in the composition of adjacent zones, singular points are present and suitable zones corresponding to these points are required.

P. Bourdet and L. Mathieu (eds.),
Geometric Product Specification and Verification: Integration of Functionality, 45-54.

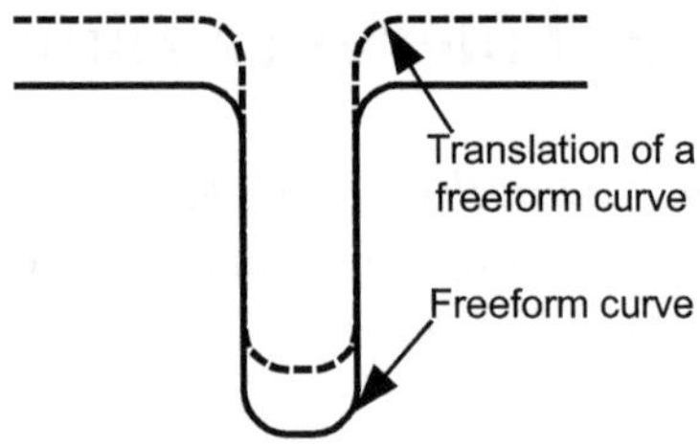

Figure 1; Translation of a freeform curve. Note that some sections have no zones

Another requirement that departs from the current specification is the need for non-uniform boundaries (tolerance zones with varying width) to specify profile tolerances on a single feature. Here again, n-offsets cannot convey the designer's intent. In the book by [Henzold, 1997], examples cases are indicated. To represent a non-uniform tolerance zone for features of aerofoil parts, [Cardew et al., 1990] use B-splines. They apply it to what they call "datum form" tolerance zones. [Utpal, 1998] suggest that the datum form tolerance is similar to a profile tolerance zone. In their paper, a concrete example that conveys the need for a varying width of the datum form tolerance zone is provided.

In section 2, details from examples are provided to capture the specification requirements. These examples further illustrate the ambiguities that arise from the composition of the zones and the need for varying the width of the tolerance zone. Next we present the applicability of parametric curves for specifying profile tolerances. The form of the parametric curve is determined by the basis function. Only certain basis functions among the available set of basis functions can meet the specification requirements. The basis functions that qualify based on their properties are listed. We also present a short review on a representation scheme that considers the singular points and another that uses a B-spline scheme for specification of tolerances. When using parametric curves to specify the zones, it might not be as easy as in the case of explicitly defined curves to classify the points [Mortenson, 1997] i.e., to determine if actual points (measured points of an actual feature) are contained in the zone. Following the review on the representation scheme, we present a conformance verification procedure for 2-D parts. Conclusions are then presented.

2. PROFILE TOLERANCE DEFINITIONS

The ANSI Y 14.5M [Standard, 1994] defines tolerance zones with respect to the geometry of features. Basically this is a division of a part into features that can be described implicitly or using free form curves. Though this division helps in the application of the syntax and realizing the semantics of regular features, singular points at the intersection of two adjacent features are not discussed. The zones corresponding to the singular points features are critical in several cases and are to be properly addressed for an unambiguous definition of the part. In the paper [Wilhem, 1996] illustrates such a case on the application of the ANSI Y 14.5M offset definition. A suspension element of a hard disk drive (small part) is used as an example to highlight the importance of providing an unambiguous specification. Further examples to illustrate the ambiguous zones due to the singular points are shown in Figure 2. In Figure 2, solid lines represent the features and the spaces between the dashed lines show the tolerance zones. Singularities are also discussed extensively in [Rossignac, 1986]. Furthermore, The boundary problem discussed in [Walker 1993] can be attributed to the singularities.

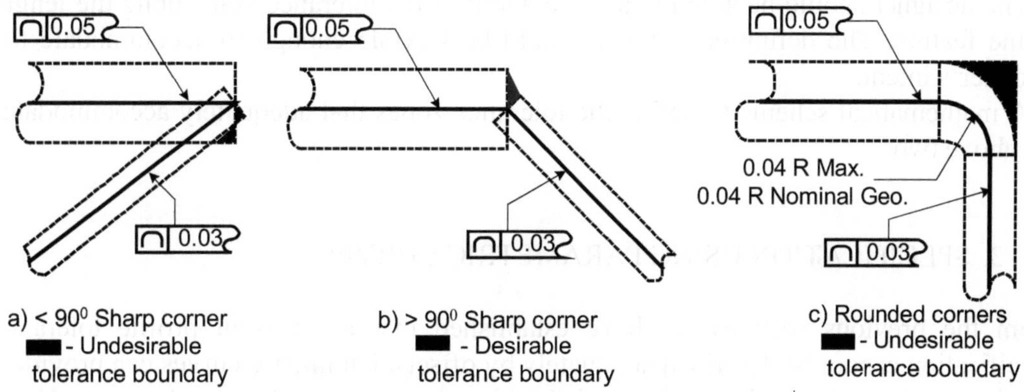

Figure 2; Ambiguity from composing adjacent zones

The paper by [Cardew et al., 1990] explains that a varying tolerance zone width on a single feature is useful for specifying the allowable variation in their example part - a complex feature on the surface of a turbine blade. While function does not always depend on the varying zone, it is economical to produce parts that exhibit varying form or to accept a range of products that obtain signatures from the manufacturing process. In the turbine blade example, some sections of a feature can have a larger tolerance width that does not alter the function while some sections need to have a smaller width, as it can be critical to the function. The authors say that such a requirement specification is implemented using B-splines, as it is necessary to have a smooth transition of outermost tolerance zone boundaries for the different sections. In their system, the designer knows the functionality and provides inputs to appropriately shape the zones.

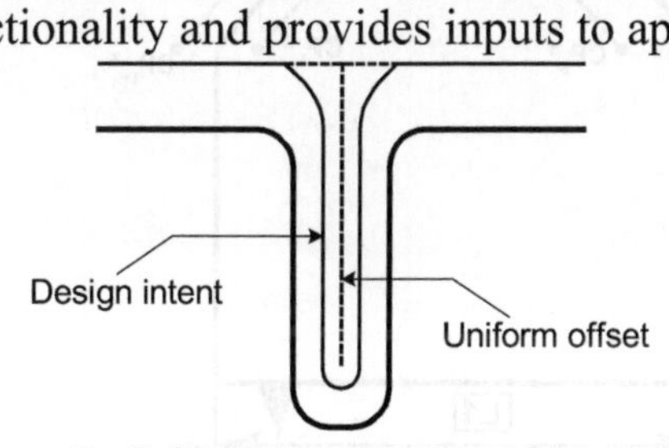

Figure 3; Self-intersection problem in concavity

Consider the concavity in Figure 3. The n-offset produces a zone as shown by the dashed line, while a possible design intent can be as shown by the continuous thin line. In the design intent, the self-intersection of the tolerance zone is not expected but a larger width at certain sections of the feature is required.

From the above discussions, a summary of requirements that the profile tolerance definition should satisfy is as follows.

2.2 Requirements of an alternate profile tolerance definition

1. A suitable specification scheme for the tolerance zone that includes zones corresponding to the singularities in the composition of adjacent zones. The designer is the best judge to decide on the zones corresponding to the singular points.
2. The designer should be able to vary the width of the tolerance zone along the length of the feature. The definition scheme should be flexible enough to accommodate the designer's intent.
3. A mathematical scheme to define the tolerance zones that adequately accommodates the above two.

3. SPECIFICATION USING PARAMETRIC CURVES

From the previous sections we have established that the current profile tolerance specification cannot be described adequately by offsets. Parametric curves can provide a suitable mathematical scheme that satisfies the requirements discussed above. These curves can be used to shape the tolerance zone boundaries.

Parametric curves [Farin, 1993] can be generated with control points and using suitable basis functions. In the profile tolerance specification, the designer can provide the control points with respect to the nominal geometry. The curve or surface that is generated using the basis function specifies the extreme boundaries. Consider Figure 4.

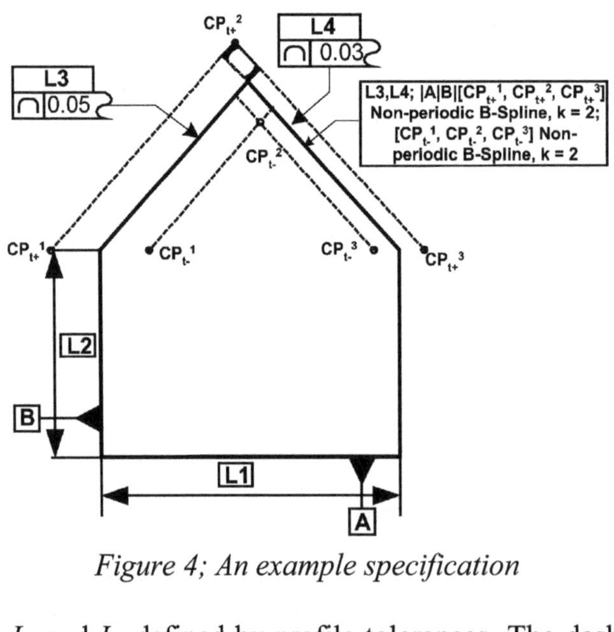

Figure 4; An example specification

It has two features L_3 and L_4 defined by profile tolerances. The dashed lines show the zone implied by the current standard. To specify the t_+ boundary, control points CP_{t+}^{1},

CP_{t+}^{2} and CP_{t+}^{3} are chosen. On applying the basis function provided in the feature control frame, a curve describing the t_+ boundary is obtained. Similarly the t_- zone can be selected with CP_{t-}^{1}, CP_{t-}^{2} and CP_{t-}^{3}. In the case of unilateral tolerance, one set of control points should specify the nominal feature. Note the shaded area that is now included in the specification. To indicate the profile tolerance specification in a drawing, one may use a scheme as shown in feature control frame supplementing the regular callouts.

The above procedure is best realized in a CAD environment, as the basis function is a recursive formulation. Further, CAD systems provide spline functions that assist in visualization. Parametric curves are economical in handling the number of points to be stored. The control points are those that are to be stored. For subsequent processing, such as in the verification of conformance of actual features, as many points as required on the tolerance boundary can be generated. As mentioned earlier, several parametric basis functions are available and a few can be used for blending. In our study of different basis functions, some appeared more applicable based on their properties. The results are shown in Table 1. The important properties considered are

a) The convex hull,
b) The variation diminishing,
c) Ease of use and,
d) Inclusion of the control points in the final (generated) curve.

The first property ensures that the curve is contained in the convex hull of the specified points. The variation diminishing property controls the number of oscillations of the generated curve. (For further information on the discussion on the properties, the readers are asked to refer to [Farin, 1993].)

Interpolating function	**Convex hull**	**Variation diminishing**	**Ease of Computing /Programming**	**Inclusion of end points in the final curve**
Lagrange	✗	✗	✗	✓
Hermite	✗	✗	✗	✓
Bezier	✓	✓	✓	✓
B-spline Any type, k = 2	✓	✓	✓	✓
B-spline Type: Periodic, k > 2	✓	✓	✓	✗
B-spline Type: Non-periodic k > 2	✓	✓	✓	✓
B-spline Type: Non-uniform k > 2	✓	✓	✓	✓

Table1; Properties of various interpolating functions.

In general, the B-spline and Bezier form were found most suitable. Since Bezier curves are special cases of B-splines, the scope can be narrowed down to B-splines. The form with B-spline and order = 1, k = 2, includes all the control points. Higher order of the curves, k > 2, can also be used to obtain a smooth transition. Multiple control points, for example [P_{t+}^{1}, P_{t+}^{2}, P_{t+}^{2}, P_{t+}^{3}, P_{t+}^{3}, and P_{t+}^{4}], pull the curve closer to those control points. Of these, the periodic form of the knot vectors does not include the first and the last points from the given set of points. Hence the non-periodic and non-uniform are well suited for k > 2. These properties are pertinent because the designer has to interactively locate the points that form the zone boundary.

Further, the boundary of the zone should be smooth i.e., should not oscillate.

4. REPRESENTATION SCHEMES

Representation schemes for tolerance zones are broadly classified as parametric and non-parametric [Requicha, 1984]. An unambiguous model is the essence of the non-parametric scheme. In such an unambiguous representation scheme [Rossignac, 1986], the singularity issue is an important factor and it has been addressed. This is the only theory that considers the zones for singular points. Due to the limitation of the offset operator, when adjacent zones are specified with different tolerance values, bulbous extensions are present as in Figure 5. The theory considers only slow variations of the actual surface. However slivers can be present in the actual part while the designer's intent can be one as shown in Figure 4 that accepts slivers within the zone.

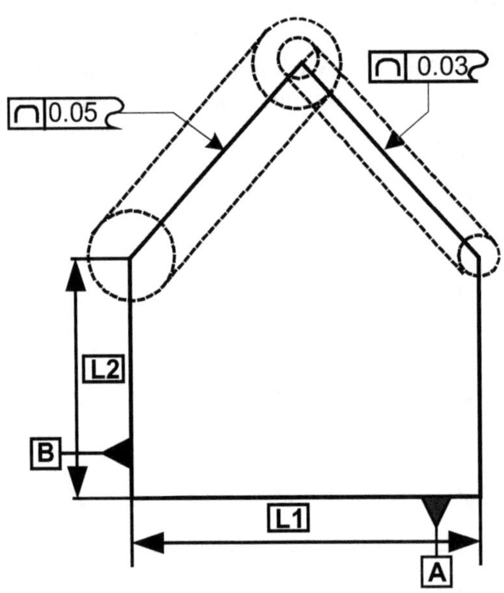

Figure 5; Zone obtained from 'non-parametrically defined zones'

[CARDEW et al., 1993], for aerospace blades, used a representation scheme based on parametric zones that implements the designer's tolerances. The input points provided are woven into a B-spline (k = 2) surface representing the zone boundary. They show that by departing from the standards, they are able to allow variations to features that have a non-uniform outer boundary.

5.CONFORMANCE VERIFICATION OF ACTUAL FEATURES

In previous sections, we have demonstrated that B-splines can provide an alternate scheme to specify profile tolerance zones. By doing so, it is also essential to show that the scheme is also useful for subsequent processing. Below, a procedure for conformance verification is presented.

As per [Standard, 1994], a feature specified by profile tolerances is checked for conformance by verifying that the actual points lie within the zone. In the case of B-splines curves, generating many points on the curve can approximate the zone boundary. Geometric algorithms can then be applied in the verification procedure.

Consider Figure 6. It shows a set of points (circles) obtained from the basis function. The points ($P_{t+} = [P_{t+}^1 \ldots P_{t+}^i \ldots. P_{t+}^{n1}]$) and ($P_{t-} = [P_{t-}^1 \ldots P_{t-}^i \ldots. P_{t-}^{n2}]$) form the inner and outer profile tolerance zone boundaries respectively. A case of the actual feature is shown inside the zones. Let $P_a = [P_a^1 \ldots\ldots P_a^n]$, be the actual data points that are collected and are to be evaluated for conformance.

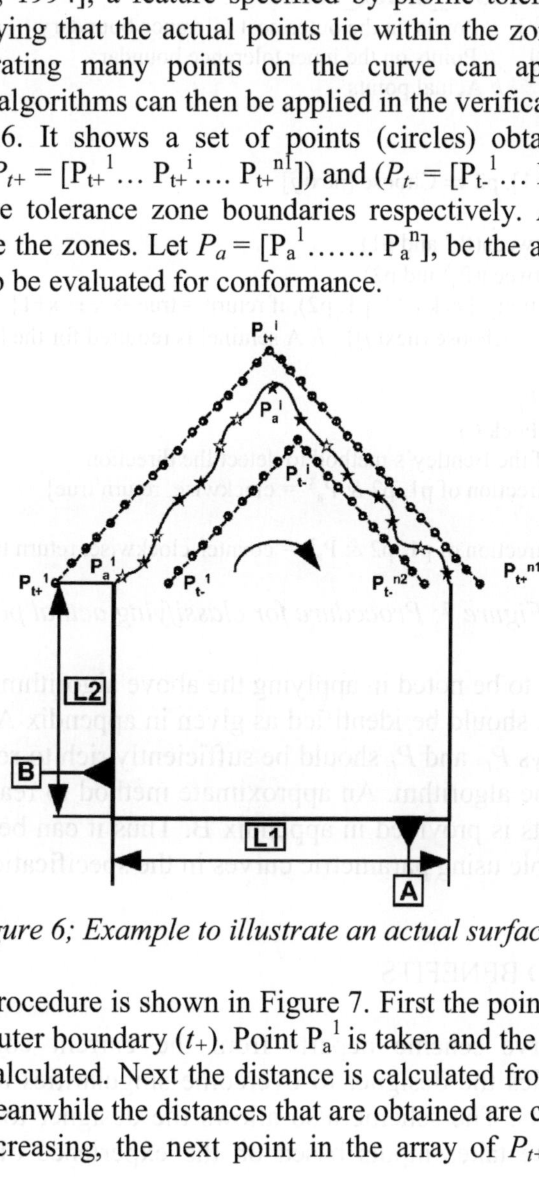

Figure 6; Example to illustrate an actual surface

The verification procedure is shown in Figure 7. First the points are checked to see if they fall below the outer boundary (t_+). Point P_a^1 is taken and the vector distance from it to the point P_{t+}^1 is calculated. Next the distance is calculated from the point P_{t+}^2. We continue doing this. Meanwhile the distances that are obtained are compared. As long as the distance keeps decreasing, the next point in the array of P_{t+} is taken up. If the distance starts to increase, say at the point P_{t+}^{x}, then Bentley's method [Sedgewick, 1989] is applied to the points P_{t+}^{x-1}, P_{t+}^{x}, and P_a^1. If the direction taken to reach P_a^1 from P_{t+}^{x-1} via P_{t+}^{x} is clockwise and then the actual point is inside the outer boundary. Otherwise the point is outside of the outer boundary and nonconformance is indicated. Next the actual data point P_a^2 is taken and the distance is calculated between P_{t+}^{x} and P_a^2 and so on until the nearest point to P_a^2in the array P_{t+} is reached. The direction test is applied again. The procedure is repeated until all of the actual data points are checked. This procedure is repeated with P_a^1 and P_{t-}^1. However the direction taken to reach P_a^1 from P_{t-}^{x-1} via P_{t-}^{x} should be counter-clockwise, where P_{t-}^{x-1} is the nearest point to P_a^1. If an actual point is on the curve, it should be resolved as shown in Bentley's method.

Procedure: **Actual_feature_inside_zone**
P_{t+} = $[P_{t+}^1 \ldots P_{t+}^i \ldots P_{t+}^{n1}]$ // Points on the outermost tolerance boundary
P_{t-} = $[P_{t-}^1 \ldots P_{t-}^i \ldots P_{t-}^{n2}]$ // Points on the inner tolerance boundary
P_a = $[P_a^1 \ldots P_a^n]$ // Actual points
Dist = d1, d2
Choose ⇐ P_{t+}
Points p1 ⇐ Choose [j = 1], p2 ⇐ Choose [next j]
For each x = 1 to n
    d1 ⇐ distance_between(P_a^x and p1)
    d2 ⇐ distance_between(P_a^x and p2)
    If (d2 > d1) {Direction_check (P_a^x, p1, p2), if return = true → x ⇐ x+1}
    Else {p1⇐ p2, p2 ⇐ Choose [next j]} // A sentinel is required for the last element in Choose
End For each
Repeat with Choose ⇐ P_{t-}
Procedure: **Direction_check ( )**
{ // Implementation of the Bentley's method to detect the direction
If {Choose = P_{t+} & if direction of p1, p2 & P_a^x = clockwise, return true}
Else {break}
If {Choose = P_{t-} & if direction of p1, p2 & P_a^x = counter-clockwise, return true} Else {break}
}

Figure 7; Procedure for classifying actual points

Two rules need to be noted in applying the above algorithm. The direction from the first to the last point should be identified as given in appendix A. Secondly the number of points in the arrays P_{t+} and P_{t-} should be sufficiently rich to represent the entire curve for the success of the algorithm. An approximate method to reach the neighborhood of the actual data points is provided in appendix B. Thus it can be shown that subsequent verification is possible using parametric curves in the specification.

6. COSTS AND BENEFITS

The parametric curve scheme departs from the current convention of specifying tolerances but enables the designer to overcome singularities and ambiguities that are critical to some parts. The scheme also allows the designer to completely specify the zones. This scheme takes inputs based on the experience of the designer and to a certain extent the skill in locating the points that best describe the design requirements. Furthermore, say for a process planner, the specification can be supplemented with conventional callouts to provide intuition in reading the blueprints. An example is provided in Figure 4. In the industrial practice, the actual points are suitably fitted and compared with templates even with the conventional specifying scheme. Using the parametric scheme can help in automating subsequent processing. However in checking for conformance, other methods, such as curve fitting techniques or converting the B-spline form to explicit equations, are available. Hence there is scope for further research.

7. CONCLUSIONS

Problems with the current profile tolerance definition were identified. They are to be addressed when function is critical. They were, associated with singularities, ambiguities in composing adjacent zones of different widths and the lack of varying the width of the tolerance zone (non-uniform zone) of a feature. A scheme using parametric curves, specifically non-periodic or the non-uniform B-splines was shown to be flexible and useful to overcome the difficulties. The scheme takes inputs from the designer to precisely meet the functional requirements. Subsequent processing such as conformance verification can be automated. To demonstrate the possibilities for subsequent processing a procedure for classifying actual points was presented.

8. ACKNOWLEDGEMENTS

The authors would like to thank Dr. Greg Hetland for his opinions and suggestions. This material is based upon work supported by the National Science Foundation under Grants EEC 9811556 and DMI-9457168. We are most appreciative of the facilities provided by the NSF IUCRC Center for Precision Metrology at UNC Charlotte.

9 REFERENCES

[Al Neu, 1995] Al Neumann; *Geometric Dimensioning and tolerancing workbook*; The American Society of Mechanical Engineers; 1995

[Cardew et. al 1993] M.J. Cardew –Hall, Lebans T, G. West and P. Dench; "A method of representing dimensions and tolerances on solid based freeform surfaces" In: *Robotics and computer –Integrated manufacturing*; Vol. 10; No. 3; pp. 223-234; 1993

[Farin 1993] Gerald Farin; Curves and Surfaces for Computer Aided Geometric Design; 3ed; 1993; ISBN 0-12-249052-5

[Henzold 1997] Henzold, G; "Tolerances of Profiles"; In: *Handbook of geometrical Tolerancing*; pp. 309-310; John Wiley & Sons; ISBN 0-471-94816-0

[Requicha 1984] Requicha A.A.G; "Representation of tolerances in solid modeling: Issues and alternative approaches", In: *Solid Modeling by Computers: from Theory to Application; M.S. Picket and Boyce, Eds*; Plenum Press; 1984; pp. 3-22

[Rossignac 1986] Rossignac, J.R and Requicha, A.A.G.; "Offsetting operations in solid modelling"; In: *Computer Aided Geometric Design*; Vol. 3; pp. 129-148; 1986
[Mortenson 1997] Mortenson M.E.; *Geometric Modeling*; 2e; Wiley Computer Publishing; 1997
[Sedgewick 1989] Sedgewick R; *Algorithms*; 2e; Addison Wesley publishers; 1989
[Standard 1994] *Mathematical Definition of Dimensioning and Tolerancing principles*; ASME Y14.5.1M-1994; January 31; 1995; The American Society of Mechanical Engineers; New York; NY; USA
[Utpal 1998] Utpal Roy and Bing Li; Representation and interpretation of Geometric Tolerances for Polyhedral Objects – 1. Form Tolerances; In: *Computer Aided Design*; Vol. 30; No. 2; pp. 151-161, 1998
[Wayne 1984] Wayne T. and Eric G Hanson; "Offsets of two-dimensional Profiles"; In: *IEEE CG&A*; pp. 36-46; Sep 1984
[Walker 1993] Walker R.K and Srinivasan V.; "Creation and evolution of the ASME Y14.5.1M standard"; In: *Proceedings of the international forum on dimensioning tolerancing and Metrology*; CRTD-Vol.27; ASME1993; pp.19-30
[Wilhelm 1996] R.G. Wilhelm; "Geometric Tolerances for Three Dimensional Profiles"; In: *Engineering Design and Automation*; Vol. 2; No.4; pp.231-242; 1996

10. APPENDIX

A. Determination of Direction:

As we incrementally traverse from the first point to the last point on the boundary, and if the nominal feature is on the right side, the actual data points should be tested for the following:

1. The direction of p_a^x should be clockwise as we move along p_{t+}^1 to p_{t+}^{n1} and counter-clockwise as we move along p_{t-}^1 to p_{t-}^{n2}.

On the other hand if the nominal feature is on the left side the actual data points should be tested for the following:

2. The direction should be p_a^x counter-clockwise as we move along p_{t+}^1 to p_{t+}^{n1} and clockwise as we move along p_{t-}^1 to p_{t-}^{n2}.

B. Determination of points in the set P_{t+} or P_{t-} :

In a verification procedure, the coordinates of the actual feature are measured at a limited number of points. Let m be the number of points dividing the nominal feature under consideration (Profile tolerances specified by B-splines) into $[m-1]$ equi-spaced distances. Later in measurements, let us suppose that the actual coordinates are measured corresponding to these points.

The B-spline curve is an affine map of the knot vector sequence. The position of a point m can be mapped to a parameter value t_i in the knot vector sequence of the defining tolerance boundary. Parameter value t_i can then be used to determine the approximate point on the boundary in the neighborhood of point m. For example say $m = 11$ and the knot vector sequence is [0,0,1,2,3,3]. Dividing $(m-1)$ with *(number of elements in the knot vector-1)*, here 10/5, yields 2. This means that between every 2 subsequent elements in the knot vector there are 2 spaces, hence 3 points. So the set t is [0, 0.25, 0.5, 0.75, 1.0, 1.5, 2.0, 2.25, 2.5, 2.75, 3.0]. Note the divisions at multiplicity. Between 0 and 1 in the knot vector sequence, 0 is repeated. So from 0 to 1 there should be 4 spaces or 5 points (0 and 1 inclusive) dividing the length. The same applies to the knot vector sequence between 2 and 3. Applying t_i in the B-spline formulation defining the tolerance boundary with gives a point in the neighborhood of the actual point m_i. More points as required in the neighborhood of t_i can be calculated in evaluating the actual point. This process assumes that the boundary does not depart considerably from the nominal curve.

Representation of tolerances using fuzzy Logic

Cédric Lelu, Marc Dahan
Laboratoire de Mécanique Appliquée R. Chaléat – Institut de Productique
24, Rue de l'Epitaphe, 25000 BESANCON (France)
Tél : +33 (0)3 81 66 60 37
Fax : +33 (0)3 81 66 67 00
e-mail : cedric.lelu@univ-fcomte.fr

Abstract: Geometric parts of mechanical assembly are not manufactured precisely. Two uncertainty measures are especially interested for the defects representation: the distribution of the uncertainties that one piece can be manufactured at dimension *X*, and the likelihood that the assembly respects the functional constraints. This paper deals with the use of fuzzy logic for the representation of tolerances. This new method is an extension of the worst case and statistical kinematic analysis. It simplifies the equations of maximum likelihood which generally uses complex Gaussian distribution and is an easy understood representation of uncertainties. We treat the case of connecting rod – crank assembly as an example.
Keywords: Tolerance analysis, Tolerance representation, fuzzy logic, cotation.

1. INTRODUCTION

A mechanical assembly is defined by many geometric subjects joined up with dimensions. These dimensions and the information of nominal geometry are necessary for the different actors to design, produce or control the product. In this paper, we are especially interested in the representation of tolerances used by the designer to validate the project, by the production office to choose the manufacturing process and by the control office to accept the different pieces.

In practice, geometric parts are not manufactured precisely. Two uncertainty measures are especially interested : the distribution of the uncertainties that one piece can be manufactured at dimension *X*, and the likelihood that the assembly respects the functional constraints.

The paper deals with the use of fuzzy logic for the representation of tolerances. Each dimension is considered as a fuzzy interval so it is defined with a 4-D vector. The proposed solution uses three fuzzy sets :"little", "correct", "great". Of course, the pieces, which satisfy the correct dimensions with the first model, have always got a maximal memberships degree for the fuzzy sets "correct". Two other limit dimensions are defined and represent the extreme dimensions which make a piece acceptable. Outside, the constraint can not be respected. New operations are defined using Zadeh's Extension Principle. This new method simplifies the equations of maximum likelihood

P. Bourdet and L. Mathieu (eds.),
Geometric Product Specification and Verification: Integration of Functionality, 55-62.

which generally uses complex Gaussian distribution and is an easy understood representation of uncertainties.

In this paper, we present the case of connecting rod – crank assembly. In first, we compute the different kinematic elements of the system and we study the influence of manufacturing uncertainties. We compare it with the same element obtained with fuzzy operators.

2. POSSIBILITY THEORY

2.1 Fuzzy sets and possibility theory

Fuzzy sets notion is an extension of Boolean sets, based on the partial belonging to a class *A* [Zadeh, 1965] :The element's degree of belonging - μ_A - is a real between *0* (the element does not belong the set) and *1* (the element fully belongs the set). Thus, the transitions between two different classes is less sudden than for Boolean logic.

$$\mu_A : \quad \Omega \to [0,1] \qquad X \mapsto \mu_A(X) \tag{1}$$

Each field Ω is divided into many fuzzy sets, generally 3 or 5. For each element *X*, the degree of possibility is calculated.

Based on fuzzy sets, the theory of possibility, developed by Dubois and Prade [1985], is an extension of the error computing and the probability. Uncertainties and non-fullments are taking into account. The belonging degree is called "possibility degree" - π_A - and is a qualitative measure of the uncertainties of an event, whereas the aim of probability is to obtain a quantitative measure of this uncertainties. Therefore, possibility degree only helps for making a choice between alternatives.

Possibility aim is to describe an inaccurate but consistent knowledge body. A coefficient between *0* and *1* is attributed to each element defined in a finite set. It evaluates how this event is possible. This calculus is quite different than the probability which takes the ratio between the number of a class element and the number of all elements. The probability theory suppose that the future events will follow the same uncertainties distribution.

The number of measure is another difference between possibility and probability. For the probability, one measure is sufficient for the evaluation of the uncertainties. As the possibility of an event has no direct correlation with the possibility of the opposite event, two measures of the uncertainties are necessary for the possibility theory : the measure of possibility, and the measure of necessity computed by $N(A) = 1 - \Pi(\neg A)$. The possibility of the event (A and B) is computed by $\Pi(A \cup B) = \max(\Pi(A), \Pi(B))$, whereas $N(A \cap B) = \min(N(A), N(B))$ [OFTA, 94].

2.2 Extension Principel

Computing with fuzzy sets require to define the operation and function for fuzzy numbers. These are generally modelled with L-R functions (2), written $\left(\underline{m},\overline{m},\alpha,\beta\right)_{LR}$ with $\alpha \geq 0$ and $\beta \geq 0$. For trapezoidal fuzzy sets, the used functions L and R are : $\forall x, L(x) = R(x) = 1 - x$.

$$\pi: \ \Omega \rightarrow \qquad [0,1]$$

$$x \mapsto \quad \mu(x) = \begin{cases} L\left(\dfrac{\underline{m}-x}{\alpha}\right) & x \leq \underline{m} \\ 1 & \underline{m} < x \leq \overline{m} \\ R\left(\dfrac{x-\overline{m}}{\beta}\right) & x > \overline{m} \end{cases} \tag{2}$$

The Extension Principel expressed by Pr Zadeh is used to define the basic operators and function. A function, noted *f*, link the fuzzy set *A* to the fuzzy set *B*. The belonging degree of each element y from B is calculated by:

$$\mu_B(y) = \sup\left\{\mu_A(x) \setminus x \in B, y = f(x)\right\}. \tag{3}$$

These possibility degrees are only quantitative measure of uncertainties [Bouchon-Meunier et al, 96]. So the result of operations or functions is approximated by fuzzy numbers. For computing the result of fuzzy function or fuzzy operators, we determine the minimal and the maximal values for which the possibility measure is *0* or *1*. For example, the result of the fuzzy numbers inverse is:

$$\mu_{1/Y} = \left(\frac{1}{\overline{m}},\frac{1}{\underline{m}},\frac{\beta}{\overline{m}\times\left(\overline{m}+\beta\right)},\frac{\alpha}{\underline{m}\times\left(\underline{m}-\alpha\right)}\right)_{LR} \tag{4}$$

and the multiplication between two strictly positive fuzzy numbers [Henn et al, 1994] is defined by:

$$\mu_{X \otimes Y} = \ \underline{m_X} \times \underline{m_Y}, \overline{m_X} \times \overline{m_Y}, \underline{m_X} \times \alpha_Y + \underline{m_Y} \times \alpha_X - \alpha_X \times \alpha_Y, \overline{\underline{m_X}} \times \ {}_Y + \overline{m_Y} \times \beta_X - \beta_X \times \beta_Y \quad {}_{LR} \tag{5}$$

We have defined the following functions : the inverse ($M^{\ominus 1}$), the opposite ($\ominus M$), the square root ($\overline{M}$), the *n*-power ($M^{\tilde{n}}$) and the multiplication by a scalar ($*$) and the operators like the addition ($\oplus$), the subtraction (~) and the multiplication $\otimes$.

3. FUZZY TOLERANCES

3.1 Representation

The representation of tolerances by fuzzy numbers is based on the expressions "the dimension is too little for respecting all constraints" – fuzzy set: "Little", "the dimension respects all constraints" – fuzzy set : "correct" - and "the dimension is too great for respecting all the constraints" – fuzzy set: "Great". Three fuzzy sets are also defined, each of them is written with fuzzy numbers (Figure 1).

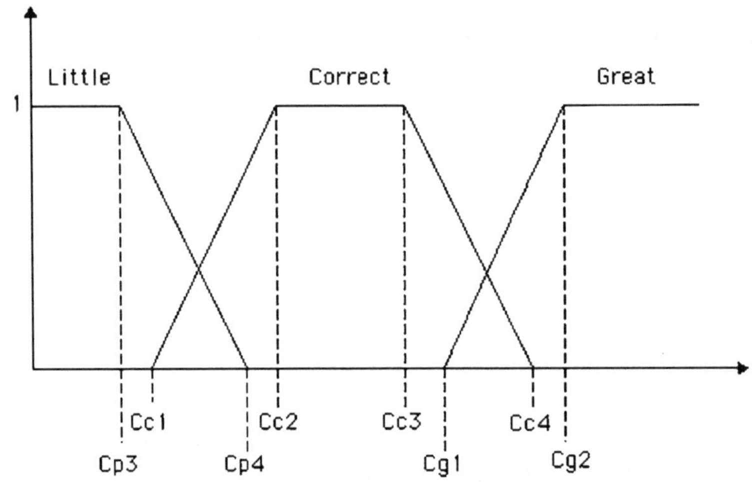

Figure 1 : Representation of fuzzy tolerances

Four parameters can be grouped : For example, C_{c1} expressed the minimal value for which a configuration of correct dimensions can be found for respecting all functional constraints. C_{p3} is the maximal value for which the dimension is too little for respecting all functional constraints if the other dimensions are correct. We obtain the simplified representation shown in figure 2.

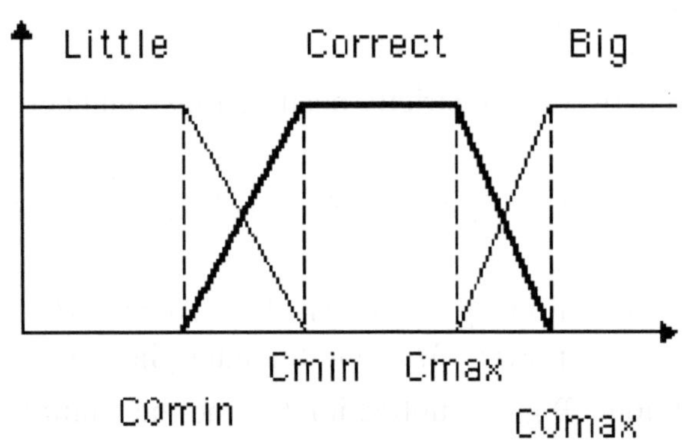

Figure 2 : simplified representation of fuzzy tolerances

So we define a fuzzy tolerance as a L-R fuzzy number, written $(C_{\min}, C_{\max}, \alpha, \beta)_{LR}$. $C_{0\min} = C_{\min} - \alpha$ is the smallest dimension for which a configuration of pieces inside their own tolerance interval respects the constraints, whereas $C_{0\max} = C_{\max} + \beta$ is the greater one. Fuzzy tolerance includes maximum and minimum likelihood representation.

3.2 Probabilistic distribution

The fuzzy set "correct" is an approximation of the logical sentence : "the dimension respects all constraints". The probabilistic measure of this event is computed by :

$$P(\text{the constraints are respected}) = \prod_k \left(\oint_{f_k(C_i, Cf_k)} \left(\prod_{i \neq j} P(C_i) \right) \right) \quad (6)$$

where $f(C_i, Cf_k)$ is

$$f(C_i, Cf_k) = \begin{cases} \left(\sum_{i=1}^{i=n} a_{i,k}.C_{i,\min} - \sum_{i=1}^{i=n} b_{i,k}.C_{i,\max} \right) - Cf_{k,\min} \\ \text{or} \\ Cf_{k,\max} - \left(\sum_{i=1}^{i=n} a_{i,k}.C_{i,\max} - \sum_{i=1}^{i=n} b_{i,k}.C_{i,\min} \right) \end{cases} \quad (7)$$

In the case of minimal constraint of two pieces, the expression of this sentence is expressed by the distribution:

$$\begin{aligned} P_1 \quad & \mathrm{IR} \to [0,1] \\ x \mapsto & \begin{cases} 1 & x \leq C_{\min} \\ \frac{1}{2}\left(\Phi\left(\frac{\overline{C_2} - C_f + x}{\sqrt{2}.\sigma_2} \right) + \Phi\left(\frac{C_{2,\max} - \overline{C_2}}{\sqrt{2}.\sigma_2} \right) \right) & x > C_{\min} \end{cases} \end{aligned} \quad (7)$$

3.3 Computing fuzzy tolerances

We consider a mechanical assembly of k functional constraints and n dimensions. $a_{i,k}$, $b_{i,k}$ can take the values :

- *0* : if the dimension vector is not used in this equation or it is not in the positive direction for $a_{i,k}$ or negative direction for $b_{i,k}$;
- *1* : if the dimension vector is in the direction of the constraint for $a_{i,k}$ or in the opposite direction for $b_{i,k}$;

The functional constraints are expressed in function of the different dimensions, using the maximum likelihood hypothesis.

$$\begin{aligned} \forall k: \quad & Cf_{k,\min} = \sum_{i=1}^{i=n} a_{i,k}.C_{i,\min} - \sum_{i=1}^{i=n} b_{i,k}.C_{i,\max} \\ \text{or} \quad & Cf_{k,\max} = \sum_{i=1}^{i=n} a_{i,k}.C_{i,\max} - \sum_{i=1}^{i=n} b_{i,k}.C_{i,\min} \end{aligned} \quad (8)$$

Using the definition, we develop the equation for computing α and β (9, 10)

$$\text{if } a_{j,k} \neq 0, \forall k: \quad Cf_{k,\min} = \sum_{i=1}^{i=n} a_{i,k}.C_{i,\min} - \sum_{i=1}^{i=n} b_{i,k}.C_{i,\min} - a_{j,k}.\alpha_j$$
$$\text{or} \quad Cf_{k,\max} = \sum_{i=1}^{i=n} a_{i,k}.C_{i,\max} - \sum_{i=1}^{i=n} b_{i,k}.C_{i,\max} + a_{j,k}.\beta_j \tag{9}$$

$$\text{if } b_{i,k} \neq 0, \forall k: \quad Cf_{k,\min} = \sum_{i=1}^{i=n} a_{i,k}.C_{i,\max} - \sum_{i=1}^{i=n} b_{i,k}.C_{i,\max} - b_{i,k}.\beta_j$$
$$\text{or} \quad Cf_{k,\max} = \sum_{i=1}^{i=n} a_{i,k}.C_{i,\min} - \sum_{i=1}^{i=n} b_{i,k}.C_{i,\min} + b_{i,k}.\alpha_j \tag{10}$$

The equations (8,9,10) show that the values of $C_{\min} - \alpha$ and $C_{\max} + \beta$ are dependent of different C_i. This values increase with the numbers of C_i. So, it is easier to obtain a configuration of correct pieces which respect the different functional constraints.

4. APPLICATION OF FUZZY TOLERANCES

4.1 Connecting rod – crank assembly

We study a connecting rod – crank assembly, shown in figure 3, and we want to precise the position of the assembly end, defined by *X*.

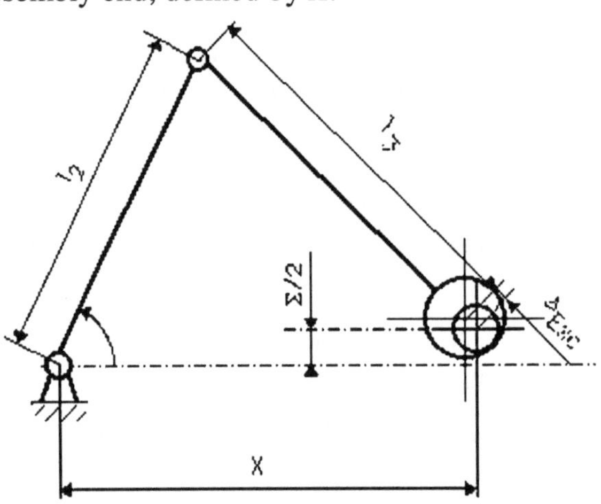

Figure 3: Connecting rod-crank assembly

The maximal defect of X is computed from the relation (11):

$$\Delta X = \max_{\varphi,\varepsilon,\Delta_{Exc},\Delta l_2,\Delta l_3} \left(\begin{array}{l} \Delta l_2.\cos(\varphi) - \sqrt{l_3^{\,2} - (l_2.\sin(\varphi))^2} + \\ + \sqrt{(l_3 + \Delta l_3 + \Delta_{Exc})^2 - ((l_2 + \Delta l_2).\sin(\varphi) - \varepsilon/2)^2} \end{array} \right) \tag{11}$$

φ is the gear orientation of the system. We use the values : $\varepsilon = 0^{\pm 0.01}$, $\Delta_{Exc} = 0^{\pm 0.1}$. Considering the minimal and maximal configuration C_{fi} (for $\varphi = \pi$ and $\varphi = 0$) of the mechanism, we obtain $l_2 = 50^{\pm 0.1}$ and $l_3 = 200^{\pm 0.1}$ from system (12).

$$\begin{cases} C_2 - C_1 = C_{f1} \\ C_2 + C_1 = C_{f2} \end{cases} \text{avec} \begin{cases} C_{f1} = 150^{\pm 0.2} \\ C_{f2} = 250^{\pm 0.2} \end{cases} \tag{12}$$

The resolution of the equations (9,10) with fuzzy tolerances gives the results :

$$\begin{cases} L_2 = (49.9, 50.1, 0.1, 0.1)_{LR} \\ L_3 = (199.1, 200.1, 0.1, 0.1)_{LR} \end{cases} \tag{13}$$

3.2 Fuzzy kinematic result

We use the equation of the position in function of the different defaults (10) and the different values are defined as fuzzy numbers, $(l_2 + \Delta l_2) = \tilde{l}_2$, $(l_3 + \Delta l_3) = \tilde{l}_2$ $\Delta_{Exc} = \varepsilon = (0, 0, 0.1, 0.1)_{LR}$. The expression (14) can not be algebraically expressed.

$$\tilde{X}^* = \tilde{l}_2 * \cos(\varphi) \oplus \sqrt{\left(\tilde{l}_3 \oplus \tilde{\Delta}_{Exc}\right)^{\tilde{2}} \sim \left(\tilde{l}_2 * \sin(\varphi) \sim \varepsilon\right)^{\tilde{2}}} \tag{14}$$

The maximal defect of the position is given by the expression $\max_{\varphi} \tilde{X}^*$.

The aim of the calculus is in first to evaluate the maximal and minimal position defect of the system. The dimensions of the two pieces are considered as : $(l_2 + \Delta l_2) = \tilde{l}_2 = (50, 50, 0.1, 0.1)_{LR}$ and $(l_3 + \Delta l_3) = \tilde{l}_2 = (200, 200, 0.1, 0.1)_{LR}$

The obtained result (15) coincides with the probabilistic method one. We observe that the nominal positions of the assembly is the interval [149.79,250.21]. This result is an information about the positions of the mechanism caused by the manufacturing defects : Considering theses defects, the minimal position is *149.58*, and the maximal one is *250.31mm.*

$$\tilde{X}^* = (149.79, 250.21, 0.21, 0.100091)_{LR} \tag{15}$$

5. CONCLUSION

We present a model of tolerances using fuzzy Logic, called Fuzzy tolerances. The applications of this new representation permits the resolution of the sizing and kinematic problems. It is an easy-understanding expression and is an extension of the maximum (the first two coordinates) and minimum likelihood method (information given by the last two coordinates). We develop the functions and operators for the computing of kinematic variations. The numerical resolution of maximal positions in function of the tolerances intervals is easier by the using of fuzzy operators. For example, the position of the connecting rod – crank assembly end is computed by fuzzy operators whereas the optimisation of the position in function of the four dimension parameters and the rotation angle is more difficult to obtain.

Another application of fuzzy tolerances is the integration of the economical constraints. As the result of equations is expressed as a membership degree of fuzzy sets, all these fuzzy sets can be added as in a fuzzy controller, and defuzzyfied. An inference motor or a genetic optimisation algorithm can use these results to obtain optimal dimensions using the economical constraints.

REFERENCES

[Bouchon-Meunier *et al*, 1996] B. Bouchon-Meunier, H. Nguyen, "Les incertitudes dans les systèmes intelligents", Presse Universitaire de France, 1996

[Dubois *et al*, 1985] D. Dubois, H. Prade, *Théorie des Possibilités*, Masson, 1985

[Henn *et al*, 1994] V. Henn, J-C Bernard "Logique Floue et Circulation", Université de Saint-Etienne, Octobre 1994

[Lelu *et al*, 2000] C. Lelu, M. Dahan, "Utilisation de la Logique Floue pour la modélisation des cotes de fabrication", In *the 3rd International Conference on Integrated Design and Manufacturing in Mechanical Engineering*, Forum 2000 SCGM/CSME, June 2001

[OFTA, 1994] Observatoire Français des Techniques~Avancées, In *Logique Floue*, Masson, 1994

[Zadeh, 1965] L. Zadeh, Fuzzy Sets, In *Information and Control*, 1965

Determination of relative situations of parts for tolerance computation

Eric Ballot, Pierre Bourdet**, François Thiébaut***
** CGS - Ecole Nationale Supérieure des Mines de Paris,*
60, bd Saint Michel 75272 PARIS cedex 06
*** LURPA - Ecole Normale Supérieure de Cachan*
61, ave du Pdt Wilson 94235 CACHAN cedex
eric.ballot@cgs.esnmp.fr

Abstract: This work is to contribute to three goals: a mathematical model of part geometric deviations, computation principles for three dimensional tolerance chains and a tolerance scheme in a mathematical form. We use an example to show present conclusions of this work according to the second goal. Then, we focus on a new problem faced in achieving the whole process of tolerancing. If the small displacement torsor is a useful tool to describe the relative situation of two simple surfaces, it's more difficult to describe the relative situation of two parts involving several different couples of surfaces, what is wanted in mechanism study. We first explain the combinatory nature of this problem and ways to reduce it.
Keywords: Computer Aided Tolerancing, small displacement torsor.

1 INTRODUCTION

1.1. Problem description and state of the art

A lot of work has been carried out in the past ten years to develop a mathematical formulation of tolerances. Two main approaches can be distinguished: by space allowed around theoretic geometry [Requicha, 1983] or by parameterization of deviation from theoretic geometry [Rivest and al., 1995], [Clément et al., 1991] and [Gupta et al., 1993].

Tolerance computation needs to take into account the relative situations of parts that are given by the relative situations of the mating surfaces. Indeed, values of each functional requirement depend on part relative position during assembly. An assembly typology and associated methods have been designed by J. Turner [Sodhi and al., 1994] to facilitate tolerance computations. The relative situations of parts may change when mechanical solicitations are applied, since they may modify contacts inside of the joints.

The values of each functional requirement depends not only on geometric deviations from the nominal part but also on relative positions of parts due to mating conditions as figure 1 illustrates.

P. Bourdet and L. Mathieu (eds.),
Geometric Product Specification and Verification: Integration of Functionality, 63-72.

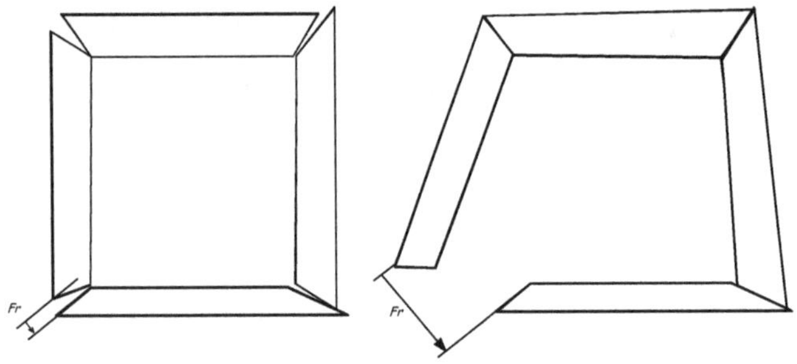

Figure 1; impact of the assembly process on a function requirement ([Sodhi et al., 1994]).

The aim of this paper is the proposal of a new concept, named configuration, in order to select, sort and compute all relative positions of parts. This concept provides an intermediary result for the computation of tolerance chains and values.

1.2. Illustration

To illustrate this paper we use the mechanism presented figure 2. This mechanism might obviously be only considered to be the support needed to reveal concepts.

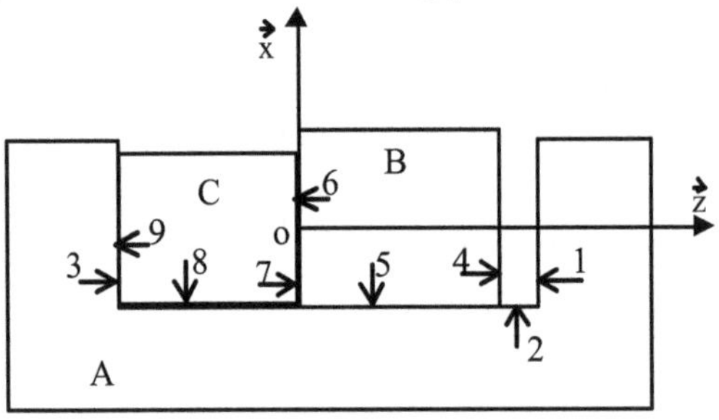

Figure 2; illustration mechanism (surfaces are numbered and part are marked by a letter)

2 THE PROPOSED MODEL

2.1. Formalization

In the following, we summarize and widely use previously published results. Earlier findings can be found in several presentations made at the CIRP seminars [Ballot, 1995] and [Ballot et al., 1996].

A geometric model of a real part is made of a nominal geometry, a model of the real surfaces and of a deviation criterion between each nominal geometric element and the model of the real surface. Computer aided design systems define the nominal geometry by theoretic surfaces, lines and points.

Three kinds of models are usually used to describe real part geometry.

First model: an unlimited set of points. This model, used by standards, describes in the most precise and conceptual manner, real part surfaces.
Second model: a limited set of points. These points are the result of a measuring process and are grouped by subset according to relevant nominal surfaces.
Third model: a geometric model is associated with each subset of points. The real part is then described by a set of substitute surfaces. Each substitute surface is an external, tangent and bonded theoretic surface.

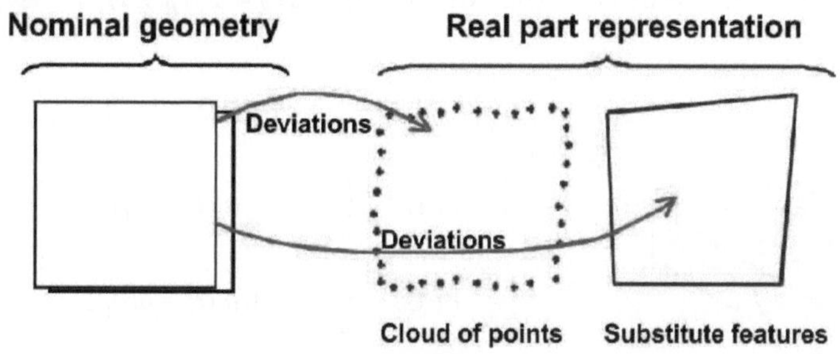

Figure 3; models of real parts

Geometry is therefore described by an allowed deviation between a model of the real part (a set of points or a substitute surface) and the appropriate nominal geometry [Bourdet and al., 1988] and [Ballot and al., 2000].

In the case of real surfaces described by a substitute surface, geometric deviations are modeled by small displacement torsors named for this matter deviation torsors [Bourdet et al., 1996]. A deviation torsor is the model of the 3D geometric deviation of a surface, a curve or a point.

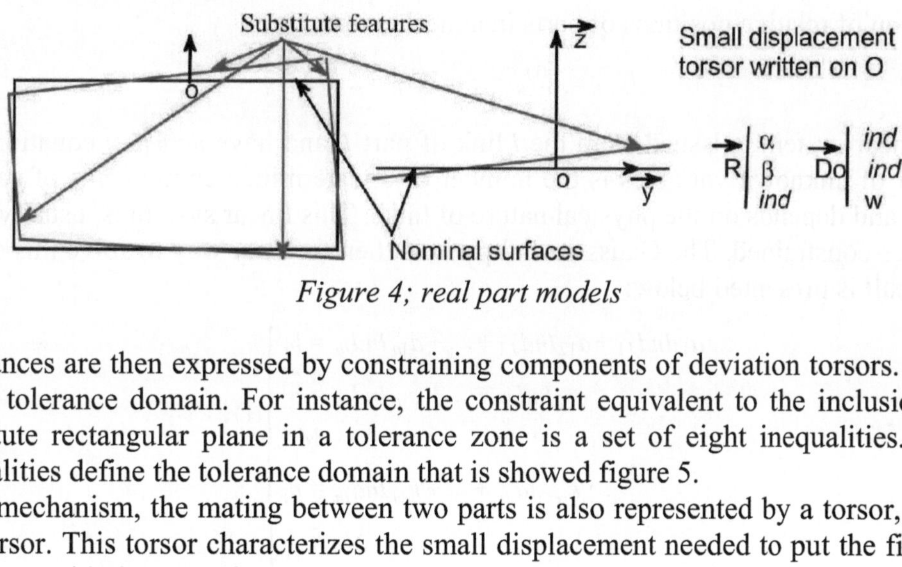

Figure 4; real part models

Tolerances are then expressed by constraining components of deviation torsors. This is called tolerance domain. For instance, the constraint equivalent to the inclusion of a substitute rectangular plane in a tolerance zone is a set of eight inequalities. These inequalities define the tolerance domain that is showed figure 5.
For a mechanism, the mating between two parts is also represented by a torsor, named gap torsor. This torsor characterizes the small displacement needed to put the first part in contact with the second.

Let us consider two substitute surfaces j and k that respectively belong to two parts A and B.

- The gap torsor is called: $T_{K/j}$ it corresponds to the small displacement from k to j.

- The deviation torsors $T_{j/A}$ and $T_{k/B}$ correspond to the deviation torsors of the two substitute surface j and k from their corresponding nominal surfaces,
- The part torsors $T_{A/R}$ and $T_{B/R}$ correspond to the parts small displacement torsor of nominal parts A and B according to a frame R.

With these notations, one can write for every link k/j between two parts A and B and furthermore for a mechanism, a set of small displacement composition. For the described link, the torsor equality is the following: $T_{B/R} = T_{A/R} + T_{j/A} - T_{k/B} + T_{k/j}$

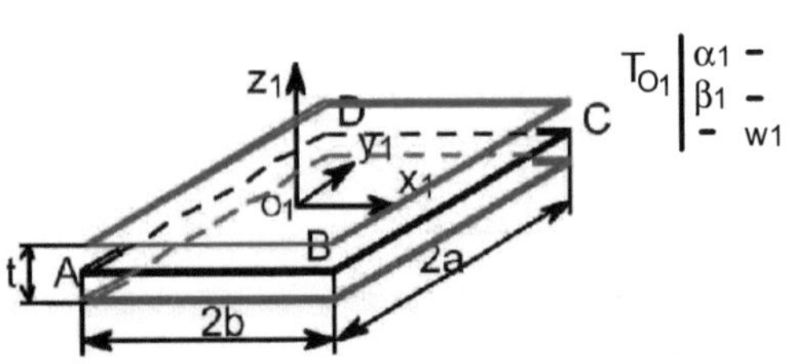

Pt A: $-\frac{t}{2} \le w_1 - a.\alpha_1 - b.\beta_1 \le \frac{t}{2}$

Pt B: $-\frac{t}{2} \le w_1 - a.\alpha_1 + b.\beta_1 \le \frac{t}{2}$

Pt C: $-\frac{t}{2} \le w_1 + a.\alpha_1 - b.\beta_1 \le \frac{t}{2}$

Pt D: $-\frac{t}{2} \le w_1 + a.\alpha_1 + b.\beta_1 \le \frac{t}{2}$

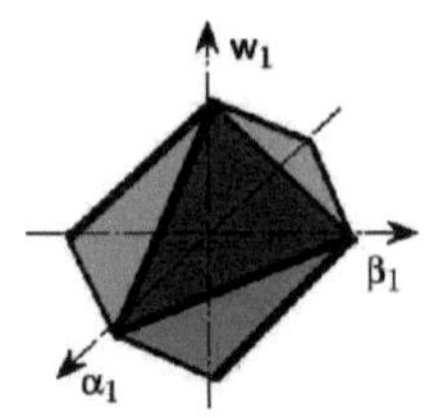

Figure 5; tolerance domain

2.2. Expression of relative part positions according to geometric deviations

To compute the complete set of relative positioning for a given part in a mechanism, we first write a composition of small displacement torsor for every link invoking this part. All these equalities define the position of the given part relative to other parts, that implies the equality of all expressions. Thus the following expression provide a definition of relative positions of parts in a mechanism.

$$T^{*}_{P/R} \equiv T^{(1)}_{P/R} = \ldots = T^{(i)}_{P/R} = \ldots = T^{(l)}_{P/R}$$

This linear system is issued from the l link of part P and have $m=6.(l-1)$ equations. The number of unknown variables is the number of undetermined components of gap "*Ind*" torsors and depends on the physical nature of links. This linear system is, usually, under and over-constrained. The Gauss partial pivot is then the only way to solve this system. The result is presented below:

$$\left.\begin{aligned} a_{11}Ind_{J1} + a_{12}Ind_{J2} + \ldots + a_{1n}Ind_{Jn} &= b_1 \\ c_{22}Ind_{J2} + \ldots + c_{1n}Ind_{Jn} &= \tilde{b}_2 \\ &\vdots \\ k_{rr}Ind_{Jr} + \ldots + k_{rn}Ind_{Jn} &= \tilde{b}_r \end{aligned}\right] (1)$$

$$\left.\begin{aligned} 0 &= \tilde{b}_{r+1} \\ &\vdots \\ 0 &= \tilde{b}_m \end{aligned}\right] (2)$$

The first part of this system settle an expression for the relative position of part P according to the linked parts. The second part of the system results from the over constrained nature of this problem. The number of remaining equations is the difference between the number m of equations and the rank r of the system. From a machine theory point of view, it is the hyperstatisum degree.

Now, we focus on the second part of this system and we give methodology and results to highlight the interest we found in using this system to compute relative part positions. The next system details variables concerned by equations (2).

$$\begin{cases} \bar{b}_{r+1} = x_{r+1,1}.\alpha_i + x_{r+1,2}.\beta_i + \ldots + x_{r+1,j}.J_{tx}(i/j) + \ldots + x_{r+1,\ldots}.J_{ty}(i/j) = 0 \\ \mathrm{M} \\ \bar{b}_m = x_{m,1}.\alpha_i + x_{m,2}.\beta_i + \ldots + x_{m,j}.J_{tx}(i/j) + \ldots + x_{m,\ldots}.J_{ty}(i/j) = 0 \end{cases}$$

In the following, this system is named: part compatibility system at part level and mechanism compatibility system at mechanism level. This system shows dependencies between gap and deviation torsor components. It means that all gap components could not be every time prescribed, as the compatibility system must be always checked without prescribing precisely values of the geometric deviation.

The next equation system is a "copy and paste" result from Mathematica® [Wolfram, 1995] of the illustration mechanism.

```
Compatibility system:
-β[2,A] + β[5,B] + γ[3,A] - γ[6,B] + γ[7,C] - γ[9,C] - J[ry,5,2] + J[ry,6,7] + J[ry,9,3] = 0

β[2,A] - β[8,C] - γ[3,A] + γ[9,C] + J[ry,8,2] - J[ry,9,3] = 0

-β[1,A] - β[3,A] - β[4,B] - β[6,B] - β[7,C] - β[9,C] + J[rx,4,1] - J[rx,6,7] - J[rx,9,3] = 0

γ[1,A] - γ[3,A] - γ[4,B] + γ[6,B] - γ[7,C] + γ[9,C] + J[ry,4,1] - J[ry,6,7] - J[ry,9,3] = 0

J[tz,4,1] - J[tz,6,7] - J[tz,9,3] - u[1,A] - u[3,A] - u[4,B] - u[6,B] - u[7,C] - u[9,C] + J[ry,4,1].x[4,1] -
J[ry,6,7] x[6,7] - J[ry,9,3] x[9,3] - J[rx,4,1] y[4,1] + J[rx,6,7] y[6,7] + J[rx,9,3] y[9,3] = 0
```

The number of equations, five in this case, is the number of hyperstatism degrees. Furthermore each equation describes the direction involved and the deviation components involved in this hyperstatism.

2.3. Rigid and non rigid mechanisms

A compatibility system is a set of linear combination of gap components and deviation components. In order for the relative position of a part to make sense, this compatibility system must be checked. Two cases have to be studied, rigid parts which means that the compatibility system can be checked whatever the values of deviation components or non rigid parts which means that some values of deviation components can be prescribed. The figure 6 illustrates the case of an assembly with non-rigid parts.

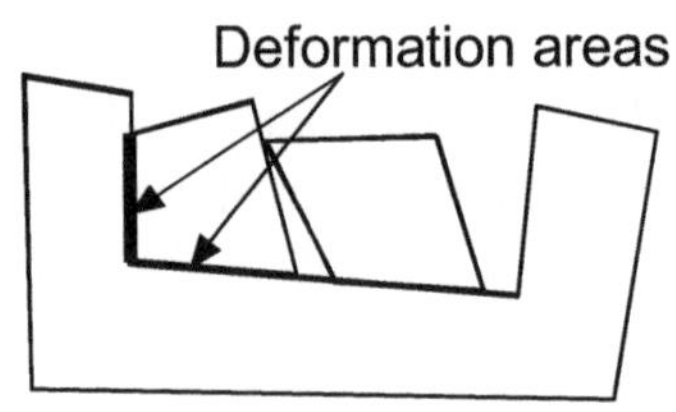

Figure 6; mechanism with non rigid parts

In the next sections we only deal with mechanism made of rigid parts.

2.4. Configuration definition

The proposed formulation allows to check if a set of gap components defines or not a mechanism configuration. But first, we have to define this original concept.

We name configuration without deformation of a mechanism a set of gap torsor components, which completely describe the relative positions of parts and where each gap component can be prescribed independently and still check the compatibility system. There are usually numerous possible configurations for a mechanism.

Figure 7; two different assembly but of the same configuration

According to this definition, a configuration is not yet an extreme position but a class of positions relative to geometric deviations and corresponding contact points as shown in figure 7. The main interest of this concept is that there is a unique tolerance chain for each configuration.

3 CHARACTERIZATION OF MECHANISM CONFIGURATIONS

Before the presentation of the computation process of configurations, we first introduce notation and give an approximation of the number of configurations for a mechanism.

3.1. Notations

We name J the set of gap torsor component of a mechanism. The set could be divided in distinct subsets: JI and JD. JI is a set of gap components that defines a configuration. JD is a set of gap components that is defined by the compatibility system and the configuration. This division of J in JI and JD is not unique and the number of divisions gives the number of configurations.

The subset JI could again be split in two subsets. In fact some of the gap components do not appear in the compatibility system and thus can be imposed whatever the compatibility system. This subset of gap components is named JL. According to the given configuration, the remainder subset is named: $JCI = JI - JL$.

At last the set of gap components related to the compatibility system is noted: JC. One then can write the compatibility system with a matrix notation: $B.JC = \Delta$ where B is the matrix of gap components coefficients JC and Δ the deviation component vector. The gaps in the set JCI are the rank of the compatibility system: $Card(JCI) = Rank(B)$.

3.2. Approximation of configuration number

The number of gap components one can choose is the same as the rank of the compatibility system.

We search for every gap component combination that possibly makes a configuration. One can choose $Card(JCI)$ in $Card(JCI) \cup Card(JD)$. The number of cases is given by the classical combinatory formula and is the maximum number of the mechanism configurations Nc:

$$Nc \leq \frac{Card(JCI \cup JD)!}{Card(JCI)!.(Card(JCI \cup JD) - Card(JCI))!} = \frac{Card(JCI \cup JD)!}{Card(JCI)!.Card(JD)!}$$

This number is over estimated because each gap vector is not necessarily a frame for this linear system. All proposed gap vectors need to be tested. It's only a necessary condition.

Using the previous formula with the numerical data of the example gives a maximum of 462 configurations to be checked.

$$Nc \leq \frac{Card(JCI \cup JD)!}{Card(JCI)!.(Card(JCI \cup JD) - Card(JCI))!} = \frac{11!}{5!.(11-5)!} = 462$$

In the case of complex mechanisms, the number of configurations increases very quickly. This is the characteristic of a combinatory problem.

4 CONFIGURATION STUDY METHODOLOGIES

Now, we describe four methods that use the compatibility system as a step to define the working domain of a mechanism and prepare the computation of extreme positions. In particular we describe how intrinsic complexity of mechanisms can be dramatically reduced and how one can avoid large combinatorial problems usually encountered.

4.1. Combinatory reduction by splitting principle

The first proposed methodology to reduce complexity of a mechanism is to find sub assemblies that could be studied independently. This fact generally occurs as designers try to control complexity for a lot of reasons. At the level of the compatibility system, this facts means that a decomposition into sub-matrix of the matrix B is feasible. The study of configurations is then made sub-matrix by sub-matrix that gives smaller combinations.

$$B = \begin{bmatrix} B_1 & \dots & 0 \\ \mathrm{M} & \mathrm{O} & \mathrm{M} \\ 0 & \dots & B_n \end{bmatrix}$$

The matrix of the example, shown bellow, can not be split into sub-matrix. This fact can be understood by the impossibility of splitting this mechanism in several independent sub mechanisms.

$$B = \begin{bmatrix} 0 & 0 & -1 & 0 & 1 & 0 & 0 & 1 & 0 & 0 & 0 \\ 0 & 0 & 0 & 0 & 0 & 1 & 0 & -1 & 0 & 0 & 0 \\ 1 & 0 & 0 & -1 & 0 & 0 & -1 & 0 & 0 & 0 & 0 \\ 0 & 1 & 0 & 0 & -1 & 0 & 0 & -1 & 0 & 0 & 0 \\ -y_{4,1} & x_{4,1} & 0 & y_{6,7} & -x_{6,7} & 0 & y_{9,3} & -x_{9,3} & 1 & -1 & -1 \end{bmatrix}$$

4.2. Working domain reduction using link selection

The simplest way to reduce the theoretical configuration number consists in limiting *a priori* the working domain. In the case of the example, if only a functional requirement has to be checked between the surfaces 1 and 4, one knows that the variables associated to the gap between 1 and 4 will never be of use in the configuration set. Then this link is suppressed of the configuration research concerning part B situation.

The new compatibility system resulting from this treatment is simpler. The maximum number of configurations in that case falls from 462 to 11 maximum cases, that is a great reduction of the complexity on large mechanical systems.

```
Compatibility system:
-β[2,A] + β[5,B] + γ[3,A] - γ[6,B] + γ[7,C] - γ[9,C] - J[ry,5,2] + J[ry,6,7] + J[ry,9,3] = 0
β[2,A] - β[8,C] - γ[3,A] + γ[9,C] + J[ry,8,2] - J[ry,9,3] = 0
```

4.3. Definition of working domain aided by configuration checking

Among the integrality of the theoretical configurations, there exist impossible cases according to the mechanism use. The assembly method and the force determination inside the mechanism make appear these impossible cases that can be suppress or obvious cases that have to be studied. The designer background permits to select the gap variable that needs to be under control.

The knowledge of the compatibility system is then of first interest since the control of the hyperstatism degree is possible. If the designer selects a set of gaps that is incompatible according to the compatibility system, he will be notified.

Concerning our illustration, if the designer selects both contacts between planes 9 and 3 and planes 8 and 2, that implies to force values of the gaps: *J[rx,9,3]*, *J[ry,9,3]*, *J[tz,9,3]*, *J[ry,8,2]*, *J[rz,8,2]* and *J[tx,8,2]*. This case is presented figure 6. The second equation of the compatibility systems is then:

```
β[2,A] - β[8,C] - γ[3,A] + γ[9,C] = constant
```

The deviations between surfaces of a part are due to the part fabrication, since the associated variables are random variables. This second equation can not be verified. The set of chosen variables is not independent with regard to the compatibility system. The chosen configuration implies deformation or is not actually usable.

4.4. Definition of mechanism operation aided by configurations enumeration

The whole set of possible configurations of a mechanism is not always obvious to describe. But the designer has to do it in order to check functional requirements in extreme positions of relative parts. In this case, the configuration system could be used to enumerate the complete set of configurations. With this information the designer can select appropriate configurations and rejects the others in the search of worst and best tolerance cases.

In that purpose we generate in a systematical way the subsets of gaps in JC with dimension $rank(B)$ and check if the result is a frame or not for the system. Application of this algorithm to the mechanism without the link between surfaces 4 and 1 gives 5 configurations. The five configurations are illustrated by the following figure.

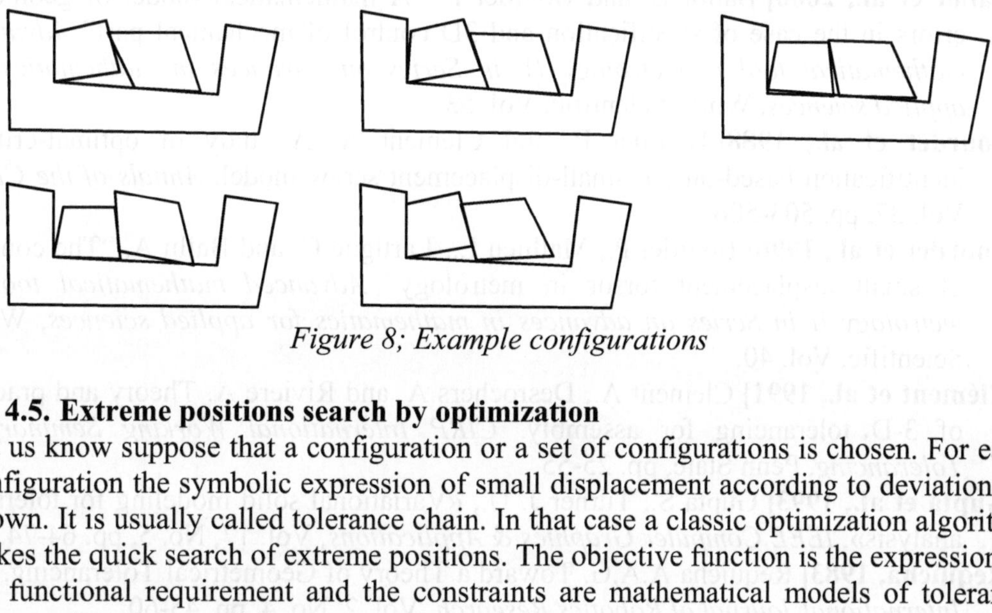

Figure 8; Example configurations

4.5. Extreme positions search by optimization

Let us know suppose that a configuration or a set of configurations is chosen. For each configuration the symbolic expression of small displacement according to deviations is known. It is usually called tolerance chain. In that case a classic optimization algorithm makes the quick search of extreme positions. The objective function is the expression of the functional requirement and the constraints are mathematical models of tolerance zones and mating conditions.

5 CONCLUSION

After presenting the major impact of relative positions on computation of tolerance, we show how the proposed model of three-dimensional tolerance chain can be used to solve this problem. In this purpose we have developed the configuration concept for a mechanism, highlighted its interest and illustrated its application. Particular applications for designers were developed in order to aid them in the description of the admissible domain of work for a mechanism.

The proposed methodology was programmed, applied and checked on various mechanisms but always in the case of rigid parts.

A possible further research is to include part deformation to this tolerance model. Finite element analysis combined with tolerance could then allow computing tolerance impact on stress for instance.

REFERENCES

[Ballot, 1995] Ballot E. "Lois de comportement géométrique des mécanismes pour le tolérancement" *Thèse de Doctorat de l'Ecole Normale Supérieure de Cachan.*

[Ballot et al., 1996] Ballot E. and Bourdet P. "A computational method for the consequences of geometrical errors in mechanisms" *Geometric Design Tolerancing: theories, standards and application, edited by El Maraghy. Chapman & Hall*, ISBN 0-412-83000-0, 488pp.

[Ballot et al., 2000] Ballot E. and Bourdet P. "A mathematical model of geometric errors in the case of specification and 3D control of mechanical part" *Advanced mathematical tool in metrology IV in Series on advances in mathematics for applied sciences*, World Scientific, Vol. 53.

[Bourdet et al., 1988] Bourdet P. and Clément A. A study of optimal-criteria identification based-on the small-displacement screw model. *Annals of the CIRP*, Vol. 37, pp. 503-506.

[Bourdet et al., 1996] Bourdet P., Mathieu L., Lartigue C. and Ballu A. "The concept of small displacement torsor in metrology" *Advanced mathematical tool in metrology II in Series on advances in mathematics for applied sciences*, World Scientific, Vol. 40.

[Clément et al., 1991] Clement A., Desrochers A. and Riviere A. Theory and practice of 3-D tolerancing for assembly. *CIRP International Working Seminar on Tolerancing*, Penn State, pp. 25-55.

[Gupta et al., 1993] Gupta S., Turner J. U., «Variational solid modeling for tolerance analysis», *IEEE Computer Graphics & Applications*, Vol. 17, No. 5, pp. 64-74.

[Requicha, 1983] Requicha A.A.G. Toward a Theory of Geometrical Tolerancing. *The International Journal of Robotics Research*, Vol. 2, No. 4, pp. 45-60.

[Rivest et al., 1995] Rivest L., Fortin C., Morel C., «Tolerancing a solid model with a kinematic formulation», *Computer-aided design*, Vol. 26, No. 6, pp. 465-485.

[Sodhi, 1994] Sodhi R. and Turner J. U. Relative positioning of variational part models for design analysis. *Computer-aided design*, Vol. 26, No. 5, pp. 366-378.

[Wolfram, 1995] Wolfram S; "Mathematica: A system for doing mathematics by computer" Addison-Wesley.

Genetic Algorithms for TTRS tolerance analysis

Jiandong Liu *and* ***Robert G. Wilhelm***
Center for Precision Metrology and Intelligent Manufacturing
Department of Mechanical Engineering and Engineering Science
The University of North Carolina at Charlotte
9201 University City Blvd., Charlotte, NC 28223-0001
Email: rgwilhel@uncc.edu

Abstract: This paper describes a computational tool for TTRS tolerance analysis that makes use of a Genetic Algorithm approach. The Technologically and Topologically Related Surfaces (TTRS) concept represents and classifies part surfaces and mechanical assemblies. We show results for examples with position and orientation tolerances that suggest that a Genetic Algorithm approach can provide an effective means for tolerance analysis.

Keywords: TTRS, Tolerance analysis, Genetic Algorithm

1. INTRODUCTION

The Technologically and Topologically Related Surfaces (TTRS) concept represents and classifies part surfaces and mechanical assemblies. TTRS can also be applied to dimensioning and tolerance analysis. Based on the TTRS concept, a tolerance zone is represented by small displacements with respect to Minimum Geometric Datum Elements (MGDE). TTRS classifies all surfaces into 7 different classes according to their invariant displacement attributes. Previous research has proposed a matrix approach based on the invariant displacement classification for tolerance analysis. In the matrix approach, the product of vectors and matrices that satisfies the specified tolerance constraint represents the tolerance zone. [Jones 1997] also provides a good introduction to the TTRS theory.

Tolerance analysis via the matrix approach can be very difficult since both the objective function and constraints are non-linear [Bank]. Currently there are no general solutions for such an optimization problem. In this paper, we develop a Genetic Algorithm approach for tolerance analysis. The complete representation scheme with operators and parameters is presented. Specifically the matrix approach is used to represent the tolerance zones. We show results for examples with position and orientation tolerances. The results from our analysis suggest that a Genetic Algorithm approach can provide an effective means for tolerance analysis.

P. Bourdet and L. Mathieu (eds.),
Geometric Product Specification and Verification: Integration of Functionality, 73-82.

2. TOLERANCE ANALYSIS MODELS AND METHODS

Tolerance analysis models can be placed in two categories: worst-case and statistical. The worst-case model assumes that all components in an assembly are at their limit position in the tolerance zone. The statistical model assumes that component variation is generally normal or Gaussian distributed; this model reflects more accurately the actual manufacturing process. The advantages and disadvantages of these two models are complementary. Worst-case analysis is accurate but the tolerance requirement for any particular part is too tight. The statistical model relaxes the part tolerance, which results in a decrease of manufacturing cost, but if the distribution estimates of the parts are not suitable, the tolerance analysis is inaccurate.

In tolerance analysis, there are two main methods of calculation. The linear programming method is applied if the analysis function of the problem is linear or may be approximated by a linear function. Tolerance analysis using linear programming can be expressed as an optimization method. If the tolerance analysis function or variables are non linear, the most common solution method is Monte Carlo simulation which is a statistical method [Gao, 1993]. Random numbers are generated for the variables in the tolerance analysis problem and these numbers are input to the analysis function. The output is then used to determine the attributes of the system. In current non-linear tolerance analysis methods, there are some limitations in that linear approximation method lack accuracy while non-linear solvers are difficult to generally apply.

3. TTRS MODEL AND TOLERANCE REPRESENTATION

The definition of TTRS proposed by [Clement 1991] is:
A TTRS is defined as an assembly formed by two surfaces (or between a surface and a TTRS or between two TTRS) belonging to the same solid (topological aspect) and located in the same kinematic loop in a given mechanism (technological aspect).
The TTRS model classifies the surfaces into 7 categories as shown in Figure 1 by the invariant displacement criteria; using this model we can represent a part or assembly by a binary tree, which makes it a very suitable data structure in an object orientated environment.

Based on the invariant displacement, a matrix approach for representing tolerance zone has been proposed by [Desrochers, 1994, 1997]. In this approach, the tolerance zones are obtained through point displacements around a coordinate system established on the theoretical surface of the part. The displacements include 3 translations and 3 rotations.

For example, Figure 2 shows a cylindrical tolerance zone obtained by translating and rotating points, on the axis, about the coordinate system. The corresponding displacement matrix is also shown. The matrix parameter reduces to 4 due to the invariant displacement property of the cylinder. The tolerance zone must satisfy the given constraints.

Surface class	Coordinate system	Matrix of invariant displacement
Any 	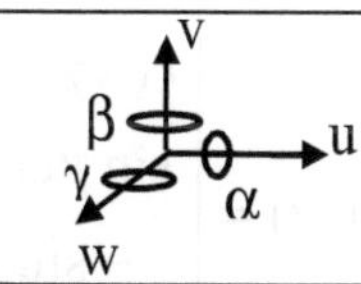	$\begin{bmatrix} C\gamma C\beta & -S\gamma C\alpha + C\gamma S\beta S\alpha & S\gamma S\alpha + C\gamma S\beta C\alpha & u \\ S\gamma C\beta & C\gamma C\alpha + S\gamma S\beta S\alpha & -C\gamma S\alpha + S\gamma S\beta C\alpha & v \\ -S\beta & C\beta S\alpha & C\beta C\alpha & w \\ 0 & 0 & 0 & 1 \end{bmatrix}$
Prismatic 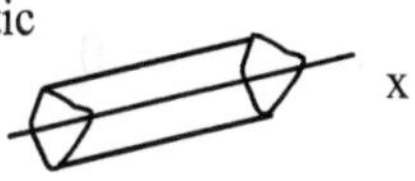	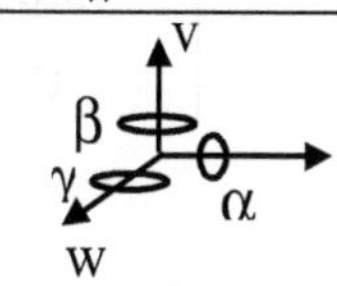	$\begin{bmatrix} C\gamma C\beta & -S\gamma C\alpha + C\gamma S\beta S\alpha & S\gamma S\alpha + C\gamma S\beta C\alpha & 0 \\ S\gamma C\beta & C\gamma C\alpha + S\gamma S\beta S\alpha & -C\gamma S\alpha + S\gamma S\beta C\alpha & v \\ -S\beta & C\beta S\alpha & C\beta C\alpha & w \\ 0 & 0 & 0 & 1 \end{bmatrix}$
Revolution 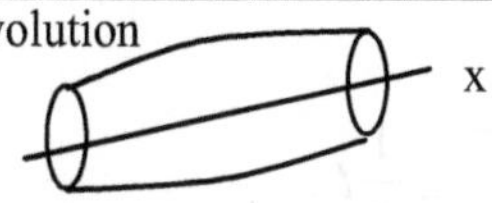	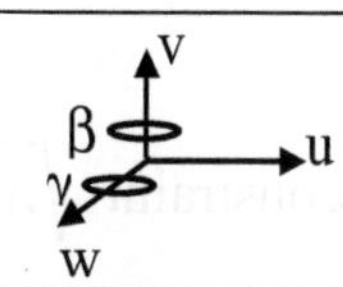	$\begin{bmatrix} C\gamma C\beta & -S\gamma & C\gamma S\beta & u \\ S\gamma C\beta & C\gamma & S\gamma S\beta & v \\ -S\beta & 0 & C\beta & w \\ 0 & 0 & 0 & 1 \end{bmatrix}$
Helical 	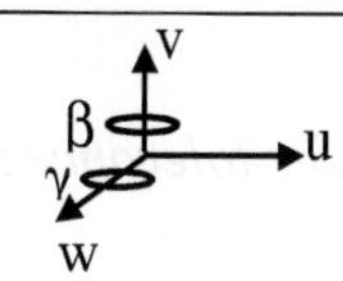	$\begin{bmatrix} C\gamma C\beta & -S\gamma & C\gamma S\beta & u \\ S\gamma C\beta & C\gamma & S\gamma S\beta & v \\ -S\beta & 0 & C\beta & w \\ 0 & 0 & 0 & 1 \end{bmatrix}$
Cylindrical 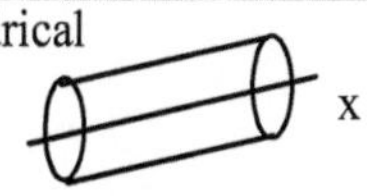	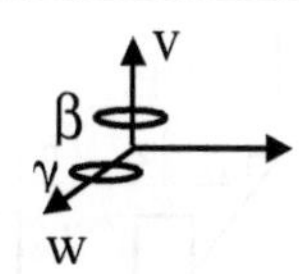	$\begin{bmatrix} C\gamma C\beta & -S\gamma & C\gamma S\beta & 0 \\ S\gamma C\beta & C\gamma & S\gamma S\beta & v \\ -S\beta & 0 & C\beta & w \\ 0 & 0 & 0 & 1 \end{bmatrix}$
planar 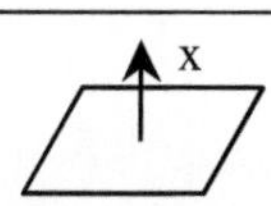	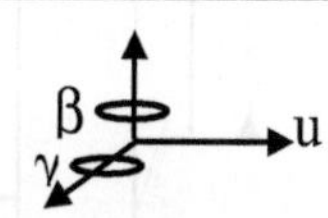	$\begin{bmatrix} C\gamma C\beta & -S\gamma & C\gamma S\beta & u \\ S\gamma C\beta & C\gamma & S\gamma S\beta & 0 \\ -S\beta & 0 & C\beta & 0 \\ 0 & 0 & 0 & 1 \end{bmatrix}$
Spherical 	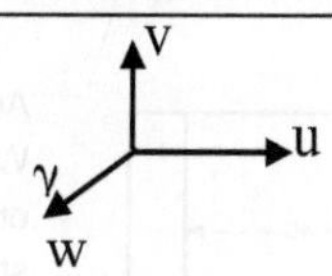	$\begin{bmatrix} 1 & 0 & 0 & u \\ 0 & 1 & 0 & v \\ 0 & 0 & 1 & w \\ 0 & 0 & 0 & 1 \end{bmatrix}$

Figure1; The 7 surface classes and their matrix representation [Desrochers 1997]

4. MATRIX APPROACH FOR TOLERANCE ANALYSIS

Following this tolerance representation, the tolerance of a part or assembly can be analyzed by building an appropriate model. [Desrochers 1997] have proposed a matrix approach for representing such tolerance zones and clearances. The example below illustrates this approach.

In this example, we evaluate the orientation tolerance of specified surfaces as indicated in Figure 3.

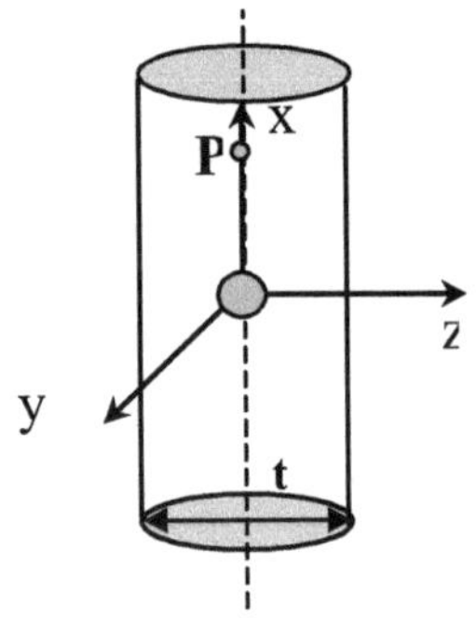

$$D(v,w,\beta,\gamma)=\begin{bmatrix}\cos\gamma\cos\beta & -\sin\gamma & \cos\gamma\sin\beta & 0\\ \sin\gamma\cos\beta & \cos\gamma & \sin\gamma\sin\beta & v\\ -\sin\beta & 0 & \cos\beta & w\\ 0 & 0 & 0 & 1\end{bmatrix}$$

$$\begin{bmatrix}X_{p'}\\ Y_{p'}\\ Z_{p'}\\ 1\end{bmatrix}=D\times P \text{ With the constraint } \sqrt{X_{p'}^2+Y_{p'}^2}\le\frac{1}{2}t$$

Figure 2; Cylinder tolerance zone

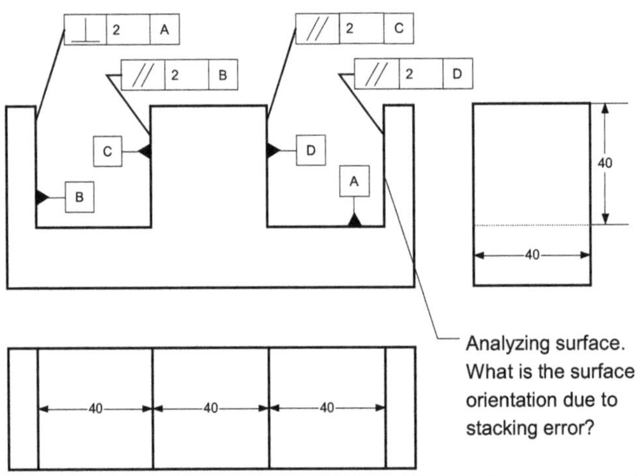

Figure 3; Example part for orientation tolerance analysis

The perpendicular and parallel tolerance zones are each volumes bounded by parallel surfaces. The part surfaces may be at any position in the corresponding tolerance zones. Due to stacking error, the orientation of the last surface is hard to analyze. We show this by finding the maximum angle of the analyzing surface with respect to the datum ***A*** surface. The procedure of finding a maximum and minimum angle for any vector in the space is the same.

The analyzing surface is indicated in Figure 3 and, as shown, the tolerance analysis problem involves a chain of surfaces. In this chain, the influence of a surface tolerance will propagate along the chain to the surfaces after it. The surface chain in the example is: $Datum\ A \rightarrow Datum\ B \rightarrow Datum\ C \rightarrow Datum\ D \rightarrow Analyzing\ \ surface$.

In the matrix approach, we establish coordinate systems on the perfect position planes of the datums *B*, *C*, *D* and the analyzing surface and we note them as R_B, R_C, R_D, and R_S. Also $T_{R_S \to R_D}$ indicates the mapping R_S to R_D. Finally we establish a coordinate system for the datum A surface. The displacements for each surface are noted as D_B, D_C, D_D, *and* D_S. Since they are all plane type displacement, they have the same transformation matrix.

When we do the tolerance analysis, we assume the analyzing surface is a flat plane obtained by translating and rotating the perfect surface plane within the tolerance zone. Under this assumption and since the tolerance zones are all convex we only need to take the four vertices of the plane as the constraint checking points. If these four points are within the tolerance zone, all of the points on the plane will satisfy the tolerance specification.

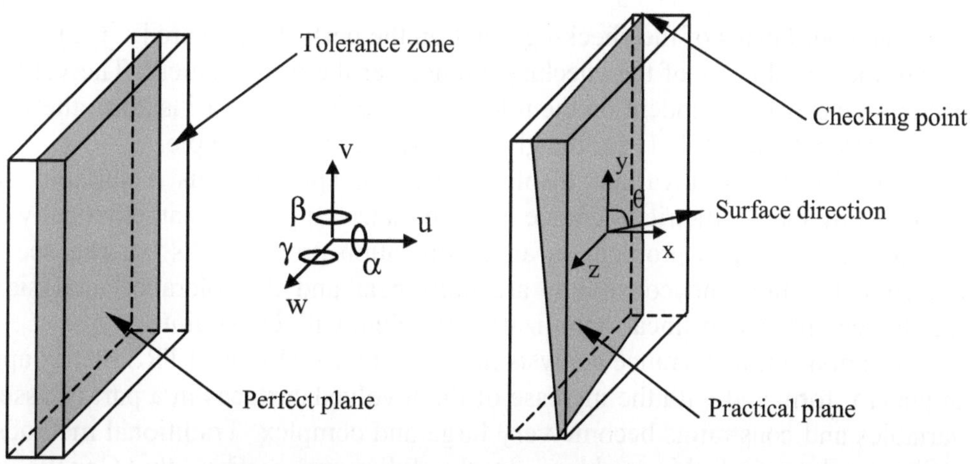

Figure 4; Tolerance zone of a plane

4.1. Objective function

To get the maximum angle between the analyzing surface and the datum A surface, we first need to express the orientation of the analyzing plane in the coordinate system established on the datum A. The perfect plane orientation of the analyzing surface in it's own coordinate system, R_S, is $m\,[1,0,0]$. By considering all of the displacements imposed on the involved datum surfaces, this vector in the coordinate system on datum ***A*** could be obtained by applying a transform chain. If we denote >> as applying a transform to the vector, the final vector m' in the datum ***A*** coordinate system will be a product of several matrices:

$$m' = m >> D_S >> T_{R_S \to R_D} >> D_D >> T_{R_D \to R_C} >> D_C >> T_{R_C \to R_B} >> D_B >> T_{R_B \to R_A}$$

The angle between m' and the orientation of the datum A $n\,[0,1,0]$ is:

$$\theta = \arccos\left(\frac{\vec{n} \cdot \vec{m'}}{|\vec{n}||\vec{m'}|}\right)$$

The solution of this optimization problem consists of the parameters' value in all of the displacement matrices. In this example, each displacement matrix has 3 parameters and the solution has 12 parameters.

4.2. Constraints

We check four points for each surface so there are 16 constraints in this problem. Each constraint requires the x coordinate of the checking point be within the tolerance zone after the translations and rotations. All of the constraints have the same format:

$$\left|x'\right| \leq \frac{1}{2}(2) \quad \text{Where,} \quad \begin{bmatrix} x' \\ y' \\ z' \\ 1 \end{bmatrix} = \begin{bmatrix} \cos\gamma \cos\beta & -\sin\gamma & \cos\gamma \sin\beta & u \\ \sin\gamma \cos\beta & \cos\gamma & \sin\gamma \sin\beta & 0 \\ -\sin\beta & 0 & \cos\beta & 0 \\ 0 & 0 & 0 & 1 \end{bmatrix} \times \begin{bmatrix} x \\ y \\ z \\ 1 \end{bmatrix}$$

The coordinates of the checking point on the perfect plane are $[x, y, z]$ while $[x', y', z']$ are the coordinates of the checking point after the displacement. The value range of each parameter is dependent on the tolerance specification; for instance, the value range of the parameter $\boldsymbol{u}$ is [-1, 1] for the given tolerance in the example.

The TTRS and invariant displacement concept gives us a suitable model for representing the geometric tolerance and the matrix approach mathematically expresses the tolerance analysis problem as an optimization problem. As we can see, both the objective function and constraints are non-linear and the tolerance analysis problem finally becomes a non-linear optimization problem with constraints.

The non-linear tolerance analysis problem induced by the TTRS matrix approach is of general form and with the increase of the involved surfaces in a part or assembly the variables and constraints become very large and complex. Traditional methods do lend well to solution of this problem. In the following section, the Genetic algorithm approach is introduced as a means for an effective and fast solution.

5. GENETIC ALGORITHM

5.1. Introduction

A Genetic Algorithm is, compared to hill climbing or simulated annealing, a multi-directional search algorithm. Genetic algorithms belong to the class of probabilistic algorithms that use stochastic processes to produce deterministic results; there is mathematical proof that genetic algorithms converge to the optimal solution in a solution space.

The main idea of GA is to generate an initial population of solutions in the solution space. These solutions can be randomly generated or by rules for better algorithm performance. Then each solution is evaluated by an evaluation function that is generally the objective function of the optimization problem; the better solutions are then given high probability to be selected to the next generation and the relatively poor solutions are rarely selected and at last killed as generations evolve. At each generation operators are applied to these solutions to produce new solutions. The two classic operators in GA are mutation and crossover. The mutation operator generates a variation version of a solution by modifying one or several parameters within a range. The crossover operator generally involves two solutions. It exchanges parts of the parameters of two solutions to result in two new solutions. After these operations, a new population having better solutions is obtained and the procedure iterates. When the termination condition is satisfied or the specified number of iteration is reached; the algorithm stops and the best result is returned.

```
Procedure GA_TTRS template
Begin
    calculate each parameter search space
    /* Generate feasible solutions randomly or by some rules,
       all parameters in the solution are set to zero in our implementation */
    initialize solution population
    set objective function
    set constraints
    set the validating algorithm        // Hill Climbing, Simulated Annealing or NULL
    t = 0
    Repeat
        evaluate each solution
        select the solutions to next generation     // high probability for better solutions
        apply the crossover operations
        validate the solution
        apply the mutation operations
        validate the solution
        adjust the validating algorithm
        t = t + 1
    Until (termination condition)
End

Procedure Validate
Begin
    If (feasible solution)     // Check all the constraints
        apply algorithm to the solution    // apply Hill Climbing or Simulated Annealing
    Else
        restore the solution to its previous value
End
```

Figure 5; Genetic Algorithm for Tolerance analysis

The GA procedure for the tolerance analysis problem is given in Figure 5. This procedure is used as a template in our implementation. It first calculates parameter value ranges, i.e. the parameters maximum and minimum value. The initial population is created with all the parameters set to zero in each solution. The objective function and constraints of the tolerance problem are set to the procedure. Also there is a validating algorithm object in the template that is used to decide whether the modified solution is accepted or not after applying the GA operators to the solution. At present, we can set this algorithm object to hill climbing, simulated annealing, or null. When we set the algorithm to null this template is a classic GA procedure. The climbing or simulated annealing algorithms produce a hybrid procedure where the algorithm is applied in each direction in the GA multi-direction search. In our tolerance analysis example, hill climbing produces a better result but takes much longer to compute.

The validating procedure decides whether a modified solution is updated or restored to its previous value by first checking all of the constraints. If any constraint is violated, which means that a tolerance specification is not satisfied, the modified solution is discarded and the solution is restored to its previous value. If all of the constraints are satisfied and the validating algorithm object is not null then the validating algorithm is applied to the modified solution and to determine acceptance.
The solution of the TTRS optimization problem consists of all the parameters in the displacement matrices. In the genetic algorithm, the operators apply to one or more parameters of the solution each time.

5.2. Evaluate function

In each generation, every solution is evaluated by a function. The objective function can be used for this purpose; the solutions with better value of the objective function are given high probability to be selected for the next generation. A simple procedure for this is as following [Michalewicz 1999]:

- Calculate the evaluate function *eval*($\boldsymbol{S}_i$) for each solution $\boldsymbol{S}_i$ (i = 1,2,..population size)
- Find the total evaluation value of the population

$$F = \sum_{i=1}^{population\_size} eval(S_i)$$

- Calculate the probability of each solution

$$p_i = \frac{eval(S_i)}{F}$$

- Calculate a cumulative probability q_i for each solution $\boldsymbol{S}_i$

$$q_i = \sum_{j=1}^{i} p_i$$

- Generate a random number $\boldsymbol{r}$ from range [0,1]
- If $r < q_1$, select the first solution, otherwise, select the i-th solution such that $q_{i-1} < r \le q_i$ to next generation.

As with spinning a roulette wheel, a better solution is selected with high probability.

5.3. Mutation operator

The mutation operator generates a variation of the selected solution. It includes two steps.
1. Randomly, or by some rules, select one parameter in the solution. We can establish some selection criteria and select the parameter that has the most influence on the

objective function. For example, the rotation parameter in TTRS matrix approach has more influence on orientation tolerance. In our implementation, we select the parameters in the invariant displacement matrix randomly.
2. Generate a variation of the parameter in its range. Assume the solution consists of two sets of TTRS matrix parameters:
$((\alpha_1,\beta_1,\gamma_1,u_1,v_1,w_1),(\alpha_2,\beta_2,\gamma_2,u_2,v_2,w_2))$
The α_1 is selected for mutation, after the mutation the solution becomes:
$((\alpha'_1,\beta_1,\gamma_1,u_1,v_1,w_1),(\alpha_2,\beta_2,\gamma_2,u_2,v_2,w_2))$, Where $\alpha'_1 = \alpha_1 + \Delta\alpha_1$ and α'_1 is in its range.

5.4. Crossover operator

The crossover operator exchanges parts of the parameters from two solutions. The exchange position can be selected randomly. For example, the crossover operator applies the following two solutions.
$V_1 = ((\alpha_{11},\beta_{11},\gamma_{11},u_{11},v_{11},w_{11}),(\alpha_{12},\beta_{12},\gamma_{12},u_{12},v_{12},w_{12}))$
$V_2 = ((\alpha_{21},\beta_{21},\gamma_{21},u_{21},v_{21},w_{21}),(\alpha_{22},\beta_{22},\gamma_{22},u_{22},v_{22},w_{22}))$
The selected crossover positions are from 2 to 3 for the first parameter set and from 3 to 6 for the second set. The two new generated solutions are:
$V'_1 = ((\alpha_{11},\beta_{21},\gamma_{21},u_{11},v_{11},w_{11}),(\alpha_{12},\beta_{12},\gamma_{22},u_{22},v_{22},w_{22}))$
$V'_2 = ((\alpha_{21},\beta_{11},\gamma_{11},u_{21},v_{21},w_{21}),(\alpha_{22},\beta_{22},\gamma_{12},u_{12},v_{12},w_{12}))$
These new solutions are inspected and the satisfied are put into the solution pool for processing in next iteration.

5.5. Results

We use the above GA template to solve the tolerance analysis problem shown in our example. The generation and population are both set to 100. The validating algorithm object is set to null and hill climbing in two different computations. The results suggest that setting the validating algorithm to hill climbing returns a better solution as shown below.
Maximum angle: 1.7708 (101.46 degree)
Parameters:

Displacement D_B	Displacement D_C	Displacement D_D	Displacement D_S
$\alpha = 0$	$\alpha = 0$	$\alpha = 0$	$\alpha = 0$
$\beta = 0$	$\beta = 0$	$\beta = 0$	$\beta = 0$
$\gamma = -0.05$	$\gamma = -0.05$	$\gamma = -0.05$	$\gamma = -0.05$
$u = 0$	$u = 0$	$u = 0$	$u = 0$
$v = 0$	$v = 0$	$v = 0$	$v = 0$
$w = 0$	$w = 0$	$w = 0$	$w = 0$

Constraints:

Datum B surface	Datum C surface	Datum D surface	Analyzing surface
Constraint # 1 = 0.9996	Constraint # 1 = 0.9996	Constraint # 1 = 0.9996	Constraint # 1 = 0.9996
Constraint # 2 = 0.9996	Constraint # 2 = 0.9996	Constraint # 2 = 0.9996	Constraint # 2 = 0.9996
Constraint # 3 = -0.9996	Constraint # 3 = -0.9996	Constraint # 3 = -0.9996	Constraint # 3 = -0.9996
Constraint # 4 = -0.9996	Constraint # 4 = -0.9996	Constraint # 4 = -0.9996	Constraint # 4 = -0.9996

The displacement parameter value means that the perfect plane rotates clockwise around the *z*-axis of its coordinate system as the Figure on the right shows. The four constraints give the x coordinates of the four checking points on each plane. All of the four displacement matrices have the same parameter values; which can be interpreted as all the planes rotate clockwise to their limit position in the tolerance zone and the objective function gets its maximum value with surface at these positions.

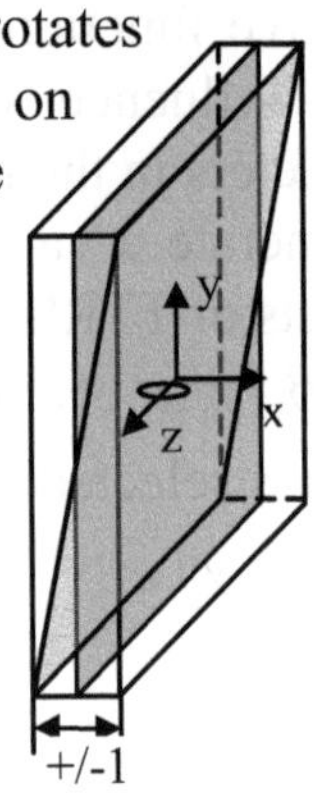

6. CONCLUSION

TTRS and the matrix approach are well suited as a tolerance representation model for parts and assemblies. When doing tolerance analysis using this model, we have to solve a non-linear optimization problem and the traditional mathematical methods do not provide an effective general solution for this type of NP hard problem. A Genetic algorithm approach has been applied in this situation and the results suggest that it could provide a general solution for such tolerance analysis problem.

7.ACKNOWLEDGEMENT

This material is based upon work supported by the National Science Foundation under Grants EEC 9811556 and DMI-9457168. We are most appreciative of the facilities provided by the NSF IUCRC Center for Precision Metrology at UNC Charlotte.

8. REFERENCES

[Clement 1991] Clement A., Desrochers A., Riviere A.; Theory and practice of 3D tolerancing for assembly; In: *2nd CIRP International Working Seminar on Computer Aided Tolerancing;* Penn State University; USA; May 1991.
[Desrochers 1994] Desrochers A, Clement A.; "A dimensioning and tolerancing assistance model for CAD/CAM system", In: *The international journal of Advanced Manufacturing Technology*; 1994; 9:352-361
[Desrochers 1997] Desrochers A, Riviere A.;"A matrix approach to the representation of tolerance zone and clearance", In: *The international journal of Advanced Manufacturing Technology*, 1997; 13:630-636
[Jones 1997] Jones A.; Keeler S.; "Mathematical sketch of the TTRS theory"; In: Boeing *technical document series ISSTECH-97-022*; 1997
[Bank] B. Bank, "Non-linear parametric optimization"; ISBN: 0817613757
[Gao 1993] Gao J.; "Nonlinear tolerance analysis of mechanical assemblies"; PhD thesis; Brigham Young University; 1993
[Michalewicz 1999] Michalewicz Z.; Genetic algorithm + data structure = evolution programs; ISBN: 3540606769

Calculation of virtual and resultant parts for variational assembly analysis

Laurent Pino, Fouad Bennis* and Clément Fortin**
ENIB, LII – Parvis Blaise Pascal, 29608 Brest – France
Laurent.Pino@enib.fr
**IRCCyN – 1 rue de la Noë, 44321 Nantes – France*
Fouad.Bennis@irccyn.ec-nantes.fr
*** Ecole Polytechnique de Montréal – Montréal – Canada*
clement.fortin@polycapp.com

Abstract: Usually, the design of mechanisms and assemblies uses the perfect model of the part. This type of model can be used within a software application to simulate and to verify the kinematics and dynamic behavior of such mechanisms. However, the analysis of the manufacturing uncertainties is also necessary to control the sensitivity and robustness of the mechanism assembly. To take into account these uncertainties, the designer deals with a class of interchangeable and functionally equivalent parts. When specifying tolerances, he essentially defines the authorized variations of the envelope of the parts and a number of classes can represent the various domains of acceptance of the parts. Since we need to take into account the infinite number of parts allowed by the variational class, the analysis of variational assembly is a hard task. In this paper, an extension of our previous work to calculate the virtual and resultant parts is proposed. The proposed method takes into account the datum chaining in the toleranced part. A kinematics model and robotic parameterisation is used to simulate the possible "motion" of the tolerance zone, allowed by MMC and LMC modifiers. Thus, the analysis of the infinite number of parts is reduced to analyse only two parts for each variational class. The analysis of the Jacobian matrix of the mechanism effectively generates the intersection and the union of all the boundaries of the features of the part. The paper presents also a significant application of this approach.
Keywords: Assembly analysis, maximum material part, virtual part, tolerance analysis.

1. INTRODUCTION

Nowadays, several models exist to take into account the effects of geometric tolerances, but the study of tolerancing for assemblies is still an open problem. The complexity of the analysis of variational assemblies comes from the simultaneous variation of the relative positions of parts in the assembly and of the part geometry. The analysis of a variational assembly must take into account the infinite number of parts in the variational class [Requicha, 1992, Robinson, 1998]. Statistical methods, as the Monte

P. Bourdet and L. Mathieu (eds.),
Geometric Product Specification and Verification: Integration of Functionality, 83-92.

Carlo method, can be used but it is time and memory consuming and one is never assured to have done enough tests to cover all limit cases of the variational assembly. Conditional tolerancing and virtual gauge methodologies have been proposed to overcome these difficulties [Srinivasan et al., 1989, Pairel 1995]. More recently, Robinson, proposes the application of an MMP (Maximum Material Part) tolerancing principle that allows a reduction of the variational class analysis to only one specific part that belongs to the class. This principle can be seen as an extension of the well-known Maximum Material Condition principle. The use of MMPs facilitates the creation of floating assemblies, the analysis of path planning and the modelling of kinematic assemblies. Finally, it simplifies the analysis and the functional specification as the conservation of a minimum material thickness [Robinson, 1997]. Also, a set of tolerancing rules is proposed by Robinson to facilitate the coverage of assemblies [Robinson, 1998].

In the following section, we recall the basic definitions of the MMP approach and introduce the virtual and resultant part search on a simple example. We point out the possible effect of datum chaining of the toleranced part. Section 3 outlines the kinematic method. Finally, an example is presented as a test case study.

2. DEFINITIONS

2.1. Maximum material part

The Maximum material part (MMP) principle was first introduced in Parratt's theory for the one-dimensional tolerancing of assemblies [Parratt, 1994]. Robinson extended this theory to geometrical tolerancing and showed the advantages of MMP application for a variational assembly analysis [Robinson, 1997]. For a given variational class, an MMP is a worst case solid, that includes all parts of the class and belongs to the variational class. Since the MMP itself is within tolerance, it does not lead to overly conservative analyses. Figure 1 shows an example of a variational class with an MMP. When each feature is at its MMC size, the part with a diameter of 40 mm and 10 mm is the MMP. It encloses all parts of the variational class.

A variational class has an MMP only if certain conditions are satisfied. To represent these, Robinson proposes a non-exhaustive list of rules to determine whether a given set of geometric tolerances leads to an MMP. An analogue list of rules is also proposed for a Least Material Part (LMP). An LMP is a part that belongs to the variational class and that is enclosed by all the parts of the variational class. For a variational assembly defined with MMP-LMP parts, the configuration space regions describing the relative motion and interference between all collections of parts are bounded by the regions derived from analysing the MMP and LMP solids alone. The list of rules proposed by Robinson points out that, datum chaining is not allowed in general. Thus, if the part is defined with datum chaining, it is not always possible to find an MMP (or LMP). In this particular case, Robinson proposes to extend the notion of MMP to virtual and resultant part and demonstrates that the analysis of an assembly with a virtual part leads to conservative results.

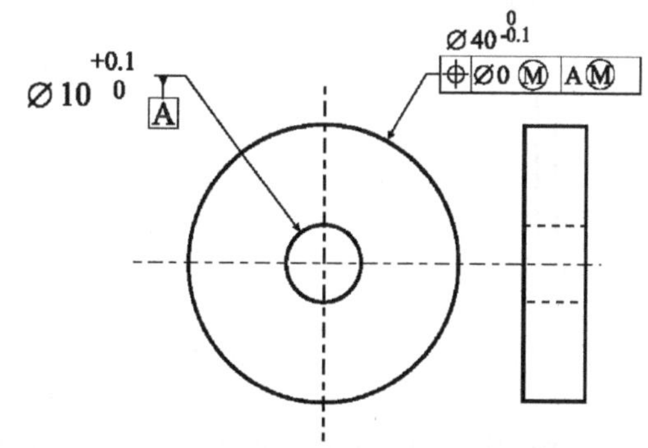

Figure 1 : Example of Maximum Material Part (from [Robinson, 1998])

Remark: Robinson uses the terms virtual solid and resultant solid instead of virtual part and resultant part which we feel are more general terms.

2.2. Virtual and resultant parts

We saw previously that it is not always possible to find an MMP (or LMP). However, for each variational class, we can find at least one particular part that encloses (and/or is enclosed in) all parts of the variational class. These parts are called virtual parts. Virtual parts are an extension of MMPs, but there exists two differences between an MMP and a virtual part [Robinson, 1998] :

- The virtual part *is not a member* of the variational class.
- For a given variational class, the virtual part is not unique.

Analogue differences exist between a LMP and a virtual part. Resultant parts are enclosed by all parts of the variational class. On the other hand, a virtual and resultant part can be found for any variational class. Robinson proposes to use sweeping tools to calculate theses two parts but he does not give any details of this calculation [Srinivasan 1993].

The major difficulty of an assembly analysis specification is taking into account the datum chaining of features. The next sections of this paper present the required tools to calculate virtual and resultant parts with datum chaining. These tools have been applied to the positional analysis and measurement applied to a given manufacturing process [Bennis et al., 1999, Pino 2000]. In order to take into account the effect of toleranced features in the tolerance chain, we analyse the sweeping movement induced by the MMC modifiers. Figure 2 shows a simple part with a hole defined with an MMC position tolerance relative to A, B and C datum reference frame (DRF).

Figure 3 depicts the calculation of the virtual and resultant conditions at MMC of this hole. The virtual condition can be obtained using the intersection of all the possible holes at MMC size. The resultant condition at MMC can be defined as the union of all possible MMC holes. In a same manner, the virtual condition and resultant at LMC condition can be calculated by sweeping LMC holes relatives to the ABC DRF. The two virtual cylinders of diameter 6.5 mm and 6.7 mm define the material boundaries of the variational class.

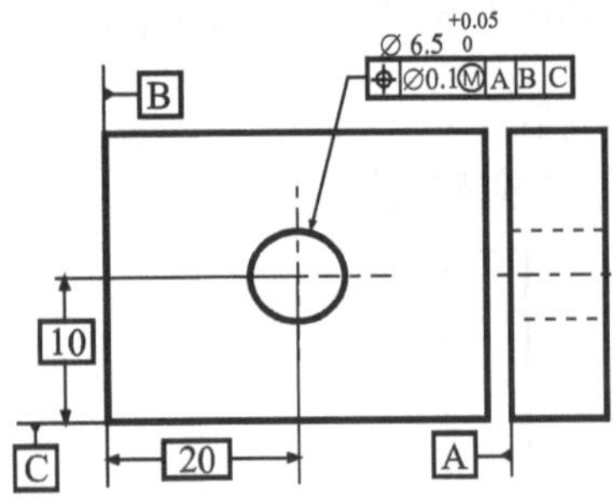

Figure 2 : Part definition

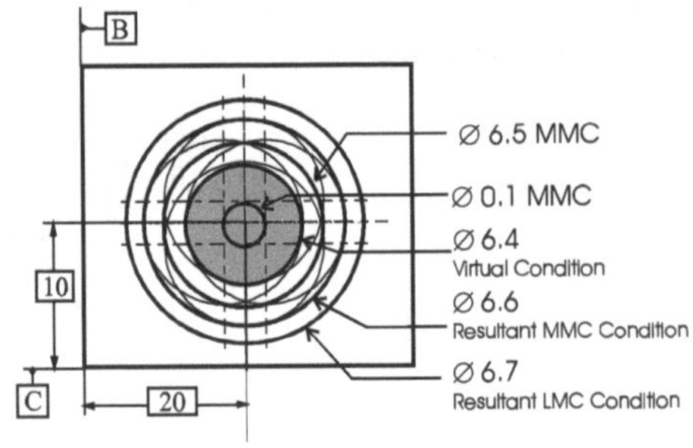

Figure 3 : Virtual and resultant MMC conditions

We will now extend these results to all the part features. Let us introduce the part defined in figure 4. It is composed of two holes. The first hole D is located at MMC with respect to datum A, B and C. The second hole E of diameter 5 mm is located by a positional tolerance with respect to datum A for orientation, datum D at MMC for location and datum B for rotation. If the actual size of D is produced away from its MMC size, the datum axis D can shift, and the relative position of the hole E with respect to the ADB DRF can also shift. The displacement of datum D must be taken into account in the analysis of all the allowed situations for hole E, relative to the ABC DRF.

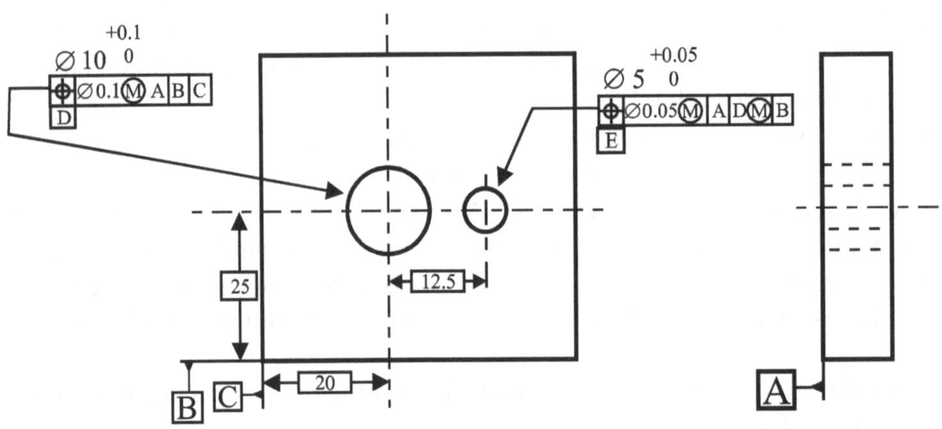

Figure 4 : Example of Datum chaining in the part definition

For the calculation of the virtual and resultant parts, one can imagine a verification protocol of hole E by using a hard gage. As shown in figure 3 of the previous example, the resultant condition of hole D at LMC is a cylinder of diameter 10.3 mm. The gage corresponding to datum D is a cylinder at virtual condition size that can be included within any location with respect to the resultant condition (diameter 10.3 mm). The left hand side of figure 5 depicts two possible locations of this *virtual gage*. For every location of the first gage, we associate one location of the second gage. For this location, we can define a resultant condition at LMC for the second axis (diameter 5.15 mm). Taking into account all axis locations of the first gage, the second axis may sweep and the resultant condition at LMC may sweep to. This external boundary of the sweeping defines an extension of the resultant condition at LMC (medium grey zone in figure 5).

The diameter of the extension of the resultant condition is equal to :

$$5.15+0.4 = 5.55 \text{ mm}$$

In a same way, we can calculate the intersection of the virtual condition of the second axis. This zone is in dark grey in figure 5. The diameter of the extension of the virtual condition is: $4.95-0.4 = 4.55$ mm.

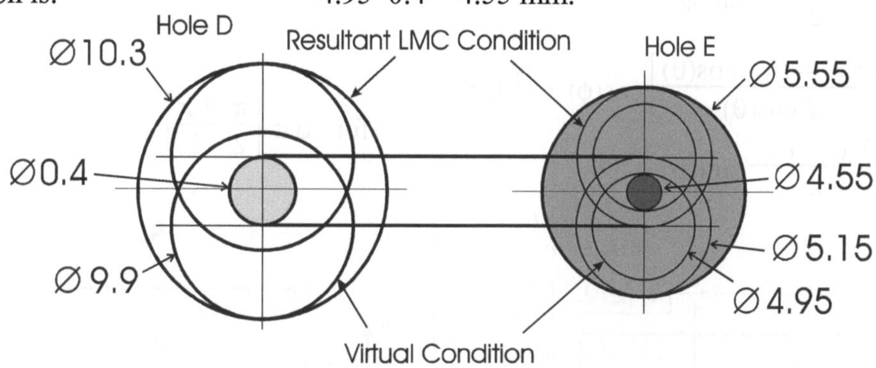

Figure 5 : Virtual and resultant condition extensions

A first hole with a diameter of 9.9 mm and the second hole with a diameter of 4.55 mm define the virtual part with respect to the ABC DRF. The two axes are in nominal locations. In a same way, a first hole with a diameter of 10.3 mm and the second hole with a diameter of 5.55 mm define the resultant part with respect to the ABC DRF.
The first part includes all the parts of the variational class with respect to the ABC DRF. The second one is included within all parts of the variational class.
This simple example shows the application of the sweeping method using intersection and union operators to calculate a virtual and a resultant part.

3. VIRTUAL AND RESULTANT PART CALCULATION

The relative situation of the DRF with respect to the actual part is the result of the composition of all locations of each datum feature that define the DRF. If some datum features are of MMC datum types and they are manufactured away from MMC size, then multiple datum feature candidate exists, and also multiple candidate DRF exist. They are called candidate datum reference frame. In the followings sections we define theoretical tools, needed for the explicit calculation of the virtual and resultant parts.

3.1. Actual Candidate Axis Set (ACAS)

For an internal feature, the ACAS represents all the possible situations of the axes of a virtual gauge cylinder in the actual feature with respect to datum reference frame [Bennis et al., 1999].
Figure 7 shows the model of an actual cylinder corresponding to the part of figure 6, and figure 8 depicts the detail of the corresponding ACAS in the plane.

The explicit relations of ACAS are given by the following equations:

$$\begin{cases} x = \left[\dfrac{D_{AM} - D_{CV}\cos(\theta)}{2\cos(\theta)}\right]\cos(\phi) - \dfrac{w}{2}\tan(\theta) \\ y = \dfrac{D_{AM} - D_{CV}}{2}\sin(\phi) \end{cases} \quad \text{for } \phi \in \left[\frac{-\pi}{2}, \frac{\pi}{2}\right] \qquad (1)$$

$$\begin{cases} x = \left[\dfrac{D_{AM} - D_{CV}\cos(\theta)}{2\cos(\theta)}\right]\cos(\phi) + \dfrac{w}{2}\tan(\theta) \\ y = \dfrac{D_{AM} - D_{CV}}{2}\sin(\phi) \end{cases} \quad \text{for } \phi \in \left[\frac{\pi}{2}, \frac{3\pi}{2}\right] \qquad (2)$$

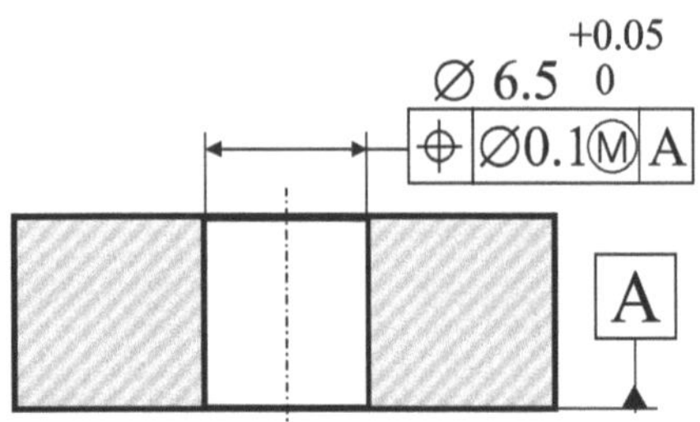

Figure 6 : Part definition

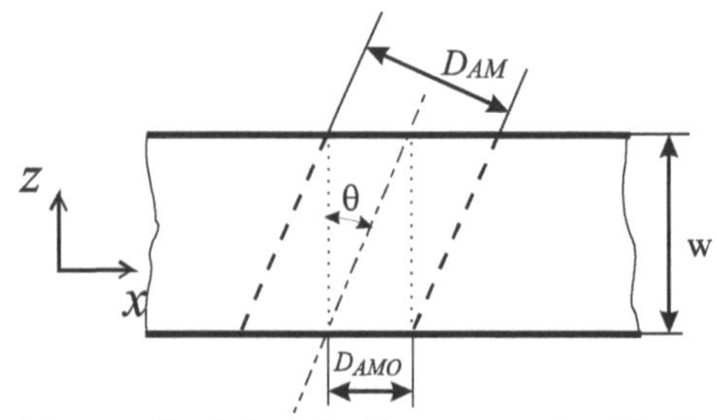

Figure 7 : Model of an actual cylinder

Generally, a simplified definition of the ACAS can be defined; it can be approximated by a cylinder of diameter D_Z with:

$$D_Z = D_{AM} - D_{CV} \qquad (3)$$

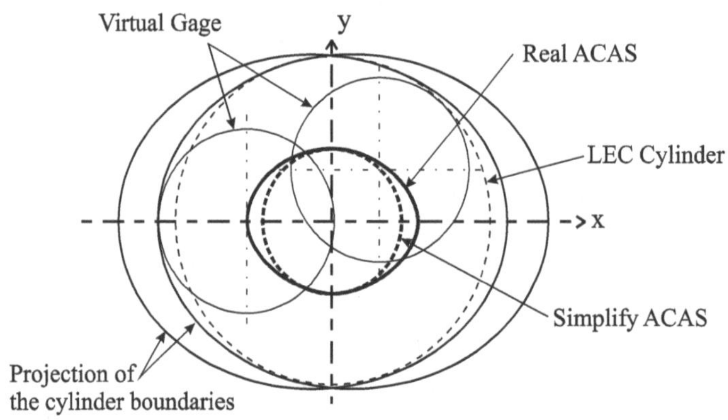

Figure 8 : Definition of the ACAS

In this paper, we use the simplified definition of the ACAS. Let's now introduce the kinematic approach to calculate the sweeping motion of the virtual gage relative to a given DRF.

3.2. Definition and study of the kinematic mechanism

The calculation of virtual and resultant parts corresponding to the part defined in figure 4 can be done manually. Indeed, the orientation of the intermediate DRF is constant. Generally, the orientation of the intermediate DRFs is not constant, and the use of a general tools is therefore necessary.

In this section, a kinematic mechanism approach is used [Bennis et al., 1999].
For an internal cylindrical datum feature, the axis motion is enclosed in the ACAS of the feature. A transformation associated to each motion defines the location of the virtual gage axis at virtual condition size with respect to the actual position axis of the hole (center of ACAS) relative to a candidate DRF. This must be done for each individual datum feature. On the other hand, the feature control frame defines an explicit constraint between each individual datum feature of the DRF (datum precedence). Each constraint can be defined by an homogenous transformation. Like in robotics, an open kinematics chain composed by joints and links can represent each homogenous transformation. The constraint transformations may close these chains. A kinematic mechanism is obtained with serial and closed chains. A similar method was proposed by Rivest to define tolerance zones for each individual feature using kinematics [Rivest, 1994].

The basic situation of the axis of each toleranced hole with respect to a candidate DRF can be defined by a constant transformation. For every candidate DRF, we associate a location of the toleranced feature axis basically located to the DRF. This is a particular configuration of the kinematic mechanism. For the mechanism point of view, the set of all true position axis of the specified feature is the workspace of the point that simulates this axis. As commonly known, the singularities for the Jacobian matrix of the mechanism for a particular point, defines the boundary of the workspace of this point. It represents all locations of the point of the mechanism. In the current case, the study is done for the end point of the mechanism. To define the Jacobian matrix, the following relations are used.

The first relation is defined by the equation of the point relative to the base DRF that is written as:

$$^{base}\boldsymbol{T}_{end}(\mathbf{q}) = \boldsymbol{U}_0 \qquad (4)$$

Where $\mathbf{U_0}$ describes the possible locations of the end effector. And $\boldsymbol{q}$ represents the parameter configuration of the mechanism.

The second relation is defined by the equation of closed chain of the mechanism. The product of the homogenous transformations between each closed-chain element forms this equation.

$$\prod {}^{a(i)}\mathbf{T}_i = \mathbf{I} \qquad (5)$$

Where $a(i)$ is the previous element of the ith element of the mechanism, and $\mathbf{I}$ is the identity matrix. There may exist more than one closed chain in the mechanism.
All the previous equations define the following system:

$$\begin{cases} \mathbf{F}(\mathbf{q}_d,\mathbf{q}_i) = 0 \\ \mathbf{\Psi}(\mathbf{q}_d,\mathbf{q}_i) = \mathbf{U}_0 \end{cases} \qquad (6)$$

Where $\mathbf{F}(\mathbf{q}_d,\mathbf{q}_i)$ is the equation of close chains, $\mathbf{\Psi}(\mathbf{q}_d,\mathbf{q}_i)$ the equation of the task, $\mathbf{q}_d$ are dependent parameters, and $\mathbf{q}_i$ are independent parameters.
The derivative of this system gives the Jacobian matrix J :

$$d\mathbf{U}_0 = \mathbf{J}\, d\mathbf{q}_i \qquad (7)$$

with $\mathbf{J} = \mathbf{K}_d(\mathbf{J}_d^{-1}\mathbf{J}_i) + \mathbf{K}_i$ and $\mathbf{J}_d = \frac{\partial \mathbf{F}}{\partial \mathbf{q}_d}$, $\mathbf{J}_i = \frac{\partial \mathbf{F}}{\partial \mathbf{q}_i}$, $\mathbf{K}_d = \frac{\partial \mathbf{\Psi}}{\mathbf{q}_d}$ and $\mathbf{K}_i = \frac{\partial \mathbf{\Psi}}{\partial \mathbf{q}_i}$

The analysis of the determinant of **J**, gives the singularities which define the boundary of the workspace of a particular point of the mechanism.

3.3. Application

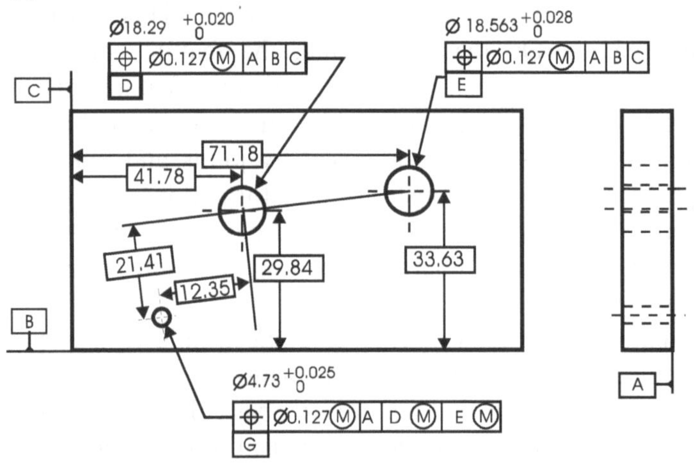

Figure 9 : Definition of a part

Several authors have studied the part defined in figure 9. It is used as a test case study for many models and methods. Hole D is located at MMC with respect to the planes A, B and C. Thus, this hole conforms to the specification when all points of the boundary of the actual hole D are not in the Virtual Cylinder of size (18.29-0.127). This virtual cylinder is basically oriented and located with respect to the DRF A, B and C. An equivalent definition is applicable to the Datum E.

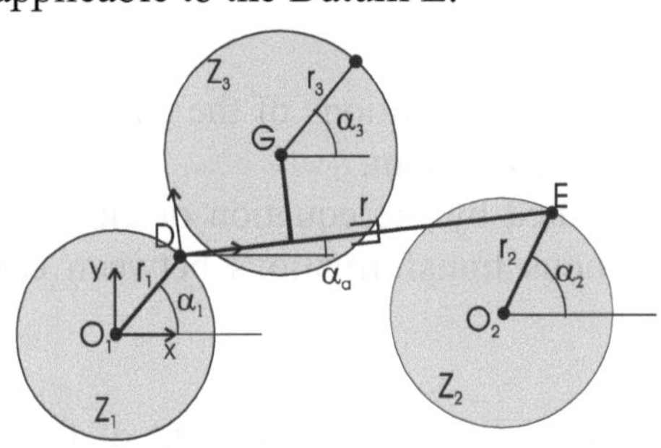

Figure 10 : Kinematic structure model

This specification is equivalent to say that if the actual mating envelope of hole D is manufactured away from its MMC virtual condition size (18.29-0.127) then the origin of the DRF may be in the circle of size (δ – (18.29 –0.127)) located at the centre of the actual mating cylinder of hole D. We note that the diameter of the cylinder D is necessarily in the interval [18.29 18.31] and the value (δ – (18.29–0.127)) belongs to the interval [0 0.147]. δ is the size of the actual mating cylindrical envelope of D, perpendicular to the primary datum A. An equivalent definition is used to obtain the set of Datum E. Datum E is used for the orientation of the DRF. The set of candidate datum D and E leads to a set of candidate DRF. The orientation of the DRF of an actual candidate part is not constant.

For the analysis of this particular example, we have to take into account the datum chaining property and the orientation of DRF. The kinematic structure of figure 10 is used to model this tolerance chain. The parameters r_1 and r_2 define the sweeping movement of the ADE datum reference frame that control the location of axis G.

The range of the parameters of the structure is calculated relative to the allowable motion of the DRF. For example, point D of figure 10 is used to simulate the motion of the axis of datum feature D. The allowable value for r_1 is calculated from the ACAS definition $T_{Z,D}$ $(0 < r_1 < T_{Z,D} / 2)$. In the same manner, datum feature E is simulated by point E and the allowable values for r_2 are $(0 < r_2 < T_{Z,E} / 2)$.

The following relations determine the tolerance zone size of each axis:

$T_{Z,D} = (\delta_D - (18.29\text{-}0.127))$, $T_{Z,E} = (\delta_E - (18.563\text{-}0.127))$, $T_{Z,G} = (\delta_G - (4.73\text{-}0.127))$,

At the simulation stage, the following values of $(\delta_D, \delta_E, \delta_G)$ or $(T_{Z,D}, T_{Z,E}, T_{Z,G})$ may be combined to give all the worst case situations of the axis of hole G (Table 1).

	Feature	Virtual condition	Actual Mating size	Tol. zone
At Minimum Material Size	D	18.163	$\delta_D = 18.31$	$T_{Z,D} = 0.147$
	E	18.436	$\delta_E = 18.591$	$T_{Z,E} = 0.155$
	G	4.603	$\delta_G = 4.755$	$T_{Z,G} = 0.152$
At Maximum Material Size	D	18.163	$\delta_D = 18.29$	$T_{Z,D} = 0.127$
	E	18.436	$\delta_E = 18.563$	$T_{Z,E} = 0.127$
	G	4.603	$\delta_G = 4.73$	$T_{Z,G} = 0.127$

Table 1: worst case situations zone of the axe of G.

Table 2 lists the size of holes D and E, and the value of parameters r_1, r_2 and r_3. These parameters are defined in figure 10, they are used to calculate the virtual and resultant parts. The left-hand side of figure 11 shows the details corresponding to the extension of the resultant condition of G with respect to datum ABC (in light grey). An extension of the virtual condition of feature G is also represented in this figure in dark grey.

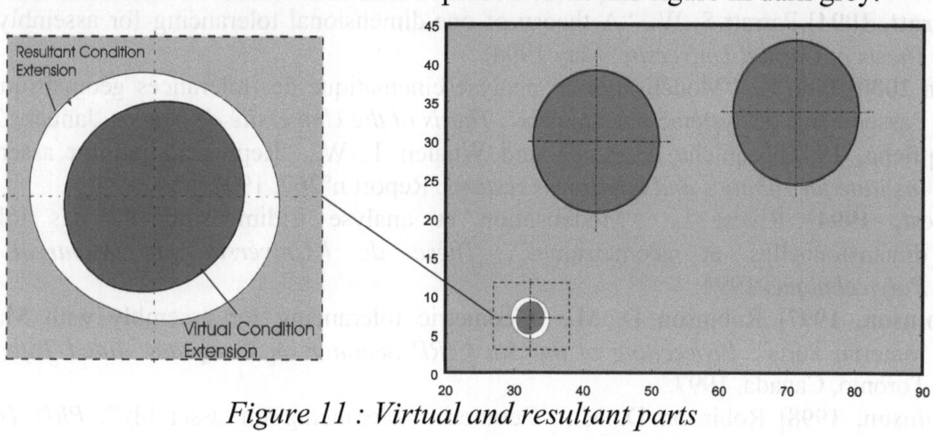

Figure 11 : Virtual and resultant parts

In figure 11, one can see that the gap between the virtual and resultant parts is 0.273 mm for D and 0.289 mm for E, but for G it was close to 2 mm.

	D	E	G			
			r_1	r_2	r_3	**operation**
Virtual part	18.163	18.436	0.147	0.155	4.603 / 2 = 2.3015	intersection
Resultant part	18.457	18.746	0.147	0.155	4.907 / 2 =2.4535	union

Table 2: Dimensions of D, E and G

4. CONCLUSION

This paper presents a direct extension of Robinson's and Rivest's works. Virtual and resultant parts are used to simplify the analysis of variational assembly. A kinematics parameterisation is used to calculate these parts. The method presented is general, it takes into account datum chaining and all the allowable orientations of the Datum reference frame controlling the features in a tolerance chain. It was shown that it is possible to draw an analogy between the sweeping domain and the Boolean union and intersection operations. These unions and intersections are deduced from the boundary of the workspace of the kinematics mechanism associated to the toleranced part. Powerful analysis tools can thus be developed from the concepts and methods presented.

REFERENCES

[**Bennis et al., 1999**] Bennis F., Pino L. and Fortin C., "Analysis of positional tolerance based on the assembly virtual state". *Proceeding of the 6th CIRP Seminar on Computer-Aided Tolerancing*, University of Twente, The Netherlands, pp. 415-424, 1999.

[**Pairel 1995**] Pairel E., "Métrologie fonctionnelle par calibre virtuel sur machine à mesurer tridimensionnelle". *Thesis of the Université de Savoie*. December 1995.

[**Parratt, 1994**] Parratt S. W., "A theory of one-dimensional tolerancing for assembly". *PhD Thesis of Cornell University*. May 1994.

[**Pino 2000**] Pino L., "Modélisation et analyse cinématique des tolérances géométriques pour l'assemblage de système mécaniques". *Thesis of the Université de Nantes*. January 2000.

[**Requicha, 1992**]Requicha A. A. G. and Whalen T. W., "Representation for assemblies". *Institute for robotics and intelligent systems*. Report n°267, 1992.

[**Rivest, 1994**] Rivest L., "Modélisation et analyse tridimensionnelle des tolérances dimensionnelles et géométriques". *Thèse de l'Université de Montréal, Ecole Polytechnique*. 1994.

[**Robinson, 1997**] Robinson D. M., "Geometric tolerancing for assembly with Maximum Material Parts". *Proceeding of the 5th CIRP Seminar on Computer Aided Tolerancing*, Toronto, Canada. 1997.

[**Robinson, 1998**] Robinson D. M., "Geometric tolerancing for assembly". *PhD Thesis of Cornell University*. May 1998.

[**Srinivasan et al., 1989**] Srinivasan V. et Jayaraman R., "Geometric tolerancing : II. Conditional tolerance". *IBM Journal of Research an Development*, 33(2) pp.105-124, 1989.

[**Srinivasan 1993**]Srinivasan V., "The Role of Sweeps in Tolerancing Semantics". *Manufacturing Review*, 6(4) pp. 275-281, 1993.

3D Tolerances Analysis, from Preliminary Study.

Paul Clozel

MECAmaster Sarl
39, chemin du Moulin Carron, 69570 Dardilly, France
mecamast@club-internet.fr

Conception Team, Ecole Centrale de Lyon,
BP 163, 69131 Ecully Cedex, France
Paul.Clozel@ec-lyon.fr

Abstract: A mechanical assembly is made up of parts and links. Its definition by solids and linkages, and the definition of tolerances for the linkages (these can be 3D position tolerances, clearances...), enables us to calculate functional conditions on the assembly.
3D tolerance analysis is performed like a unidirectional chain of dimensions, with computed coefficients (sensitivities) that fully take into account the 3D effects. Results are presented as calculation tables, and are fully explained. A 3D graphical representation of the sensitivities and contributions enables us to understand immediately where greater precision is (or is not) required.
The matching of a car door, positioned by an assembly tool on the "body in white" will be presented, in order to compute the clearances and the surfaces alignments.
These calculations only require the definition of the linkages (or functional surfaces) between parts and can be done on the basis of preliminary studies (choice of the mechanical assembly architecture, of the tolerancing type, of references... , choice of the assembly process...) until the final tolerance distribution is optimized. The whole assembly process of an airplane is currently being studied by a major aeronautics company, before the parts are designed, using the wireframe architecture.
MECAmaster software, standalone (Windows) or integrated in Cad systems (Cadds5, Euclid3, Catia) works this way and is used daily by major automobile and aeronautics companies.
Keywords: 3D tolerances, chain of dimensions, sensitivities, preliminary study, product and process

1. 3D CHAINS OF DIMENSIONS

1.1. Introduction

The tolerancing of a three-dimensional mechanical system is done in order to obtain certain functional characteristics: tolerance of positioning, functional gaps...
These characteristics depend on the parts and the assemblies which make up the system.
The assemblies consist of surfaces in contact and can be described by linkages (indeed, the

P. Bourdet and L. Mathieu (eds.),
Geometric Product Specification and Verification: Integration of Functionality, 93-104.

analysis of all the possible contacts between two parts has led to the definition of standardized linkages).

From the definition of the mechanical system by linkages, a kinematic simulation is possible for any isostatic 3D mechanical system, in order to know the sensitivity that the local tolerances of the parts and assemblies have on the functional tolerances chosen, and reciprocally.

Then, the definition, for the various linkages of tolerance values (which can correspond to tolerances of 3D positions, gaps...) makes it possible to evaluate the selected functional tolerance. One obtains the equivalent of a "3D chain of dimensions".

The following model represents an exhaust system (simplified...) of a motor vehicle, held by two assemblies. These are modeled by perfect linkages to which one will assign a tolerance to each component of positioning.

The problem is to determine which is the most important factor in this mechanical system for the vertical positioning of the exhaust pipe exit.

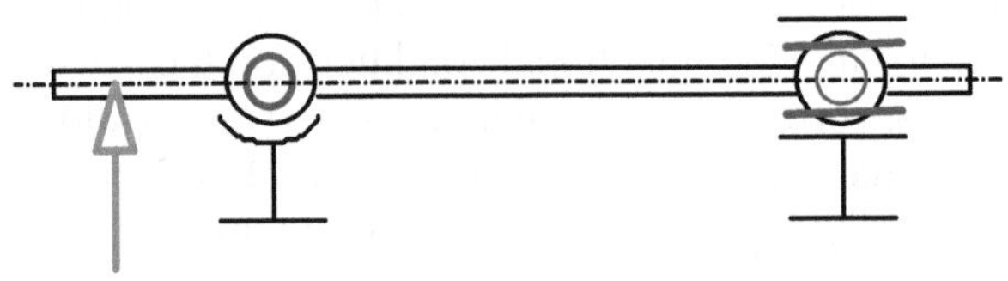

1.2. Definition of the data:

The place where one wishes to control the tolerance is defined:

Tolerance type	POSITIONAL TOLERANCE
between	body
and	pipe
Name	pipe exit
Position.	600,-3200,-200
Direction.	0,0,1

The assemblies are defined in MECAmaster by linkages which ensure a positioning corresponding to reality. For example, for the front fastener:

Linkage type	CIRCULAR_CONTACT
between	body
and	pipe
Name.	front
Center.	600,-2000,-200
Translation direction	0,1,0
Radial tolerance	3

This is the only sort of information necessary for the calculation.

1.3. Principle of calculation:

The software then simulates a small displacement for each component of each linkage, as one can do it on a simple case:

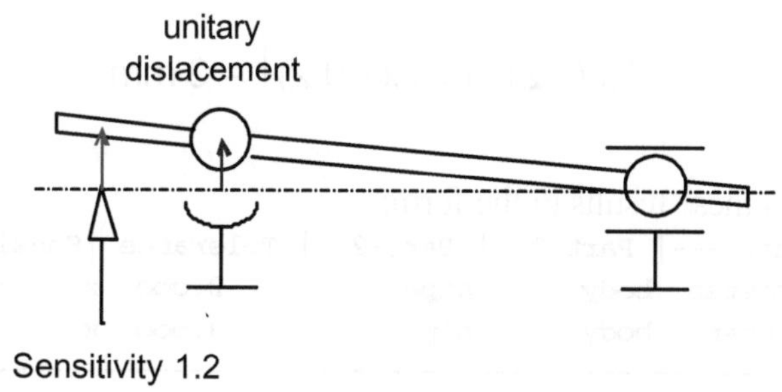

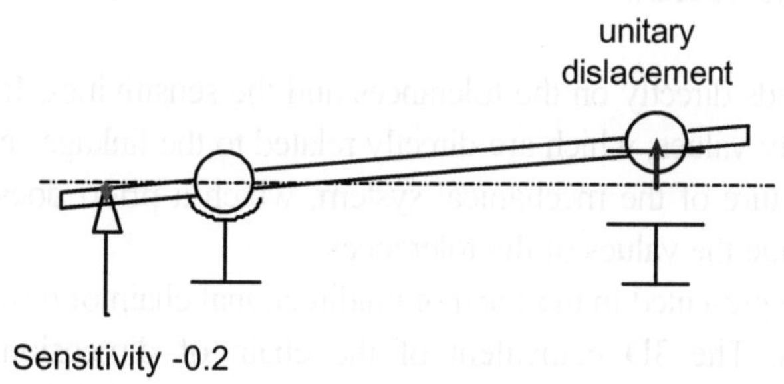

For small displacements on the horizontal components of the linkages, the resulting vertical displacement for the pipe exit are null.

1.4. Results: "3D chains of dimensions"

The sensitivities (or influences), i.e. the ratios induced displacement / initial displacement are calculated and displayed, at their place of origin, with lengths proportional to their intensity:

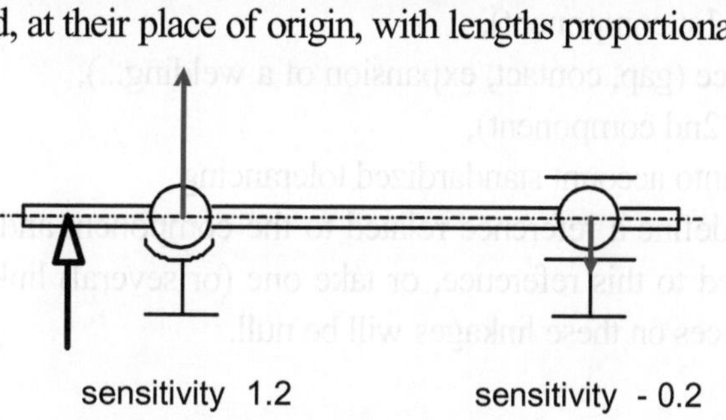

The simultaneous display of all the sensitivities makes it possible to see immediately which are the dominant tolerances for the system.

In the same way, the absence of sensitivity for certain components is very interesting information...

The relation between the vertical tolerance of the pipe and the tolerances of the assemblies is:

$$\Sigma \left| \text{Tolerance x Sensitivity} \right| = \text{Resulting Tolerance}$$

If the vertical tolerance of the rear fastener is 2 mm and that of the front fastener 3 mm, one obtains:

$$\left| 2 \text{ x } (1.2) \right| + \left| 3 \text{ x } (-0,2) \right| = 3 \text{ mm}$$

MECAmaster presents these results in the form:

```
--------|----Names----| Part-1   | Part-2   | Tolerance |Sensitivity| Contribut
SPHERIC  front fasten  body       pipe            2.000   x     1.200   =      2.400
CIRC CON back fasten   body       pipe            3.000   x      .200   =       .600
------------------------------------------------------------------------------------
ARITHMETICAL calculation:                     Value of the Tolerance    =      3.000
```

The final value depends directly on the tolerances and the sensitivities. It is of primary interest to act on the sensitivity values, which are directly related to the linkages and their positions, i.e. to act on the architecture of the mechanical system, which a priori does not "cost" anything. Then, one could change the values of the tolerances.

These results are presented in the form of unidirectional chain of dimensions, in which one adds the coefficients. The 3D equivalent of the chain of dimensions is obtained. Many concepts (statistical distributions of tolerances...) are directly applicable.

1.5. Tolerance / interface / tolerance

Beyond the principle of calculation explained previously, it is interesting to be able to take into account various tolerances for a linkage.
The possibility of defining for each linkage three tolerances,
- a tolerance (for the 1st component),
- an interface tolerance (gap, contact, expansion of a welding...),
- a tolerance (for the 2nd component),

allows to easily take into account standardized tolerancing.

One can either define a reference related to the component and define the tolerances of the linkages compared to this reference, or take one (or several) linkages as references. The corresponding tolerances on these linkages will be null.

1.6. Note:

The limits of the approach are related to the hypotheses, which are not restrictive:
- perfect linkages (but taking into account gaps is possible, and the definition of the assembly on the reality of the contacts by combination of linkages).
- rigid solids (but the taking into account of known deformations or local flexibility is possible).
- small displacements, which is generally very well checked in tolerancing.

2. STUDY OF A CAR DOOR MATCHING

2.1. The door matching

In production process, the assembly of the doors of a car on the "body-in-white" is an important operation, for it determines the quality of the finished product. The objective is to have gaps (for example, between the door and the body, or between the front door and the rear door) as small as possible (reduction of the wind noises, esthetics) and the most regular possible (esthetics, to avoid a "billiard cue"). In the same way, the surfaces alignments (for example, door/body or front door/rear door must be perfectly controlled (reduction of the wind noises, esthetics).

The assemblage of the doors and the "body-in-white" can be done in various ways. MECAmaster is used to determine the best method by several car manufacturers, by means of simulation of the various assembly techniques.

The study hereafter describes an industrial technique very much used: the back door is taken, i.e. is positioned in the "back assembly tool", then the "back assembly tool " is positioned relative to the body.

The hinges of the back door are then welded. Then, the front door is taken in the "front assembly tool", then the "front assembly tool" is positioned relative to the back door and on the body. The hinges of the front door are then welded.

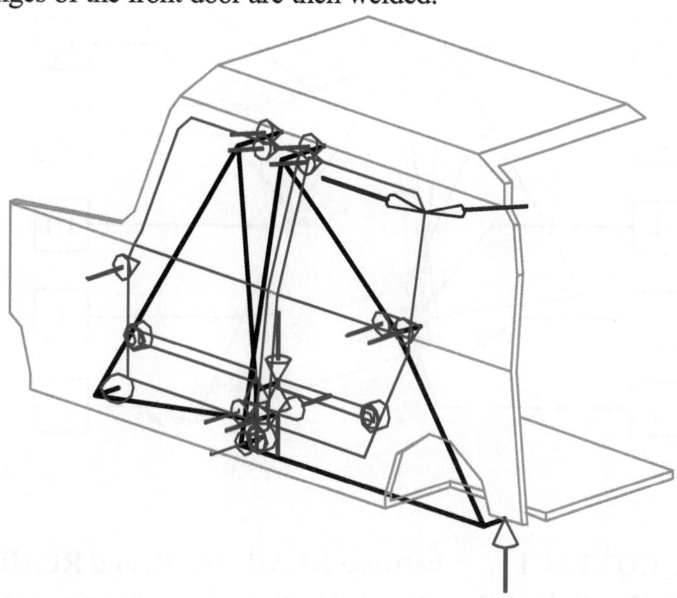

These various positionings use zones of contact and various holes (side-impact bar, fixing of bumper, dedicated holes).

The questions which arise are of the type: "where must the support of the front tool on the body be?", "are the high zones of positioning the same for the front and back doors?", "where must they be?", "must one position the front tool exclusively on the body or also on the back door?"... and without rigorous quantification it is difficult to make the choices.

2.2. Definition of the Linkages

We shall now present the positioning of the back door. The MECAmaster model is defined by using existing CAD models. The linkages are chosen according to the positionings they ensure. All the values are given in± x mm.

We will take reasonable values for the tolerances of the various parts, that are:

- Body in the plane of the metal sheet: 1,5
- Body in direction perpendicular to the metal sheet: 1,75
- Door in the plane of the metal sheet: 1
- Door in direction perpendicular to the metal sheet: 1,25
- Assembly tool (relatively precise because mechanical): 0,2
- Gap between Assembly tool and Body Case or Doors: 0,1

One could make different choices of references for the various parts. For example, for the door, all is defined relative to a "stamping and crimping" reference, but one could also have chosen the side-impact bar holes as references (the tolerances defined above for the door would then be different).

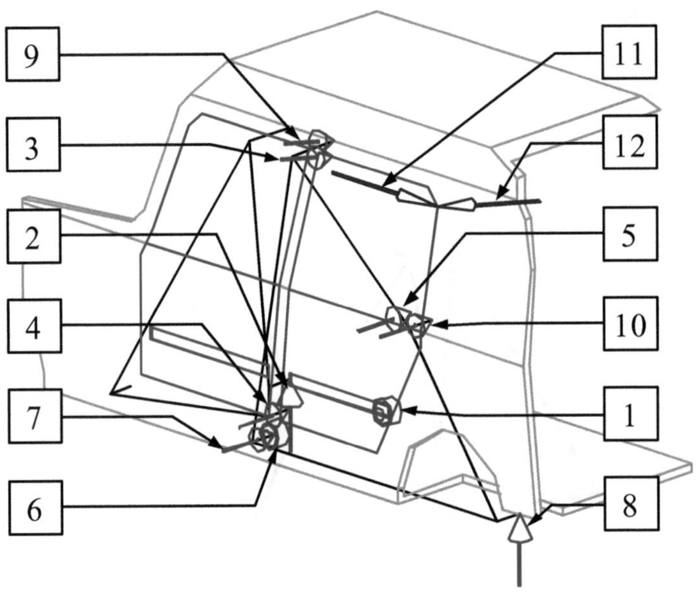

1 : CIRCULAR_CONTACT between REAR_TOOL and REAR_DOOR

Positioning of REAR_DOOR in REAR_TOOL by the back hole of the side-impact bar. The hole and pin are localized in circular zones and are circular (the gap is thus radially uniform). Positioning on X and Z axis are dependent.

tolerance:	0,2	REAR_TOOL tolerance
interface:	0,1	Gap between Assembly Tool and Door
tolerance:	1	REAR_DOOR tolerance

2 : POINT_CONTACT between REAR_TOOL and REAR_DOOR

Positioning of REAR_DOOR in REAR_TOOL by the front hole of the side-impact bar. The hole is rectangular and ensures only positioning on Z axis.

tolerance:	0,2	REAR_TOOL tolerance
interface:	0,1	Gap between Assembly Tool and Door
tolerance:	1	REAR_DOOR tolerance

3 4 5 : POINT_CONTACT between REAR_TOOL and REAR_DOOR

Positionings of REAR_DOOR on REAR_TOOL (3 surfaces alignments on Y axis).

tolerance:	0,2	REAR_TOOL tolerance
interface:	0	REAR_TOOL is in contact with REAR_DOOR
tolerance:	1,25	REAR_DOOR tolerance

6 : CIRCULAR_CONTACT between BODY and REAR_TOOL

Positioning of REAR_TOOL on the BODY by a hole which ensures positioning X, Y and Z One could be tempted to put a Spherical Contact, but technologically, what is done in the X-Z plane is independent of what is done on Y axis (the positioning of the pin on Y axis does not depend on the position in X-Z in the circular hole, the localization of the hole in X-Z is independent of its position on Y axis. The zones of positioning and gap are cylindrical rather than spherical. One will thus use this CIRCULAR_CONTACT (n°6) and the POINT_CONTACT (n°7) for this hole.

tolerance:	1,5	BODY tolerance
interface:	0,1	Gap between Assembly Tool and Body
tolerance:	0,2	REAR_TOOL tolerance

...

2.3. Choice of the functional conditions

When the linkages are defined, one just has to define the functional conditions to be analyzed. The same model makes it possible to analyze the various functional conditions.

One is interested here in "the" question which always interests the car manufacturers: the positions on X, Y, Z axes of the higher back corner of the door.

11 : POSITIONAL TOLERANCE between BODY and REAR_DOOR

One analyzes the dimension condition "Gap on X axis" between BODY and REAR_DOOR.

tolerance:	1,5	BODY Form Tolerance
tolerance:	1	REAR_DOOR Form Tolerance

12 : POSITIONAL TOLERANCE between BODY and REAR_DOOR

One analyzes the dimension condition "Surfaces alignment on Y direction" between BODY and REAR_DOOR.

tolerance:	1,75	BODY Form Tolerance
tolerance:	1,25	REAR_DOOR Form Tolerance

2.4. Results

The values of the arithmetic (worse case) and statistical tolerances are then calculated by MECAmaster. For the tolerance number 11, the "Gap on X axis", one obtains:

```
--|------ Names -------|------ Parts -----| Tolerance |Sensitivity| Contribut
• • •
                         rear_tool          0.200000 \
PC  5  y rear_door middle                      0.000  x 0.162709  =  0.235928
                         rear_door             1.250 /                   2.0%
-----------------------------------------------------------------------------
                         body                  1.750 \
PC  10  y body middle                          0.000  x 0.128226  =  0.250041
                         rear_tool          0.200000 /                   2.1%
-----------------------------------------------------------------------------
                         body                  1.500 \
PC  8  pin z body side                      0.100000  x 0.791524  =     1.425
                         rear_tool          0.200000 /                  12.2%
-----------------------------------------------------------------------------
                         rear_tool          0.200000 \
PC  2  pin z rear_door                      0.100000  x    1.568  =     2.039
                         rear_door             1.000 /                  17.5%
-----------------------------------------------------------------------------
                         body                  1.500 \
CI  6  pin xz body side                     0.100000  x    1.284  =     2.311
                         rear_tool          0.200000 /                  19.8%
-----------------------------------------------------------------------------
                         rear_tool          0.200000 \
CI  1  pin xz rear_door                     0.100000  x    1.846  =     2.399
                         rear_door             1.000 /                  20.6%
-----------------------------------------------------------------------------
                         body                  1.500 \
PT  11  Gap on X axis                                 x    1.000  =     2.500
                         rear_door             1.000 /                  21.5%
=============================================================================
ARITHMETICAL calculation:                  Value of the Tolerance  =   11.651

STATISTICAL calculation:                   Value of the Tolerance  =    3.845
```

The result value depends directly on the values of the tolerances and the sensitivities. It is of primary interest to act on the sensitivity values, which are directly related to the linkage and to their positions, i.e. to act on the architecture of the mechanical system, which a priori does not "cost" anything. Then, if necessary one could change the value of certain tolerances.

The graphic results, with lengths proportional to their intensity, are:

- sensitivities. The sensitivities depend only on the geometry.
- contributions (contribution = tolerance x sensitivities). They allow a total perception of the contribution of each linkage (or component) to the functional condition. The contributions depend on the geometry and the tolerances.

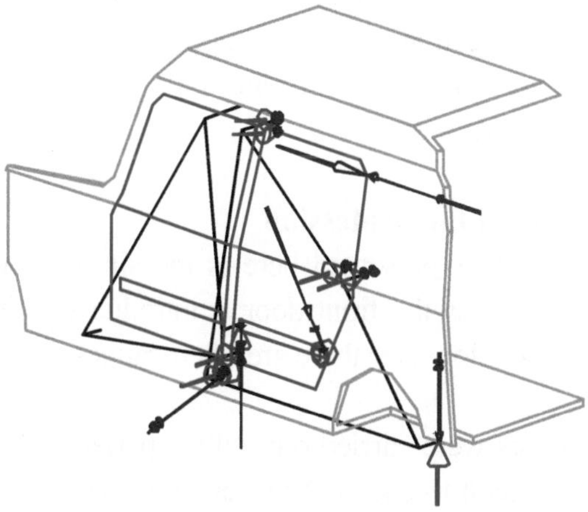

Display of the contributions to gap size in the X direction

It is to be noted (as one might have expected) that for the gap between BODY and REAR_DOOR in the X direction, the most important contributions are located in XZ plane (in fact, the directions of the linkages are not rigorously X Y and Z, but those at a tangent to or perpendicular to the body).

With the same model, one determines the value of the tolerance for the "Surfaces alignment in the Y direction":

```
• • •
=========================================================================
ARITHMETICAL calculation:                 Value of the Tolerance =   9.279

STATISTICAL calculation:                  Value of the Tolerance =   3.112
```

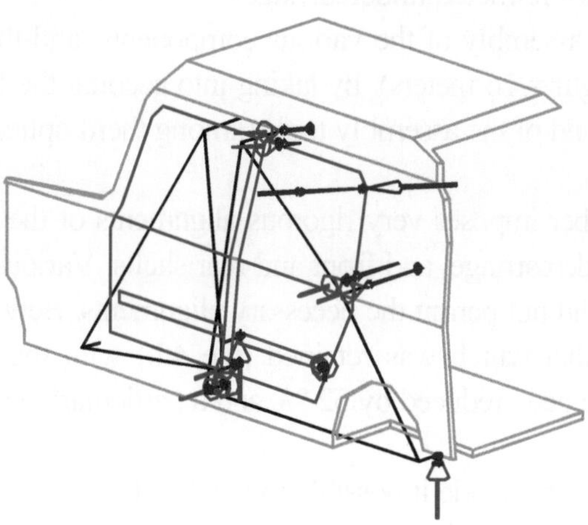

Display of the contributions, Surfaces alignment on Y direction

It is to be noted (as one might have expected) that for the surface alignment between BODY and REAR_DOOR in the Y direction, the most important contributions are located in the Y direction (in fact, the directions of the linkages are not rigorously X Y and Z, but those at a tangent to or perpendicular to the body).

2.5. An evolution of the profession

The door-fitting of the back door presented here is the simplest and most comprehensible situation that we meet. As soon as the front door is introduced, with common references, it becomes much more complex. Moreover there are often gauges for the assembly of the hinges and handles…

In the past, these analyses were carried out with experience, knowledge, and 1D chains of dimensions. But it was difficult to choose between architectures, and the quantification was not very precise. The MECAmaster analysis brings a rigorous answer. The time saving is enormous (analyses which required 4 to 5 weeks are carried out in 1 day), and several solutions can be evaluated. However, a technical limitation remains, because the solids (body and doors) are far from being infinitely rigid.

3. AIRBUS A340-600 KEAL BEAM ASSEMBLY

We now present the analysis of the assembly of the central 15-21 section of the AIRBUS A340-600 (lengthened version of A340).

The increase in capacity of this plane imposes a very important reengineering, and the use of innovative technologies, specially for the central 15-21 section (major section which is very highly solicited because it includes the wing fixings, the middle undercarriage, and has to provide space for 3 sets of retracted undercarriage).
The study related to the assembly of the various components, and the integration of the Keal Beam (carbon fiber, length = 16 meters), by taking into account the tolerances of the different parts, of the assemblies and of the assembly tools (among them optical beam).

The rigidity of carbon fiber imposes very rigorous alignments of the beam, particularly on the fittings of the central undercarriage, and front and rear shells. Various assembly methods were studied, some of which did not permit the necessary alignments. Between the initially imagined assembly process, and that which was retained (the 4th) with the two separated sides, the alignment distances have been reduced by 52 %, and a particularly critical value by 72%.

The study with MECAmaster made it possible to validate the constructive principles, the tools for assembly, the process of assembly, more than one year before the realization of the beam.

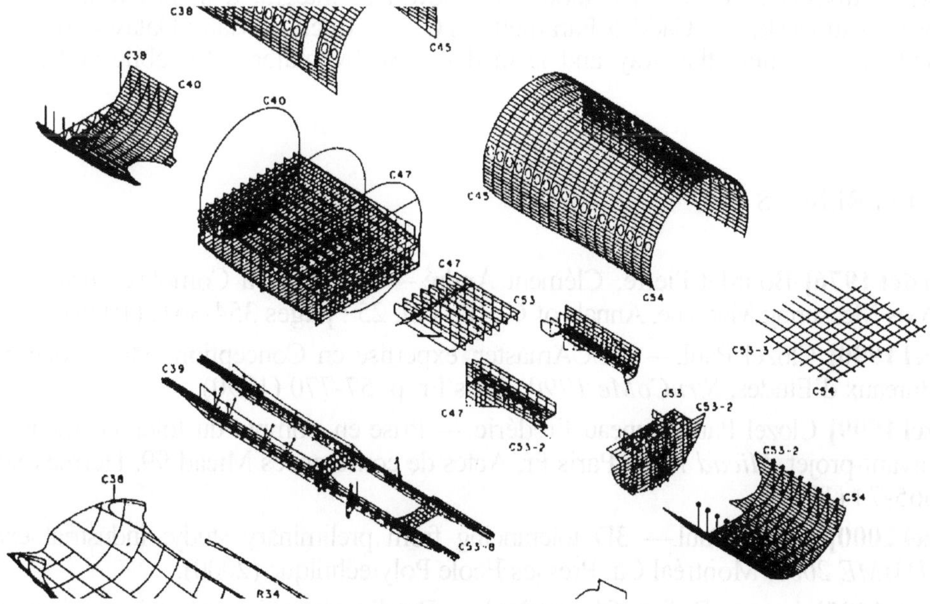

Choice of the assembly process
of the central 15-21 section / keal beam of the A340-600 AIRBUS

4. 3D TOLERANCES ANALYSIS, FROM PRELIMINARY STUDY

This 3D Tolerances analysis only requires the definition of linkages (or functional surfaces) between parts and can be done from preliminary study. It enables:

- the choice of the mechanical system architecture, choice of the assembly process, and, even when one only knows the definition of the linkages (or functional surfaces), the minimization of the coefficients of sensitivity,
- the choice of the tolerancing types, of the references, in order to structurally reduce the contributions.

These quantifications and comparisons of the various solutions enable the choice of the best solution at minimal cost, and thus generates important economies.

Then, once the major design choices are made, one can do detailed studies to distribute and optimize the values of the tolerances, to obtain the desired functional conditions at lower cost, until the end of the design phase.

The whole assembly process of an aircraft is actually under study by a major aeronautics company, before the parts are designed, on the basis of the preliminary wireframe architecture.

Other major advantages are

- With the calculation table, one can understand the results.
- The use is possible on any isostatic system, even with many internal chains. The study of an hyperstatic system without gaps, and isostatic with gaps, is possible.
- The calculations are fast (1 minute for an assembly with one hundred 3D parts), with a usual computer.

MECAmaster software, standalone (Windows) or integrated in different Cad systems (Autocad–Autodesk, Cadds5-Parametric TC, Euclid3-Matra Datavision, Catia-Dassault Syst)) works this way and is in daily use by major automobile and aeronautics companies.

REFERENCES

[Bourdet 1976] Bourdet Pierre, Clément André.— Controlly a Complex Surface with a 3 Axis Measuring Machine, Annals of CIRP, Vol. 25 : pages 354-361, (1976).

[Clozel 1990] Clozel Paul.— MECAmaster, expertise en Conception Mécanique pour les Bureaux d'Etudes, *StruCoMe 1990,* Paris Fr, p757-770 (1990).

[Clozel 1999] Clozel Paul, Luneau Frédéric.— Prise en compte du tolérancement 3D dès l'avant-projet, *Micad 1999,* Paris Fr, Actes de conférences Micad 99, Hermes Sciences, p65-74 (1999).

[Clozel 2000] Clozel Paul.— 3D tolerancing from preliminary study: industrial examples, *IDMME 2000*, Montréal Ca, Presses Ecole Polytechnique (2000).

[Lacour 1995] Lacour Didier, Clozel Paul.— De l'analyse des mécanismes à la cotation I.S.O., *Colloque PRIMECA*, La Plagne Fr, (1995).

[Le Borzec 1974] Le Borzec, Lotterie.— Théorie des mécanismes, *Editions Dunod* Fr, (1974).

[Marguet 2000] Marguet Benoît.— The Key Charasteristic Process at Aerospatiale-Matra Airbus, *Key Characteristics Symposium*, Long Beach, California, January 19-21 (2000).

[Mecamaster 1996] MECAmaster.— Logiciel calcul efforts 3D, hyperstatismes, tolérances 3D. *Notice d'utilisation*, Fr, MECAmaster Sarl, 39, ch moulin Carron, 69570 Dardilly, Fr (1996).

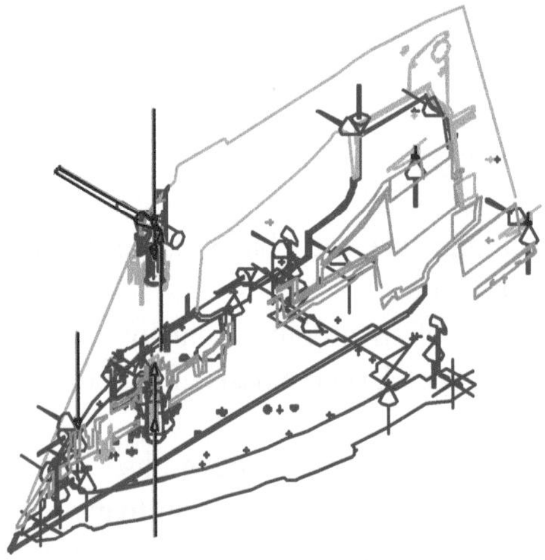

Domestic iron, whose architecture and values of tolerances were optimized

Statistical Tolerance Analysis Using GapSpace

Zhihua Zou,   ***Edward Morse***
The Center for Precision Metrology
Department of Mechanical Engineering & Engineering Science
The University of North Carolina at Charlotte
NC 28223, USA | zzou@uncc.edu

Abstract: This paper describes a new method for statistical tolerance analysis using the GapSpace model. We show how, after modeling the distribution of manufactured dimensions with a joint probability density function, assembly analysis is performed by integrating the density function over a tolerance region or assembly region described in GapSpace. Through an example, this paper addresses how statistical tolerance analysis should mimic the "manufacturing scheme" of the parts and how the GapSpace model can be used to identify optimal results.
Keywords: statistical tolerancing, assembly modeling, assembly analysis, GapSpace

1. INTRODUCTION

Tolerance analysis is a continued focus of the manufacturing industry as it seeks to both improve the quality of its products and produce them with lower cost. The appropriate allocation of tolerances in an assembly may result in lower costs per assembly and a higher probability of good fit, reducing the number of rejects or the amount of rework. Statistical tolerance analysis is a powerful analytical method because it not only predicts the effect of manufacturing variation on design performance and production cost, but also allows designers and manufacturing personal to take advantage of statistical averaging to relax the component tolerances without sacrificing quality.

The “GapSpace Model” proposed by Morse [Morse 99] can be used to capture necessary and sufficient conditions for the satisfaction of various assembly criteria. The problem of interchangeable assembly is represented by partitioning the GapSpace into regions where assembly is possible and regions where it is not. The model of an individual assembly or a population of potential assemblies can be located in the GapSpace and be analyzed by comparison to these regions.

In this paper we distinguish between several types of parameter assignments, or dimension schemes. The design scheme describes how dimensions and tolerances are specified for the parts, the manufacturing scheme captures the dimensions that vary due to the manufacturing process, and the inspection scheme identifies the dimensions that are measured on the finished components. The manufacturing scheme will provide the best information for statistical assembly analysis, although it may not be available during the initial design phase. Different manufacturing schemes with the same design

P. Bourdet and L. Mathieu (eds.),
Geometric Product Specification and Verification: Integration of Functionality, 105-114.

scheme may generate huge differences in statistical tolerance analyses. In this paper, this difference is quantitatively and qualitatively shown using the GapSpace model.

In the next section we identify a few selected works in assembly analysis, and relate these works to our research. The GapSpace model is described briefly using an example, as it is the basis for later analysis. We then present the execution of statistical tolerance analysis in GapSpace and show how the optimization results are realized.

2. RELATED WORK

The linear "stack-up" method is a fundamental tolerance analysis technique, as described in [Bjørke 89] and others. This analysis requires that assembly components be in contact at mating faces and closed loops are generated through these faces. Every loop is composed by chain of dimensions and a single gap. The "worst case" tolerance analysis is performed by accumulation of the variability in the chain. In 1-D analysis, all sensitivities of dimensions are ± 1. The RSS (Root Sum Squared) model and its complex forms [Bender 62] are widely used for statistical tolerance analysis. We adopt the "RSS" model in the paper, in which the distributions of part dimensions are assumed to be normal.

The group led by K.W. Chase at Brigham Young University has worked to extend this stack-up analysis to 2- and 3-dimensional assemblies [Chase 97][Gao 95]. They use vector loop-based assembly models for these analyses, in which closed vector loops describe the small kinematic adjustments and open vector loops describe critical clearances or other assembly features. A commercial software package based on this work is available and can perform worst-case and statistical analyses.

Monte Carlo simulation is commonly used for statistical tolerance analysis [Grossman 76]. A random number generator is used to simulate variability of each component's size and form. These values are combined through the assembly function to determine the resulting influence on some clearance or gap dimension. This simulation method can be quite time-consuming.

Mullins and Anderson [Mullins 98] have proposed a graph-based approach to tolerance analysis. By identifying constraints that exist among components of an assembly, the Physically Constraining Face Set (PCFS) is generated. Loops are constructed by working on the assembly graph and PCFS. In the GapSpace model used in this paper, the step for identifying PCFS is not necessary as these constrains are implicitly included in the rules of searching loops in the assembly graph.

With the exception of the PCFS model, the methods described above require that there is contact between assembly components and at most one clearance exists in every loop. The GapSpace model does not have these restrictions. In the examples that follow, we show how assemblies without required contact conditions can be analyzed for both worst case and statistical criteria.

3. GAPSPACE MODEL

The GapSpace model allows us to capture the physical requirements in one-dimensional tolerancing analysis independently of how the parts are dimensioned. It can be used to induce whether particular dimension schemes impose sufficient conditions for assembly. This section introduces GapSpace model, as it is the basis of other sections of the paper.

Two basic concepts, *directed dimension tree* and *liaison*, are used throughout this paper and need to be explained. The directed dimension tree is a data structure that represents the dimension scheme for parts in a single direction. A directed dimension tree describes a unique set of feature relationships, although a single part may have multiple equivalent dimension trees, as shown in Figure 1.

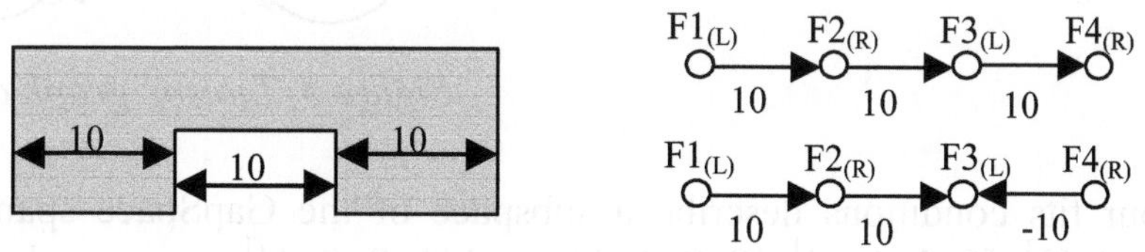

Figure 1; A nominal part and two equivalent directed dimension trees.

A liaison is defined as the specification of adjacency for a pair of opposing features, which are from different parts and may potentially interfere when assembling. The value of a liaison is the signed distance between the two features. The hinge example in Figure 2 is used in this paper to show how the GapSpace model works. There are four liaisons in Figure 2, which are named g_1, g_2, g_3 and g_4. Treating the gaps as independent variables, we can construct a four-dimensional linear vector space with basis vectors $[1\ 0\ 0\ 0]^T$, $[0\ 1\ 0\ 0]^T$, $[0\ 0\ 1\ 0]^T$, and $[0\ 0\ 0\ 1]^T$. This linear vector space is called the GapSpace for the assembly.

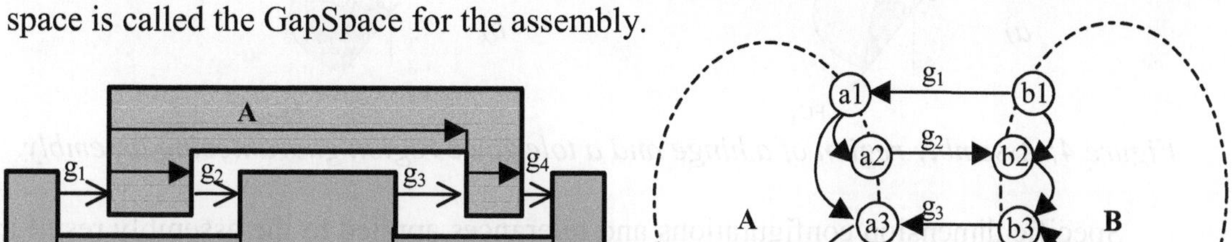

Figure 2; A hinge assembly and its assembly graph.

After identifying all liaisons and dimension trees, an assembly graph can be constructed. In the assembly graph, liaisons are represented with directed arcs between the parts which, in turn, are represented using dimension trees. Assembly graphs can have different levels of detail. For instance, the low-level graph in Figure 3, called a liaison graph, has each part regarded as a node and the arcs in the graph are the liaisons between parts. Higher-level graphs explicitly show the directed dimension trees for each part. Certain analyses can be thought of as "dimensioning independent" and only require the information captured in the liaison graph.

Based on principles detailed in [Morse 99], assembly cycles are determined by following the directed arcs in the liaison graph. *Fits Conditions* (FCs) correspond to these assembly cycles and represent the physical requirements for assembly. Four assembly cycles exist in the hinge example, so four FCs are generated. These FCs must be non-negative to satisfy the assembly requirement; this is shown in equations (1).

$$\begin{cases} FC1 = g1 + g2 \geq 0 \\ FC2 = g2 + g3 \geq 0 \\ FC3 = g3 + g4 \geq 0 \\ FC4 = g1 + g4 \geq 0 \end{cases} \quad (1)$$

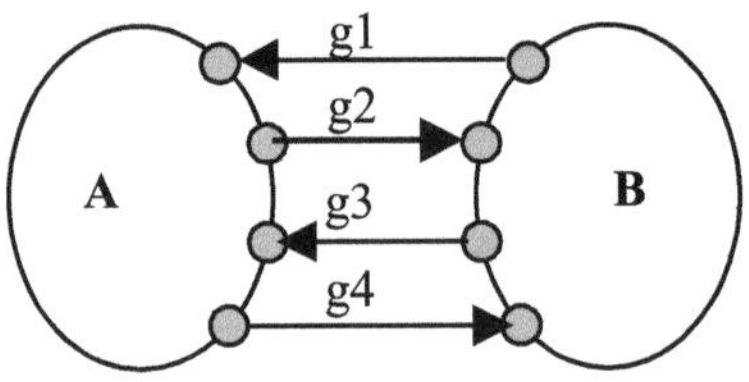

Figure 3; Liaison graph of Figure 2.

The four fits conditions describe a subspace of the GapSpace spanned by the vectors $[1\ 1\ 0\ 0]^T$, $[0\ 1\ 1\ 0]^T$, $[0\ 0\ 1\ 1]^T$, and $[1\ 0\ 0\ 1]^T$ corresponding to the fits condition gap pairs. This subspace is named the assembly space. The assembly space for the hinge example is shown in Figure 4a. The assembly region corresponds to the portion of this subspace in which the inequalities from equation (1) are satisfied; it is independent of part representation and captures the necessary and sufficient conditions required for assembly.

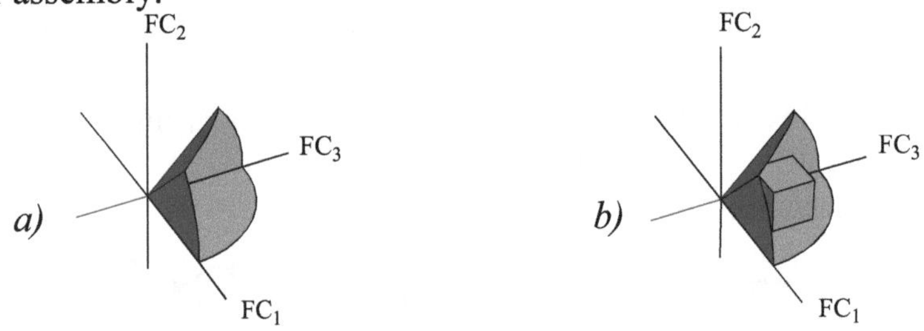

Figure 4; Assembly region of a hinge and a tolerance region guaranteeing assembly.

Specific dimension configurations and tolerances applied to the assembly result in a region in the assembly space representing the possible dimension combinations permitted by the design specification. This region is called the tolerance region of the assembly. Figure 4b shows a tolerance region contained within the assembly region. The tolerance region for a single part can be generated by applying tolerances to that part while maintaining fixed nominal dimensions for all other parts in the assembly.

4. STATISTICAL TOLERANCE ANALYSIS

Statistical tolerance analysis can be used by designers and manufacturing personnel to take advantage of statistical averaging over assemblies of parts, allowing the use of less restrictive tolerances in exchange for admitting the small probability of non-assembly. In this section we show how statistical analysis is facilitated by the GapSpace model. We discuss first the manufacturing dimension scheme and its representation of the statistical variability of dimensions. Then we describe the role of the tolerance regions and assembly regions in statistical analysis. We examine statistical independence of the dimensions and the errors incurred by an incorrect assumption of independence. An optimization result is obtained by assigning dimensions effectively. Finally, we compare different manufacturing schemes to determine which predicts the highest assembly yield.

4.1. Manufacturing dimension scheme

The manufacturing dimension scheme of a single part is a parametric representation of the variability imposed by the manufacturing process. In the scheme, certain dimensions vary as a function of process parameters while other dimensions are dependent on these 'primary' dimensions.

Because of manufacturing process limitations, the manufacturing dimension scheme may not be fully consistent with the design scheme. For example – the geometric position tolerance applied at MMC in Figure 5 allows additional position variability as the feature sizes depart from MMC. However, many processes that could be used to manufacture the features have independent factors which affect the size and the position of the features. It is difficult to envision a realistic manufacturing process that will automatically admit more positional variation as the features depart from MMC.

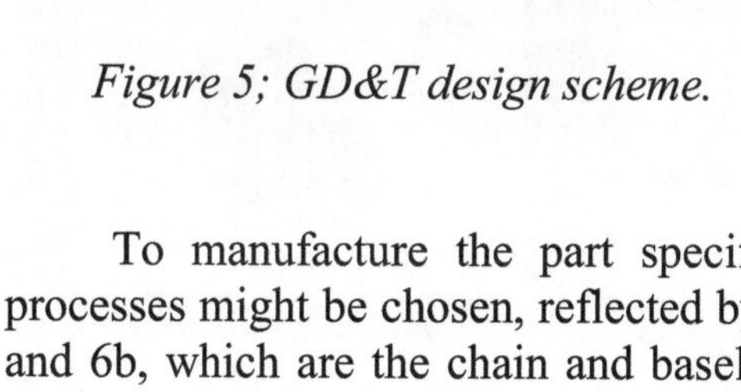

Figure 5; GD&T design scheme.

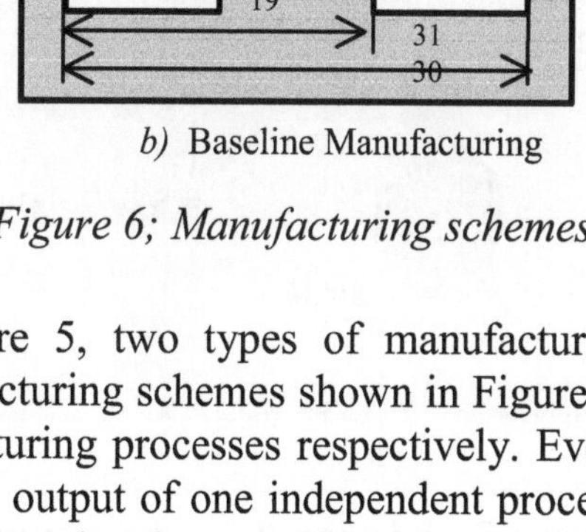

a) Chain Manufacturing

b) Baseline Manufacturing

Figure 6; Manufacturing schemes.

To manufacture the part specified in Figure 5, two types of manufacturing processes might be chosen, reflected by the manufacturing schemes shown in Figure 6a and 6b, which are the chain and baseline manufacturing processes respectively. Every dimension labeled in Figure 6 is assumed to be the output of one independent process. The range of the dimensions reflects an initial setup by the manufacturing engineer corresponding (he hopes) to $\pm 3\sigma$ of the process for creating that dimension.

4.2. Representation

After assuming process parameters for each manufactured dimension, we can construct a joint probability density function (joint pdf) that represents the probability of every combination of dimension values. Once the joint pdf has been constructed, the probability of a part meeting its tolerance specification or an assembly satisfying a set of assembly criteria can be calculated by integrating the pdf over its tolerance region or assembly region respectively.

The probability that an individual part **x** satisfies its tolerance specification and the probability that an assembly **y** satisfies its assembly criteria are shown in equations (2) and (3) respectively. In these equations, TR represents the tolerance region, AR represents the assembly region, the B_i are the basis vectors for the assembly space, and the number of integration symbols is the dimension of the assembly space. The joint pdfs p_x and p_y depend on the distributions of the component dimensions in the assembly.

$$P(\text{x meets tolerance spec}) = \int \cdots \iint_{TR} p_x(B1, B2, \cdots) d\text{TR} \tag{2}$$

$$P(\text{y meets assembly criteria}) = \int \cdots \iint_{AR} p_y(B1, B2, \cdots) d\text{A}R \tag{3}$$

If the three dimensions for chain manufacturing process in Figure 6a exhibit independent normal statistics with the same variance, the joint pdf can be represented with spherical iso-probability surface as shown in Figure 7a. The assembly for this example is comprised of the part specified in Figure 5 and the nominal complement part from Figure 1. Because the basis vectors for the assembly space are each influenced by only one of the chain dimensions, the chain dimensions are orthogonal in this representation, hence the spherical surface. The tolerance region for the assembly is the polyhedron shown on Figure 7a. The fractional containment of the joint pdf within the tolerance region represents the probability that the part with the manufacturing process will satisfy the tolerance specifications.

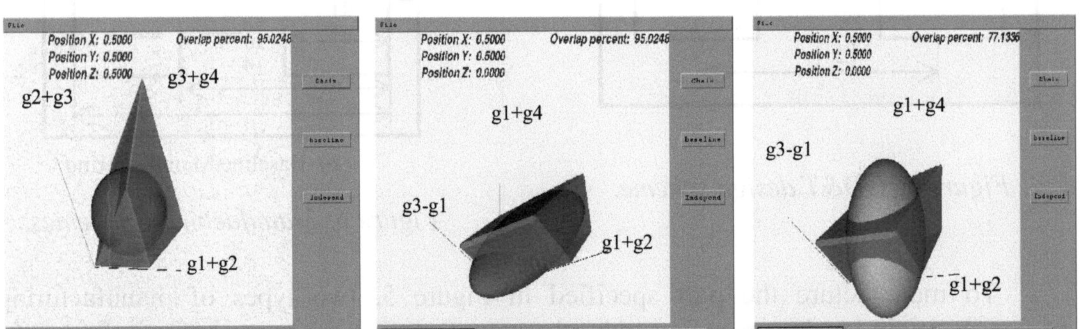

a) *b)* *c)*

Figure 7; a) Joint pdf for chain manufacturing process contained in tolerance region.
b) Joint pdf for baseline manufacturing process contained in tolerance region.
c) Joint pdf when misunderstanding the independence of the dimensions.

4.3. Assessment

In order to determine if the manufacturing process follows the manufacturing dimension scheme, we would perform measurements on a subset of population to estimate the mean and variance of the population dimensions according to an inspection scheme. If we assume the inspection scheme is the same as the manufacturing scheme shown in Figure 6a, a sufficient random sample of parts will allow us to estimate that the dimensions have approximately the same variance, and correlation between them is near zero. The estimated probability of conformance to the tolerance specification will be the same as we determined from the analysis of the manufacturing scheme.

But what if the inspection process is performed per the baseline inspection scheme in Figure 6b, but the parts are manufactured as the chain scheme in Figure 6a? If we estimate the covariance terms between the baseline dimensions in addition to the means and variances of the individual dimensions, we will find these terms are no longer near zero. The joint pdf will not be spherical when we plot it in a coordinate system where the baseline dimensions are orthogonal. However, the proportion of the pdf containing in the tolerance region will keep the same because the tolerance zone is distorted also as shown in Figure 7b.

The distortion of the tolerance region and the joint pdf comes from the linear transformation from chain dimensions to baseline dimensions. The matrix C in equation (4) is used to implement the transformation in the hinge example, where **csy_{chain}** and **csy_{Baseline}** are the basis vectors for the different coordinate systems.

$$\begin{pmatrix} g1+g2 \\ g1+g4 \\ g3-g1 \end{pmatrix} = \begin{pmatrix} 1 & 0 & 0 \\ 1 & 1 & -1 \\ -1 & 0 & 1 \end{pmatrix} \bullet \begin{pmatrix} g1+g2 \\ g3+g4 \\ g2+g3 \end{pmatrix}, \text{ or } \quad csy_{Baseline} = C \bullet csy_{chain} \qquad (4)$$

If the distribution of the chain dimensions is a multivariate normal distribution, we can apply the C matrix as shown in Equation (5) to transform the distribution into the baseline coordinate system.

$$\text{If pdf}_{chain} \sim \text{N}[\mu, \Sigma] \text{and pdf}_{Baseline} = C \bullet pdf_{chain}, \text{ then pdf}_{Baseline} \sim \text{N}\left[C\mu, \text{C}\,\Sigma C^T\right] \qquad (5)$$

If, during inspection, we ignore the correlation between dimensions of the baseline inspection scheme, then we incorrectly estimate $\text{pdf}_{Baseline}$ ~N [Cμ, diag(CΣ CT)]. This distribution corresponds to treating each of the baseline dimensions manufactured independently as shown in Figure 7c. Note that far less of the pdf is contained in the tolerance region, and we would conclude (incorrectly) that fewer parts assemble.

4.4. Example

To quantify the statements about the containment of the distribution in the tolerance region, we provide the numerical values used to generate these figures. Let's use the same example in Figure 5 for GD&T design scheme, which will assemble with the nominal part in Figure 1. The manufacturing scheme of the part is the chain scheme in Figure 6a.

The joint distribution is shown on equation (6) and the iso-probability surface shown in Figure 7.a is drawn at the corresponding to the pdf value 7.05×10^{-6}. The variance 0.028 equals to $(range/6)^2$, where range is 1 from Figure 6a.

$$pdf_{chain} = \begin{pmatrix} g1+g2 \\ g3+g4 \\ g2+g3 \end{pmatrix} = \begin{pmatrix} C1 \\ C2 \\ C3 \end{pmatrix} \sim N\left[\begin{pmatrix} 0.5 \\ 0.5 \\ 0.5 \end{pmatrix}, \begin{pmatrix} .028 & 0 & 0 \\ 0 & .028 & 0 \\ 0 & 0 & .028 \end{pmatrix}\right] \quad (6)$$

To find the probability that the part will satisfy the specified tolerance, we simply integrate the equation (2) using pdf_{chain} from equation (6). The resulting value is 95.0%.

We now consider the baseline representation of the tolerance region and joint pdf as shown in Figure 7.b. The transformed distribution $pdf_{Baseline}$ is listed in equation (7). The integration of the pdf over the tolerance region in baseline coordinates is also 95.0%. As we expected, the integration result is independent of the coordinate system.

$$pdf_{Baseline} = \begin{pmatrix} g1+g2 \\ g1+g4 \\ g3-g1 \end{pmatrix} = \begin{pmatrix} B1 \\ B2 \\ B3 \end{pmatrix} \sim N\left[\begin{pmatrix} 0.5 \\ 0.5 \\ 0 \end{pmatrix}, \begin{pmatrix} .028 & .028 & -.028 \\ .028 & .083 & -.056 \\ -.028 & -.028 & .056 \end{pmatrix}\right] \quad (7)$$

If the inspection process uses baseline schemes, and the dimensions are unwittingly assumed to be independent. The incorrect baseline distribution behaves as equation (8). The integration of $pdf'_{Baseline}$ over the baseline tolerance region is 77.1%. The value equals the containment indicated in Figure 7c, which is much less than 95.0%.

$$pdf'_{Baseline} = \begin{pmatrix} g1+g2 \\ g1+g4 \\ g3-g1 \end{pmatrix} = \begin{pmatrix} B1' \\ B2' \\ B3' \end{pmatrix} \sim N\left[\begin{pmatrix} 0.5 \\ 0.5 \\ 0 \end{pmatrix}, \begin{pmatrix} .028 & 0 & 0 \\ 0 & .083 & 0 \\ 0 & 0 & .056 \end{pmatrix}\right] \quad (8)$$

The difference between the two integration values comes from the accumulation of different independent dimensions for statistical analysis, so setting up the independence of dimensions correctly is critical for analysis.

4.5. Optimization result

Designers will always attempt to have higher containment, as that means the probability of a failed assembly will be lower. There are two ways to increase the containment for the above example. One is to decrease the variances of the manufacturing process. However, a side effect of this method is that the cost for the manufacturing process will increase. Another way is to choose the mean, or target, dimensions effectively. In the example given above, the means are assumed to be at the center of the dimension range. Changing the means can improve the containment without increasing the variances.

Because the equation (2) and (3) are not differentiable with respect to the means, we designed a program to search the optimization result by testing the increasing trend of probability related to the means. The optimization result is 98.29% in the example when the means becomes 10.57, 9.57 and 10.57 rather than 10.5, 9.5, and 10.5

respectively. Only the mean vector of the joint distribution from equation (6) has been changed to obtain this result; the variances are unchanged.

4.6. Comparison by different manufacturing schemes

We discussed two manufacturing schemes as Figure 6 above. But the designer of manufacturing process may use different schemes to manufacture the part. In the table 1 below, we compare different schemes after assuming the standard deviation of each manufactured dimension is independent of the dimensional size and has the value 1/6 (using the normal distribution again). Dimensions labeled in the figures of Table I are the means, or “target” dimensions for each case.

The containment for each scheme is listed in Table I. The optimization result is calculated as described in section 4.5. The first scheme has the highest probability, which may be the best choice for the part designed in Figure 5. In high volume applications, a few percent increase in the yield can result in significant cost savings.

Table I; Assembly yield for different manufacturing schemes and optimization results.

Manufacturing scheme	**Probability in Tolerance region ($P_{(X \in TR)}$)**	**Optimization Result**	
		Dimensions changed	**Probability**
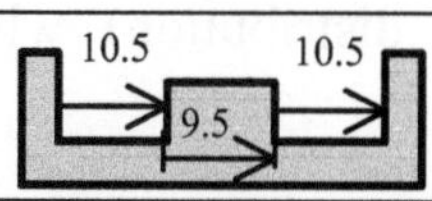	95.0%	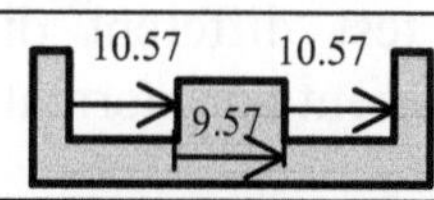	98.3%
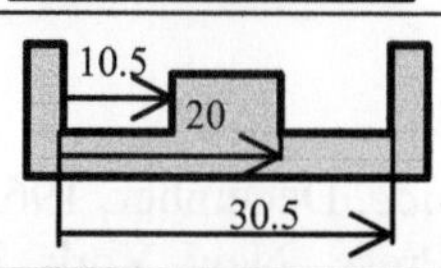	94.6%	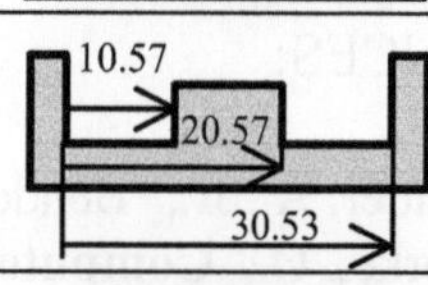	96.2%
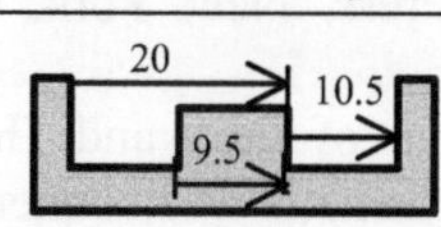	94.6%	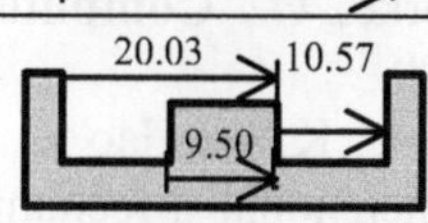	96.1%
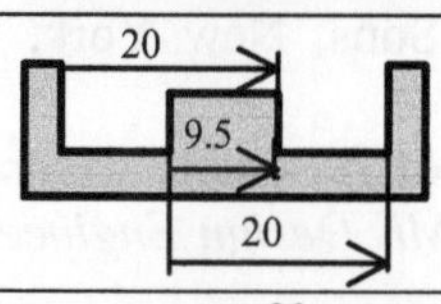	91.1%	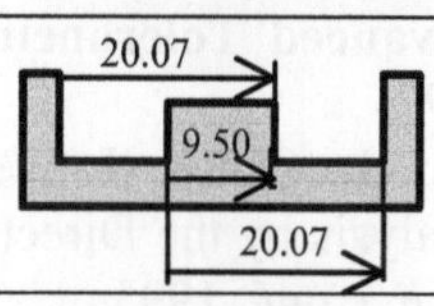	92.8%
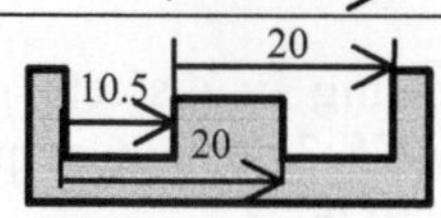	89.9%	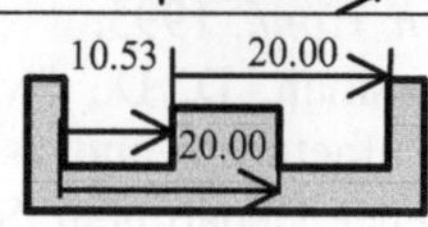	91.8%

5. VTK IMPLEMENTATION

Using Visualization Toolkit (VTK) [Schroeder 97], we developed an application program for assembly analysis that generated the pictures in Figure 7. We can zoom, move and rotate the 3D objects in the pictures. We can also pick up and move the joint pdf object with respect to the tolerance region. The containment percentage and the center (mean) location are shown in the window and updated continuously with the movement of the joint pdf location. The optimization result above can also be found "manually" by moving the pdf object until finding the maximal containment.

6. SUMMARY

This paper begins by introducing GapSpace for one-dimensional analysis. We then show that the probability of satisfying assembly criteria or tolerance requirements can be calculated by integrating a joint pdf (representing the statistical variability of the parts) over the assembly region or tolerance region. Through a hinge example, we also show that the statistical analysis should correspond to the manufacturing scheme of the parts. Misunderstanding the independence of the dimensions in the manufacturing method may generate very large differences in the statistical analysis result. An optimization result shows that thoughtful dimension assignment can achieve higher assemblebility without having to improve the manufacturing process. An application using VTK is implemented to visualize the analysis process.

Future work is underway to extend the GapSpace model to 2-D and 3-D assembly analyses and to test different distributions (not only normal distribution) where appropriate to represent non-normal processes.

7. REFERENCES:

[Bender 62] Bender, A., Jr., "Benderizing Tolerances", *Graphic Science*, December, 1962.

[Bjørke 89] Bjørke, Ø., **Computer Aided Tolerancing**, ASME Press, New York, NY, 1989.

[Chase 97] Chase, K.W.; Gao J.; Magleby S P.; "Tolerance Analysis of Two- and Three-Dimensional Mechanical Assemblies with Small Kinematic Adjustments." **Advanced Tolerancing Techniques**, John Wiley & Sons, New York, NY, 1997.

[Gao 95] Gao, J.; Chase, K. W.; Magleby, S. P.; "Comparison of Assembly Tolerance Analysis by the Direct Methods", *Procedure of the ASME Design Engineering Tech. Conf.,* 1995.

[Grossman 76] Grossman, D. D., "Monte Carlo simulation of tolerancing in discrete parts manufacturing and assembly", *Research Report STAN-CS-76-555*, Computer Science Department, Stanford University, Stanford, CA, 1976.

[Morse 99] Morse, E. P.; "Models, Representations, and Analyses of Toleranced One-Dimensional Assemblies", Ph.D. Thesis, Sibley School of Mechanical and Aerospace Engineering, Cornell University, Ithaca, NY, 1999.

[Mullins 98] Mullins, S. H., Anderson, D. C.; "Automatic identification of geometric constraints in mechanical assemblies", *Computer-Aided Design*, vol. 30, no. 9, pp. 715-726, August, 1998.

[Schroeder 97] Schroeder, W., Martin, H., **The Visualization Toolkit, 2nd Edition**, Prentice Hall, NY, 1997.

Analysis of functional geometrical specification

Philippe SERRE and Alain RIVIERE
Laboratoire d'Ingénierie des Systèmes Mécaniques et des MAtériaux
3, rue Fernand Hainaut, 93407 Saint-Ouen Cedex, France
Tel: 33 1 49 45 29 22 Fax: 33 1 49 45 29 29
e. mail: philippe.serre@ismcm-cesti.fr
e. mail: ariviere@ismcm-cesti.fr

André CLEMENT
Dassault Systèmes
e.mail: andre_clement@ds-fr.com

Abstract: "Good" functional tolerancing can only be obtained once the geometrical specification of the mechanism being studied has been analysed from a technological and geometrical standpoint. A geometrical analysis based on a vector approach has been developed. It enables existing tolerancing relations to be written between the design parameters relative to each part and the relative position parameters between two parts of the mechanism in question.
Keywords: tolerancing analysis, overconstraint, geometrical specification

While we can consider that standardised ISO specification for the tolerancing of an isolated mechanical object {with reference elements (point, straight-line and plane) and tolerance zones}is currently satisfactory, partly due to work undertaken in the last 6 CIRP seminars, the same does not apply to the functional specification of an assembly forming a mechanism.

1. POSITION OF THE STUDY

A "designer" using a CAD/CAM console is confronted with two problems which he deals with in succession but which are indissociable:

1. To construct each geometrical object at its nominal value, using only dimensional specifications that are consistant with one another. To do this, he has only one method: to use the minimum number of specifications! This requires great expertise, even for relatively commonplace problems since a system of equations always has to be resolved (for example, to build a sphere of a given radius, tangent to any two cylinder axes)

P. Bourdet and L. Mathieu (eds.),
Geometric Product Specification and Verification: Integration of Functionality, 115-125.

2. To assemble and specify the tolerance for a mechanism using only those elements which he controls, i.e.
   - the dimensional parameters (angles and lengths and their potential variations) with their parametric tolerances;
   - geometrical constraints (parallelism and point coincidence, straight line and plane [Clément et al., 1998], [Clément et al., 1999]), defined by the 7 classes of the TTRS, which he knows to be invariable; and, finally
   - any gap that exists whose maximum value he must specify (to ensure functional quality) and the minimum value (to ensure assemblability). ISO does not cover this issue at all.

The elements of the solution that we propose to implement in this instance on two examples, one simple, the other more complex, consists of determining:

- a specifications model (independently of any Cartesian reference point);
- the conditions of existence of the object or assembly (incorrectly termed the overconstraint conditions);
- the relations between the variations of these specifications and any gap that exists.

The "designer" only needs tolerancing for the relations between specifications which are generally called "overconstraint relations". These relations must first be determined and even when initially verified, they are bound to become inconsistent as a result of manufacturing variations if gap or deformation is not introduced into the specifications chain.

The theoretical contribution of the TTRS concept to define the object classes using the group of displacements seems particularly significant. However, the group structure does not allow the notion of "distance" to be introduced which is vital to establish these conditions of existence and their variations. To do this, it has to be completed by a vectorial space structure which can be equipped with a distance concept.

It is this analysis that justifies the new description of these classes of objects by a metric tensor $\overline{\overline{G}}$. This metric tensor will integrate all of an object's dimensional and geometrical properties with n dependent or independent dimensions, independently of any reference point since it is a vector. The advantage of the model lies in the non-independence of the metric tensor components since this will enable the operator to modify the object without necessarily using the specifications that have served for its construction. A metric tensor $\overline{\overline{G}}$ will be a specification model for a unique solid as well as for a mechanism. From this specifications model, we will be able to generate both the conditions of existence of the mechanism (which will enable assembly) and the relations between the specifications variations (which will enable the functional conditions to be specified).

In the first example below, we can model the set of specifications using a metric tensor while, in the second example, we will use the angles between planes in addition and model all the specifications using a metric tensor and a metric bitensor.

2. EXAMPLE 1

This example relates to the assembly of a mechanical structure composed of 4 bars and 2 pivot links whose axes are carried by the bars and two spherical links.

2.1. Vectorial modeling of the parametric geometrical specification

The structure in question is composed of 4 bars with lengths L_1, L_2, L_3 and L_4. Unit vectors $\overrightarrow{e_1}$, $\overrightarrow{e_2}$, $\overrightarrow{e_3}$, and $\overrightarrow{e_4}$ carried by each of these bars have been indicated on the Figure 1.

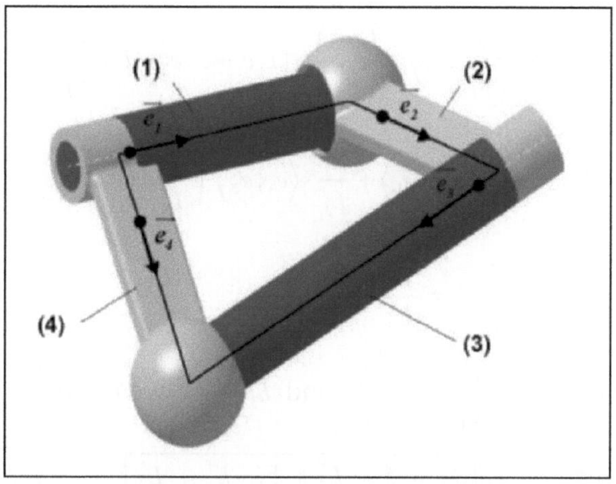

Figure 1: Mechanical structure being studied and associated vectorial modeling

The specification model ([Serré, 2000], [Lesage, 2000]) that we will use is the metric tensor, as follows:

$$\overline{\overline{G}} = \left(\begin{array}{c|cccc} & L_1\overrightarrow{e_1} & L_2\overrightarrow{e_2} & L_3\overrightarrow{e_3} & L_4\overrightarrow{e_4} \\ \hline L_1\overrightarrow{e_1} & L_1^2 & L_1 \cdot L_2 \cdot \left\langle \overrightarrow{e_1}, \overrightarrow{e_2} \right\rangle & L_1 \cdot L_3 \cdot \left\langle \overrightarrow{e_1}, \overrightarrow{e_3} \right\rangle & L_1 \cdot L_4 \cdot \left\langle \overrightarrow{e_1}, \overrightarrow{e_4} \right\rangle \\ L_2\overrightarrow{e_2} & - & L_2^2 & L_2 \cdot L_3 \cdot \left\langle \overrightarrow{e_2}, \overrightarrow{e_3} \right\rangle & L_2 \cdot L_4 \cdot \left\langle \overrightarrow{e_2}, \overrightarrow{e_4} \right\rangle \\ L_3\overrightarrow{e_3} & - & - & L_3^2 & L_3 \cdot L_4 \cdot \left\langle \overrightarrow{e_3}, \overrightarrow{e_4} \right\rangle \\ L_4\overrightarrow{e_4} & - & - & - & L_4^2 \end{array} \right)$$

Let us suppose, for example, that we have specified the 4 lengths L_1, L_2, L_3 and L_4 and the 2 angles $\left(\overrightarrow{e_1}, \overrightarrow{e_2}\right)$ and $\left(\overrightarrow{e_3}, \overrightarrow{e_4}\right)$. We thus obtain the following metric tensor:

$$\overline{\overline{G}} = \left(\begin{array}{c|cccc} & \overrightarrow{e_1} & \overrightarrow{e_2} & \overrightarrow{e_3} & \overrightarrow{e_4} \\ \hline \overrightarrow{e_1} & 1 & S_1 & \xi_2 & L_1^* + S_1 \cdot L_2^* + \xi_2 \cdot L_3^* \\ \overrightarrow{e_2} & - & 1 & \xi_1 & S_1 \cdot L_1^* + L_2^* + \xi_1 \cdot L_3^* \\ \overrightarrow{e_3} & - & - & 1 & S_2 = \xi_2 \cdot L_1^* + \xi_1 \cdot L_2^* + L_3^* \\ \overrightarrow{e_4} & - & - & - & 1 = L_1^* \cdot \left(L_1^* + S_1 \cdot L_2^* + \xi_2 \cdot L_3^* \right) + L_2^* \cdot \left(S_1 \cdot L_1^* + L_2^* + \xi_1 \cdot L_3^* \right) + L_3^* \cdot S_2 \end{array} \right)$$

The 2 angle specifications S_1 and S_2 appear directly; the closed-loop topological specification is expressed in vectorial form by the relation $L_4 \vec{e_4} = L_1 \vec{e_1} + L_2 \vec{e_2} + L_3 \vec{e_3}$, and the unknown scalar products $\langle \vec{e_1}, \vec{e_3} \rangle$ and $\langle \vec{e_2}, \vec{e_3} \rangle$ are expressed by the variables ξ_1 and ξ_2 .

Note. we can easily find the expressions in the last column by expressing the scalar products in the following manner:
Let us consider, for example $\langle \vec{e_3}, \vec{e_4} \rangle$

$$S_2 = \langle \vec{e_3}, \vec{e_4} \rangle = \left\langle \vec{e_3}, \frac{1}{L_4} \cdot \left(L_1 \vec{e_1} + L_2 \vec{e_2} + L_3 \vec{e_3} \right) \right\rangle$$

$$.. = \frac{L_1}{L_4} \cdot \langle \vec{e_3}, \vec{e_1} \rangle + \frac{L_2}{L_4} \cdot \langle \vec{e_3}, \vec{e_2} \rangle + \frac{L_3}{L_4} \cdot \langle \vec{e_3}, \vec{e_3} \rangle$$

$$.. = \frac{L_1}{L_4} \cdot \xi_2 + \frac{L_2}{L_4} \cdot \xi_1 + \frac{L_3}{L_4}$$

By changing the notation $L_1^* = \frac{L_1}{L_4}$, $L_2^* = \frac{L_2}{L_4}$ and $L_3^* = \frac{L_3}{L_4}$, we obtain

$$\boxed{S_2 = \xi_2 \cdot L_1^* + \xi_1 \cdot L_2^* + L_3^*}$$

2.2. Analysis of the geometrical specification

An initial level of analysis consists of analysing the specification of the first three vectors which constitute an open loop.

We know the three lengths but the 2 angle specifications, corresponding to the variables ξ_1 and ξ_2 are missing. The last column of the tensor gives us 2 equations (corresponding to the specification S_2 and the unit vector $\vec{e_4}$) with the 2 variables ξ_1 and ξ_2 as unknowns. Simple enumeration could lead us to the conclusion that the problem is iso-constrained. We will see that this is not the case.

The system will be iso-constrained if, and only if, the system of 2 equations with 2 variables ξ_1 and ξ_2

$$\begin{pmatrix} 2 \cdot L_2^* \cdot L_3^* & 2 \cdot L_1^* \cdot L_3^* \\ L_2^* & L_1^* \end{pmatrix} \cdot \begin{pmatrix} \xi_1 \\ \xi_2 \end{pmatrix} = \begin{pmatrix} 1 - \left(L_1^{*2} + L_2^{*2} + L_3^{*2} + 2 \cdot S_1 \cdot L_1^* \cdot L_2^* \right) \\ S_2 - L_3^* \end{pmatrix}$$

is rank 2. However, the determinant is visibly nil so there is only one independent equation.

By applying Rouché-Fontené's theorem [Lichnerowicz, 1983], we easily obtain the mathematical condition of existence which is generally termed a "hyperconstrained relation":

$$\boxed{1-L_1^{*2}-L_2^{*2}-L_3^{*2}-2\cdot S_1\cdot L_1^*\cdot L_2^*-2\cdot S_2\cdot L_3^*=0}$$

This is the condition that the geometrical object specifications must satisfy in order to achieve it.

Note. This algebraic relation translates the geometrical existence of two rigid triangles which must share the same third side.

If we wish to make the specification iso-constrained, the angle specification S_2 of the scalar product $\langle \vec{e_3}, \vec{e_4} \rangle$ has to be removed and, for example, the angle specification S_3 of the scalar product $\langle \vec{e_2}, \vec{e_4} \rangle$ added.

We can easily check that the system of equations then obtained is a rank 2.

Consequence on the parametric tolerancing

The variations due to manufacturing uncertainties can no longer be independent, they have to satisfy the following relation, obtained by differentiation of the hyperconstrained relation.

$$\boxed{-\delta L_1^*\cdot\left(L_1^*-S_1\cdot L_2^*\right)-\delta L_2^*\cdot\left(L_2^*-S_1\cdot L_1^*\right)-\delta L_3^*\cdot S_2-\delta S_1\cdot L_1^*\cdot L_2^*-\delta S_2\cdot L_3^*=0}$$

This equation can be directly verified by deformation or the adjustment of one of the offending parameters or indirectly by adding a gap vector to the modeling.

3. EXAMPLE 2

This example relates to the assembly of a mechanical structure, composed of 4 bars and 4 pivot links whose axes are perpendicular to two consecutive bars. The 4 bars all have the same morphology (see Figure 2). In point of fact, a bar i is specified by two design parameters: the angle α_i and the distance L_i between the axes of the two cylinders.

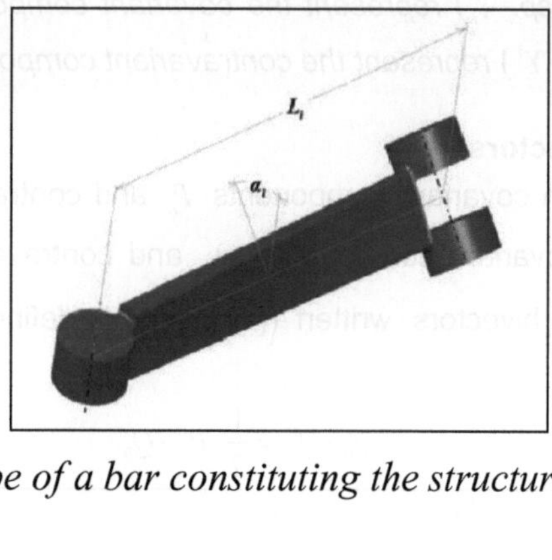

Figure 2: Shape of a bar constituting the structure being studied

These 4 parts are assembled by ensuring that the cylinders coincide, as shown in Figure 3. This assembly is similar to a Bennett mechanism since the geometry of the bars which compose it is the same. However, it is different in that the Bennett mechanism [Perez et al., 2000] has a degree of mobility whereas the system being studied does not take this into consideration and constitutes a rigid assembly for which only the conditions of existence are being sought.

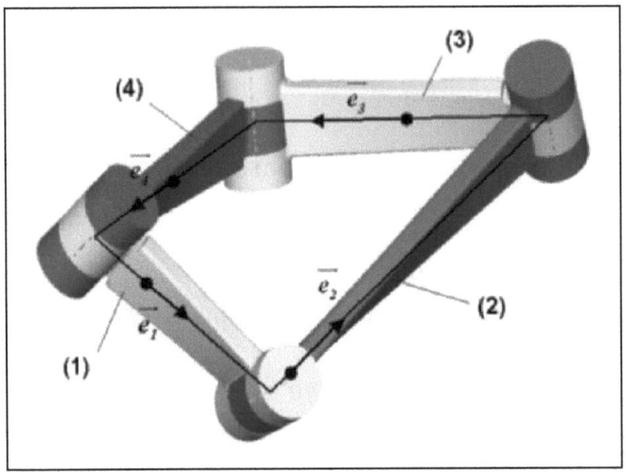

Figure 3: Mechanical structure being studied and associated vectorial modeling

3.1. Vectorial modeling of the structure and the parametric geometrical specification

A vector is associated with each part as indicated in the above figure. We have chosen to model the geometrical specification by a metric tensor, in order to specify all the angles between vectors and the vectorial sums, and by a metric *bitensor*, in order to specify all the angles between vectorial planes that we will be modeling by *bivectors*.

Definition of a *bivector* [Cartan, 1946]

[Cartan] We call a *bivector*, written $[\overrightarrow{xy}]$, the figure formed by two vectors $\vec{x}$ and $\vec{y}$ ranked in a certain order. The covariant components of this bivector are defined by the quantity $(X_i \cdot Y_j - X_j \cdot Y_i)$ and the contravariant components by the quantity $(X^i \cdot Y^j - X^j \cdot Y^i)$.

Note: the quantities X_i (resp. Y_i) represent the covariant components of vector $\vec{x}$ (resp. $\vec{y}$) and the quantities X^i (resp. Y^i) represent the contravariant components of vector $\vec{x}$ (resp. $\vec{y}$).

Scalar product of two bivectors

Given the bivector $[\overrightarrow{xy}]$ with covariant components P_{ij} and contravariant components P^{ij}, and given bivector $[\overrightarrow{uv}]$ with covariant components Q_{ij} and contravariant components Q^{ij}. The scalar product of these two bivectors, written $\langle[\overrightarrow{xy}],[\overrightarrow{uv}]\rangle$, is defined by either one of two equal expressions

$$\frac{1}{2} \cdot P_{ij} \cdot Q^{ij} = \frac{1}{2} \cdot P^{ij} \cdot Q_{ij}$$

Which, when developed, gives,

$$\frac{1}{2}\cdot P_{ij}\cdot Q^{ij}=\frac{1}{2}\cdot\left(X_i\cdot Y_j-X_j\cdot Y_i\right)\cdot\left(U^i\cdot V^j-U^j\cdot V^i\right)$$

Finally, the expression of the scalar product between bivectors in relation to the scalar product between vectors is,

$$\boxed{\left\langle\overrightarrow{[xy]},\overrightarrow{[uv]}\right\rangle=\left\langle\vec{x},\vec{u}\right\rangle\cdot\left\langle\vec{y},\vec{v}\right\rangle-\left\langle\vec{x},\vec{v}\right\rangle\cdot\left\langle\vec{y},\vec{u}\right\rangle}$$

As the scalar product between bivectors has been defined, we define a new metric tensor, called a *bitensor*, which is constructed from a set of bivectors. For example:

$$\overline{\overline{GG}}=\left(\begin{array}{c|cc} & \overrightarrow{[xy]} & \overrightarrow{[uv]}\\ \hline \overrightarrow{[xy]} & \left\langle\overrightarrow{[xy]},\overrightarrow{[xy]}\right\rangle & \left\langle\overrightarrow{[xy]},\overrightarrow{[uv]}\right\rangle\\ \overrightarrow{[uv]} & \left\langle\overrightarrow{[uv]},\overrightarrow{[xy]}\right\rangle & \left\langle\overrightarrow{[uv]},\overrightarrow{[uv]}\right\rangle \end{array}\right)$$

As the definitions have been given, we will now look at geometrical specification.

- The length of the bars will be written L_i,
- The specification of the closed loop is expressed by the normal vectorial relation,

$$\sum_{i=1}^{4}L_i\vec{e_i}=\vec{0},$$

- The angular orientation of the links in relation to one another is given by the angle formed between two bivectors. For example, angle α_2 corresponds to the angle between the two bivectors $\overrightarrow{[e_1e_2]}$ and $\overrightarrow{[e_2e_3]}$.

Specification of the closed loop

As in the previous example, we use the metric tensor constructed from vectors $\vec{e_i}$, and express the fact that vector $\vec{e_4}$ is a linear combination of the three vectors $\vec{e_1}$, $\vec{e_2}$ and $\vec{e_3}$. Thus, the metric tensor $\overline{\overline{G}}$ is written,

$$\overline{\overline{G}}=\left(\begin{array}{c|cccc} & \vec{e_1} & \vec{e_2} & \vec{e_3} & \vec{e_4}\\ \hline \vec{e_1} & 1 & \cos x & \cos t & \cos v=-\frac{1}{L_4}\cdot\left(L_1+L_2\cdot\cos x+L_3\cdot\cos t\right)\\ \vec{e_2} & - & 1 & \cos y & \cos u=-\frac{1}{L_4}\cdot\left(L_1\cdot\cos x+L_2+L_3\cdot\cos y\right)\\ \vec{e_3} & - & - & 1 & \cos z=-\frac{1}{L_4}\cdot\left(L_1\cdot\cos t+L_2\cdot\cos y+L_3\right)\\ \vec{e_4} & - & - & - & L_1^2+L_2^2+L_3^2+2\cdot L_1\cdot L_2\cdot\cos x+2\cdot L_2\cdot L_3\cdot\cos y+2\cdot L_3\cdot L_1\cdot\cos t=L_4^2 \end{array}\right)$$

We thus obtain 4 vectorial closure relations

Specification of the relative orientation of the links

The relative orientation angular specifications are represented in the bitensor $\overline{\overline{GG}}$. As we know the expression of each case of the bitensor, in relation to the cases of the tensor, we thus have 4 new relations.

The metric bitensor $\overline{\overline{GG}}$ is written:

$$\overline{\overline{GG}} = \left(\begin{array}{c|cccc} & \overrightarrow{e_{[e_1e_2]}} & \overrightarrow{e_{[e_2e_3]}} & \overrightarrow{e_{[e_3e_4]}} & \overrightarrow{e_{[e_4e_1]}} \\ \hline \overrightarrow{e_{[e_1e_2]}} & 1 & \cos\alpha_2 = \left\langle \overrightarrow{e_{[e_1e_2]}}, \overrightarrow{e_{[e_2e_3]}} \right\rangle & \left\langle \overrightarrow{e_{[e_1e_2]}}, \overrightarrow{e_{[e_3e_4]}} \right\rangle & \cos\alpha_1 = \left\langle \overrightarrow{e_{[e_1e_2]}}, \overrightarrow{e_{[e_4e_1]}} \right\rangle \\ \overrightarrow{e_{[e_2e_3]}} & - & 1 & \cos\alpha_3 = \left\langle \overrightarrow{e_{[e_2e_3]}}, \overrightarrow{e_{[e_3e_4]}} \right\rangle & \left\langle \overrightarrow{e_{[e_2e_3]}}, \overrightarrow{e_{[e_4e_1]}} \right\rangle \\ \overrightarrow{e_{[e_3e_4]}} & - & - & 1 & \cos\alpha_4 = \left\langle \overrightarrow{e_{[e_3e_4]}}, \overrightarrow{e_{[e_4e_1]}} \right\rangle \\ \overrightarrow{e_{[e_4e_1]}} & - & - & - & 1 \end{array} \right)$$

Finally, after transforming the equations to evidence the symmetric form of the problem to be solved, we obtain this system of equations:

$$\begin{cases} P + L_1 \cdot L_2 \cdot \cos x + L_2 \cdot L_3 \cdot \cos y + L_3 \cdot L_1 \cdot \cos t = L_4^2 \\ P + L_1 \cdot L_2 \cdot \cos x + L_2 \cdot L_4 \cdot \cos u + L_4 \cdot L_1 \cdot \cos v = L_3^2 \\ P + L_1 \cdot L_3 \cdot \cos t + L_3 \cdot L_4 \cdot \cos z + L_4 \cdot L_1 \cdot \cos v = L_2^2 \\ P + L_2 \cdot L_3 \cdot \cos y + L_3 \cdot L_4 \cdot \cos z + L_4 \cdot L_2 \cdot \cos u = L_1^2 \\ \cos x \cdot \cos y - \sin x \cdot \sin y \cdot \cos\alpha_2 = \cos t \\ \cos y \cdot \cos z - \sin y \cdot \sin z \cdot \cos\alpha_3 = \cos u \\ \cos z \cdot \cos v - \sin z \cdot \sin v \cdot \cos\alpha_4 = \cos t \\ \cos v \cdot \cos x - \sin v \cdot \sin x \cdot \cos\alpha_1 = \cos u \end{cases} \quad \text{with } P = \frac{1}{2} \cdot \left(L_1^2 + L_2^2 + L_3^2 + L_4^2 \right) \quad \text{(Eq.1)}$$

This set of relations represents all the geometrical specifications covering the structure being studied. As a result, to construct an example of the structure involves finding a set of parameters which satisfy these 8 relations.

3.2. Analysis of the specification and generation of tolerancing relations

Unlike the example presented in the previous chapter, the formal resolution of the previous set of non-linear equations is complex and, as a result, it is quite difficult to formally generate the relations of existence by a computer system. The solution proposed is as follows: Firstly, we generate an instance of the geometrical structure being studied by the numeric resolution of the system of equations. Then, by a first order approximation, we write the relations between the geometrical parameter uncertainties.

In the remaining part of this paragraph, we will describe this approach as applied to the example. First of all, the numeric resolution was obtained by using a Newton-Raphson algorithm, starting from an initial situation [Bouma et al., 1995]. The number of parameters that the designer must specify is 6 since we have a system of equations comprising 14 parameters and 8 relations. As seen above in the previous chapter, the choice of the 6 parameters cannot be made at random since, in certain cases, the system of equations will not be of the maximum rank. For the mechanism being studied, we made the choice detailed below:

Example of solution

The following table presents a solution of the mechanical structure being studied. The six parameters specified are shaded and we determine the initial value and final value of each parameter calculated.

	Data
γ (°)	30
δ (°)	50
L_a (mm)	100
L_b (mm)	180
L_c (mm)	130
x (°)	80

	Initial value	Final value
L_d (mm)	100	143.641676
α (°)	28.647889	32.433214
β (°)	40.107045	44.124522
y (°)	-17.188733	125.788930
z (°)	5.729577	72.680086
t (°)	-45.836623	132.453306
u (°)	-51.566201	147.644701
v (°)	-17.188733	107.631457

Expression of uncertainties

Once the numeric solution is obtained, we write the relations between the uncertainties of the geometrical parameters. By expressing the Jacobian matrix of the system of equations (Eq. 1) in relation to the 14 variables of the mechanism being studied as $T_{L_1,L_2,L_3,L_4,\alpha_1,\alpha_2,\alpha_3,\alpha_4,x,y,z,t,u,v}$, we can write the following equality:

$$T_{,L_1,L_2,L_3,L_4,\alpha_1,\alpha_2,\alpha_3,\alpha_4,x,y,z,t,u,v} \cdot \begin{pmatrix}\delta L_1 & \delta L_2 & \delta L_3 & \delta L_4 & \delta\alpha_1 & \delta\alpha_2 & \delta\alpha_3 & \delta\alpha_4 & \delta x & \delta y & \delta z & \delta t & \delta u & \delta v\end{pmatrix}^T = 0$$

To express the relations between the uncertainties of the geometrical parameters, we simply need to solve the previous linear equation. This system comprises 8 equations for 14 parameters and we can give the expression of 8 parameters in relation to 6 others if the system is of maximum rank. This choice is indifferent and we can then transfer the tolerancing specification into any space on the geometrical parameters under consideration.

We have chosen to write uncertainties δL_1, δL_2, δx, δy, δz, δt, δu and δv in relation to uncertainties δL_3, δL_4, $\delta\alpha_1$, $\delta\alpha_2$, $\delta\alpha_3$, and $\delta\alpha_4$. To do this, we transform the linear system, as follows:

$$T_{,L_1,L_2,x,y,z,t,u,v} \cdot \begin{pmatrix}\delta L_1 & \delta L_2 & \delta x & \delta y & \delta z & \delta t & \delta u & \delta v\end{pmatrix}^T + T_{,L_3,L_4,\alpha_1,\alpha_2,\alpha_3,\alpha_4} \cdot \begin{pmatrix}\delta L_3 & \delta L_4 & \delta\alpha_1 & \delta\alpha_2 & \delta\alpha_3 & \delta\alpha_4\end{pmatrix}^T = 0$$

and are thus able to deduce

$$\begin{pmatrix}\delta L_1 & \delta L_2 & \delta x & \delta y & \delta z & \delta t & \delta u & \delta v\end{pmatrix}^T = -\left(T_{,L_1,L_2,x,y,z,t,u,v}\right)^{-1} \cdot T_{,L_3,L_4,\alpha_1,\alpha_2,\alpha_3,\alpha_4} \cdot \begin{pmatrix}\delta L_1 & \delta L_1 & \delta\alpha_1 & \delta\alpha_2 & \delta\alpha_3 & \delta\alpha_4\end{pmatrix}^T$$

Numeric application

Numeric application is achieved around the position previously calculated. By separating the 8 design parameters from the 6 relative position parameters, we obtain two families of tolerancing relations.

The first family gives the assemblability conditions of the mechanical system being studied; here we obtain two relations between the design parameters:

$$\begin{pmatrix} \delta L_1 \\ \delta L_2 \end{pmatrix} = \begin{pmatrix} -0.28393 & 1.0372 & 0.94278 & -1.7197 & 1.4519 & 0.11235 \\ 0.87345 & 0.50339 & -1.1251 & -0.94287 & -0.50404 & 1.7125 \end{pmatrix} \cdot \begin{pmatrix} \delta L_3 \\ \delta L_4 \\ \delta\alpha_1 \\ \delta\alpha_2 \\ \delta\alpha_3 \\ \delta\alpha_4 \end{pmatrix}$$

The second family gives the expression of the position uncertainties in relation to the machining uncertainties; here we obtain six relations between the bar position parameters and the design parameters:

$$\begin{pmatrix} \delta x \\ \delta y \\ \delta z \\ \delta t \\ \delta u \\ \delta v \end{pmatrix} = \begin{pmatrix} -1.2659 & 1.2466 & 3.0072 & -3.9031 & -0.76921 & 2.2046 \\ 1.3907 & -1.3696 & -3.0588 & 2.5654 & 0.36088 & -0.36431 \\ -1.1729 & 1.1551 & 3.8946 & -2.1637 & -1.1567 & 0.46383 \\ 0.13972 & -0.13759 & -0.4639 & 0.64666 & 0.34571 & -1.3518 \\ 0.10639 & -0.10478 & -1.0689 & 0.19627 & -0.18186 & -0.1273 \\ 0.92172 & -0.90769 & -3.0604 & 3.56 & 1.9032 & -1.9213 \end{pmatrix} \cdot \begin{pmatrix} \delta L_3 \\ \delta L_4 \\ \delta\alpha_1 \\ \delta\alpha_2 \\ \delta\alpha_3 \\ \delta\alpha_4 \end{pmatrix}$$

4. CONCLUSION

A mechanism must be toleranced when its geometrical specification is hyperconstrained. Any tolerancing approach for a mechanical structure must, therefore, start by the analysis of its geometrical specification.

Two levels of analysis were identified in this study: a first level, relative to the enumeration of the angular parameters and a second level, relative to the generation of equations translating the conditions of existence of an object or an assembly.

From these relations of existence, we can deduce the tolerancing relations between the uncertainties of the set of geometrical parameters concerned. This is achieved directly by differentiation, or indirectly after a numeric resolution stage. We are then able to solve the problem of the transfer of the tolerancing specification in the case of parametric tolerancing.

REFERENCES

[Bouma et al., 1995] Bouma W.; Fudos I.; Hoffmann C.M.; Cai J.; Paige R.; « Geometric constraint solver »; *Computer-Aided Design;* Vol. 27.6, pp. 487-501 (1995).

[Cartan, 1946] Cartan E.; « *Leçons sur la géométrie des espaces de Riemann » (Lessons on Riemann spaces geometry);* 2nd Edition Gauthier-Villars, Paris (1946), réimpression (reprint). Jacques Gabay, Paris (1988); ISBN 2-87647-008-X.

[Clément et al., 1998] Clément A.; Rivière A.; Serré P.; Valade C.; « The TTRS: 13 Constraints for Dimensioning and Tolerancing », in *Geometric design tolerancing: theories, standards and applications*, Chapman et Hall, pp. 122-129, (1998).

[Clément et al., 1999] Clément A.; Rivière A.; Serré P.; « Global Consistency of Dimensioning and Tolerancing », *Keynote paper of CIRP Computer Aided Tolerancing, 6th Seminar*, Enschede, The Netherlands, March 22-24, (1999).

[Lesage, 2000] Lesage D.; Léon J.C.; Serré P.; « A declarative approach to a 2D variational modeler »; *3ème conférence internationale sur la conception et la fabrication intégrées en mécanique* (3rd international conference on the integrated design and manufacturing in mechanical engineering) Montréal, Canada, May 16-19 (2000).

[Lichnerowicz, 1983] Lichnerowicz A.; « *Algèbre et analyse linéaires » (Linear algebra and analysis);* Ed. Masson (1983).

[Perez et al., 2000] Perez A.; McCarthy J.M.; « dimensional synthesis of bennett linkages »; In: *Proceedings of DETC'00*, Baltimore, Maryland, USA, September 10-13 (2000).

[Serré, 2000] Serré P.; « *Cohérence de la spécification d'un objet de l'espace euclidien à n dimensions »(Consistency of the specification of an object in n-dimensional Euclidian space);* Thèse de Doctorat (Doctorate Thesis), Ecole Centrale de Paris (2000).

Simultaneous analysis method for tolerancing flexible mechanisms

Serge Samper, Max Giordano
LMécA ESIA 41 avenue de la Plaine BP 806 74016 ANNECY Cedex
nom@esia.univ-savoie.fr

Abstract: We propose a simultaneous method to make a tolerancing analysis of a deformable system. This method solves effective configurations on an assembly, deformations of mechanical parts and joints, and internal forces. It can be used in the aim of testing tolerancing and its technological consequences as life cycles or load levels (maximum static equivalent stress, fatigue stress,...). We have to discretize the assembly in elastic parts and joints (a joint can be a couple of surfaces or a component).

In order to make this analysis, we must program the component behaviour. This is the relationship between displacements and forces within or without the clearance domain for a joint component. A functioning point has got up to thirteen parameters maximum : six for displacements plus six for forces plus one for technological limitation. The later can be the life cycle or the maximum equivalent stress or maximum force.
Keywords: "tolerancing","elastic mechanism","life cycle","finite element analysis"

1. MECHANISM MODEL

Mechanisms are flexible, their precision increases so it becomes interesting to analyse both elastic behaviour and tolerancing.

They are deformed under the action of external forces such as reducers or mechanical power links. They can also be deformed under internal forces especially in the case of hyperstatisms. We are only interested in elastic strains because a mechanism must often work inside its elastic domain and we are not concerned about the history of loading. In order to analyse the behaviour of a mechanism in tolerancing it is necessary to know all the displacements of joint surfaces. So the minimal discretisation of an assembly is to divide it at every joint surface. We create part elements and joint elements. Degrees of freedom are unauthorised kinematic movements. We can also use components which whichare submechanisms.

Remarks:

R1: A joint element always has two link surfaces, thus two nodes (in general merged).

R2: Joint elements often have a non linear elastic behaviour (e.g. : small contact surfaces in Hertz theory).

P. Bourdet and L. Mathieu (eds.),
Geometric Product Specification and Verification: Integration of Functionality, 127-134.

R3: Part elements always have a linear elastic behaviour.
R4: Components (such as ball bearings, ...) may have an elastic non linear behaviour with clearances.

When the mechanism is built in parts and joints elements, the analysis can be done. This analysis takes clearances in joints (functional tolerancing on assembly), geometrical deviations of surfaces of each part (geometrical specifications in a part), and elastic laws into account. We present here two methods, first sequential then simultaneous, in order to reach this aim.

2. SEQUENTIAL MODELS

We previously worked on building sequential models [Samper 98]. Our concern in this study is to build simultaneous models but it is interesting to first consider the former. This method is the logical way to extend the LMecA's models of clearance and deviation spaces. Here, we complete that analysis by following the frame of model presented figure 1.

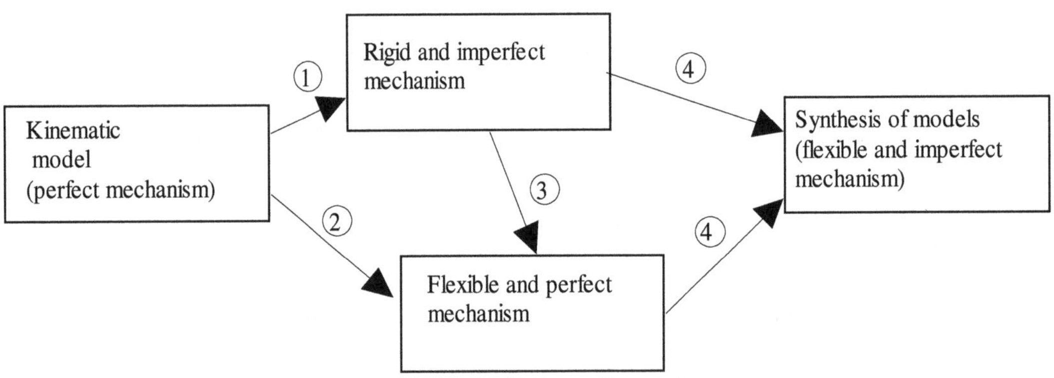

Figure 1 : Frame of models

① input of geometry and clearances (and tolerances in a verification process)
② input of external forces
③ input of contact configurations
④ input of displacement results to make the synthesis

The kinematic model is the mechanism defined by its parts and joints in nominal geometry. The rigid and imperfect mechanism is a model with its treatment which allows to make the tolerancing of each part by knowing clearances [Giordano 1999]. In the flexible and perfect mechanism model there is an elastic solving of parts and joints in order to know elastic displacements of joint surfaces due to external and internal forces. Those displacements are divided into elastic clearances for joints and elastic deviations for parts.

In fact this method cannot simply be used in one way. Step ③ supposes to input external forces to know where there will be contacts in a joint. Then we can make the elastic analysis without clearance and then verify that contacts are not modified. If contacts are modified the analysis must run another time with new input.

We can see here the main problem of this sequential analysis, its efficiency will depend on the good contact configuration inputs.

3. SIMULTANEOUS MODELS

We have then consider to making the analysis simultaneously. It allows to minimise the hypotheses and thus be close to reality. Here, the mechanism is imperfect, with clearances, and flexible. The first step is to discretize the whole assembly and identify each component behaviour.

Then, the problem can be solved simultaneously by merging the "rigid imperfect mechanism" with the "perfect elastic mechanism" analyses (figure 2).

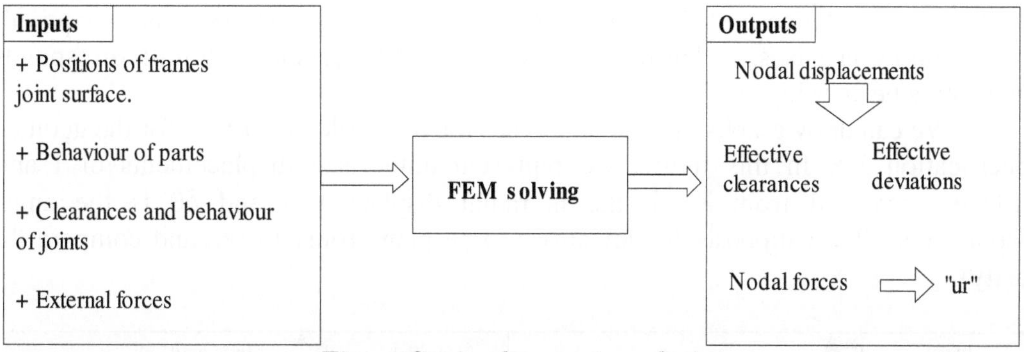

Figure 2: simultaneous analysis

3.1. Part behaviour

Parts are supposed to be perfectly elastic with linear laws. We input the frame positions and degrees of freedom of every joint surface. Then it is possible to build the stiffness matrix of this element.

The behaviour of a part is expressed by its tolerancing and its stiffness, but forces and therefore displacements must be limited. We input a scalar of use rate “ur”which expresses a level of use. This scalar can vary from 0 to 1. When it is 0, then the use of the part has no limit and when it is 1 the limit of use is reached.

For a static part, this limiting scalar will be the ratio of maximum Von Mises stress on allowed stress for example. For a rotating part we can calculate the relative fatigue stress. For a shaft, it will be the relative life cycle.

With this rate, it is possible to analyse a whole assembly with different limits and then know which part of it will be closest to its limit. Then "ur" can be translated in the technological limit of each part.

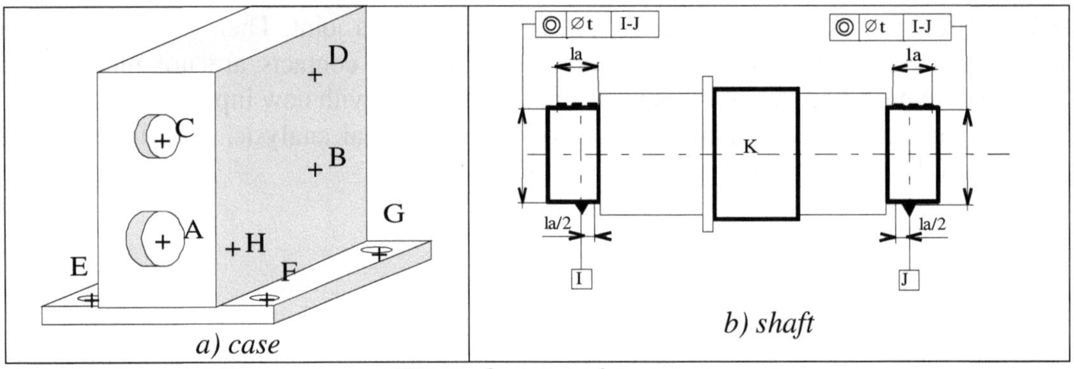

Figure 3: parts elements

This case will be a finite element in which nodes are surface links of bearings (A B C D) , screw and plane (E F G H) for example. The eight nodes will have six degrees of freedom.

The shaft will be an element with three joints I J K, one for each bearing and one for the gear. In the case of the shaft, "ur" can be the fatigue stress limit or equivalent stress limit (Von Mises,...), but for the case, "ur" will be the elastic limit (corresponding forces may be very high).

We can draw graphs in figures 4 to see the available deviations for the geometric specification 3.b. In this figure, we represent authorised displacements of I and J surfaces measured from K, δr the is radial displacement and δθ is the angular displacement if we suppose the deviations to be plane (only to see and compare them easily).

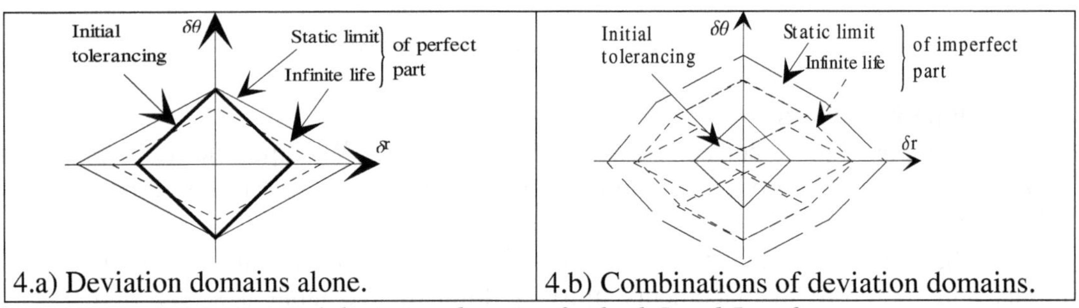

Figure 4 deviation domains for both I and J surfaces.

The initial tolerancing results from an analysis which consist in fitting deviation domain to clearance domain (of bearings in this case). When no forces are input (ur = 0), we have to verify this initial tolerancing so that every part has deviations in itsallowed domain. Allowable displacements can be buid with a maximum fixed stress (static or infinite life,...) like in figure 4.a). In figure 4.b) tolerance defects in the part are input and allowable elastic displacements are added. A Minkowski sum of polyhedra is used to obtain the resulting deviation domains.

As we can see, the initial tolerancing domain (rigid deviation space) is extended by taking elastic limits (ur between 0 and 1) of the beam element (using the beam theory in this case but FEM can also be used) into account . Two limits are shown, first the infinite life rate (Wölher theory here) then elastic stress limit. So it is interesting to see that initial tolerancing can be extended by knowing its consequences in the whole mechanism.

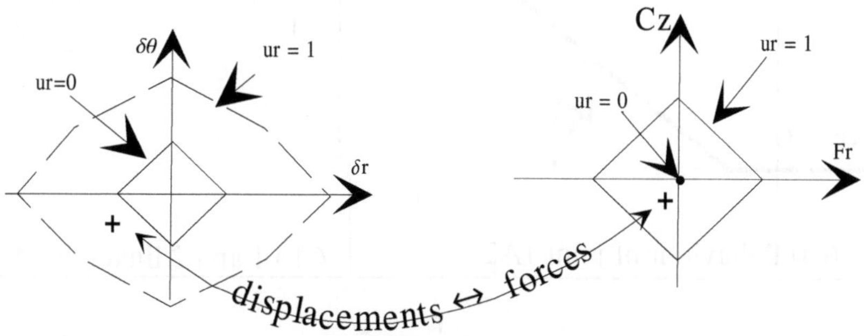

Figure 5

We can also see in figure 5 that there is a duality between displacements and forces outside of the rigid deviation domain. In the analysis, a part will have a fixed imperfect geometry. The relative positions of each surface of a part will be inside the deviation space. This will be all the data input of the elastic resolution. For every part of the assembly, each geometrical specification allows combinations of deviations. If the solution domain is explored completely, the only option will be to analyse small assemblies. If this method is to be used with conventional assembly, we must input a limited set of input deviations. This set will be that of critical configurations of deviations.

3.2. Joint behaviour

In the case of joints, the behaviour is more difficult to build because of clearances. Joints without clearances (axial or pre-loaded bearings,...) can be treated like parts. The behaviour will be expressed by relative positions of the two link frames of the joint. These allowed relative displacements can be represented by the clearance rigid domain of the joint. If relative forces are input into the joint, this domain is extended to a bigger one, which size depend on the use rate (ur) [samper 2001].

Case of a uniaxial joint

A relation between a force F and a displacement δ (with ur) for a joint with clearance is built.

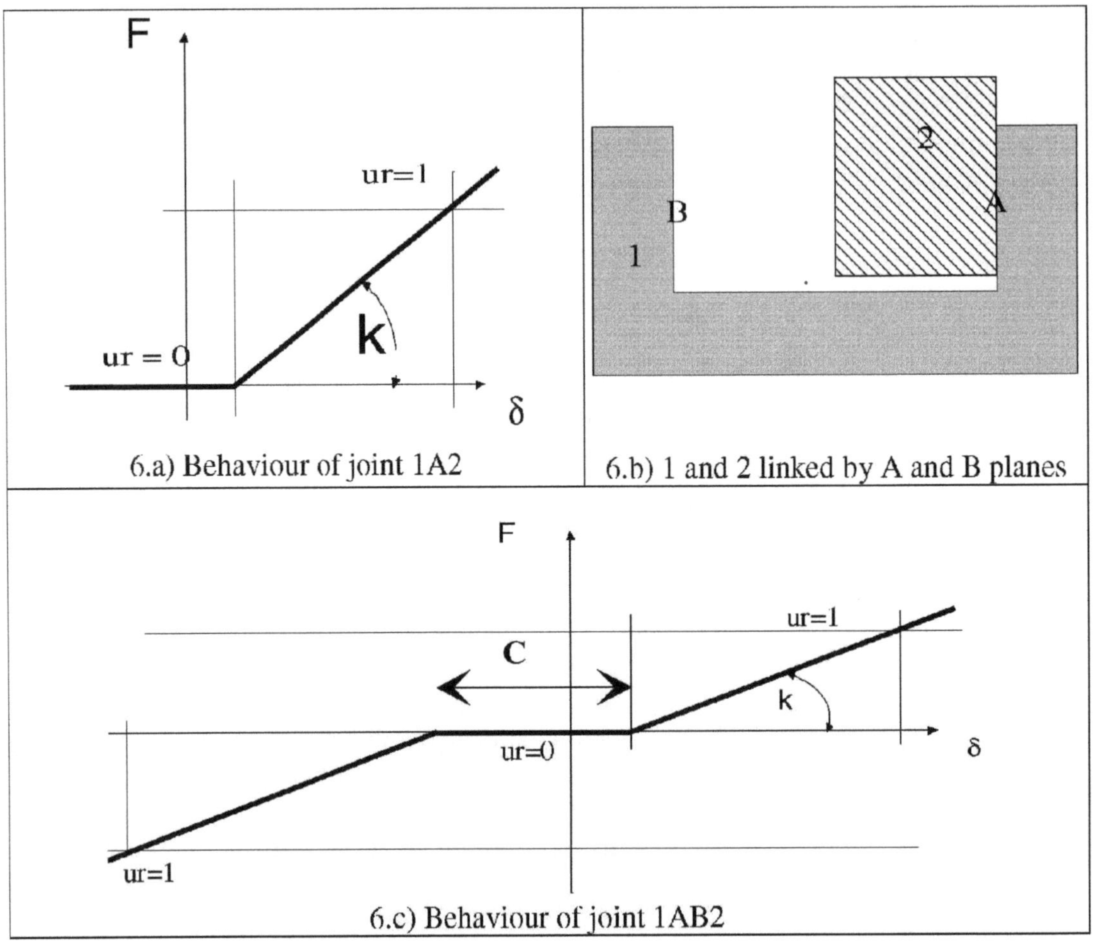

Figure 6 Examples of uniaxial joints

Here, we can see in 6.b) a two planes joint between two parts. If the stiffness of each part is known, we can say that parts are rigid and joints are flexible. Then the behaviour of a joint is shown in 6.a) and when we want to analyse the assembly, we can use the behaviour of 6.c).

Case of a multi axial joint

We can extend the method above to a general joint (two surfaces with no mobility and clearances). Thus this joint has six degrees of freedom for the finite element analysis. We have to know relations between displacements and forces and "ur". Inside the rigid clearance domain, forces and "ur" are nil and outside of the rigid clearance domain, forces depend on displacements and a functioning point is associated with "ur".

$$\text{Inside}\begin{Bmatrix} u \\ v \\ w \\ \beta \\ \gamma \end{Bmatrix} \Rightarrow \begin{Bmatrix} Fx=0 \\ Fy=0 \\ Fz=0 \\ My=0 \\ Mz=0 \\ ur=0 \end{Bmatrix}$$

$$\text{and outside the rigid domain} \begin{bmatrix} \begin{Bmatrix} u \\ v \\ w \\ \beta \\ \gamma \end{Bmatrix} \Leftrightarrow \begin{Bmatrix} Fx \\ Fy \\ Fz \\ My \\ Mz \end{Bmatrix} \end{bmatrix} \Rightarrow ur \qquad (1)$$

So like in figure 5 in the case of the shaft, equation 1 for a joint links displacements with forces and "ur" outside of the rigid clearance domain.

Case of a ball bearing

We have studied ball bearings [Perotto 1999] and some results are presented here.

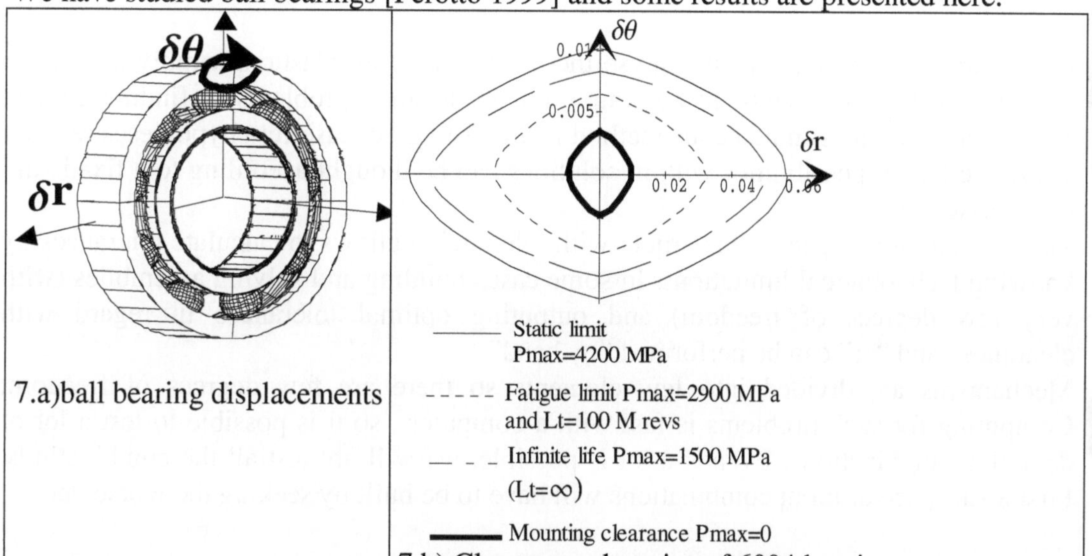

7.a)ball bearing displacements

7.b) Clearances domains of 6004 bearing

Figure 7 : Model of the ball bearing 6004

This ball bearing has got 1 movement so 5 (elastic) degrees of freedom. This joint is calculated from of the relatives position and forces of inner and outer diameters and its life cycle. In fact there are many ways to code this behaviour. A space of 6 dimensions with several clearance domains associated with "ur" growing from 0 to 1 (domain with ur=0 inside one with 0.1 and so on to ur=1) can be built.

To simplify representation, two displacements are shown here (radial displacement and in plane rotation).

In figure 7.b, we can see four clearances domains of a ball bearing from the rigid one (smallest with ur=0) to the static limit (ur=1). But we can specify ur=1 for the third (one million revolutions).

It is very interesting to link the life cycle with tolerancing as done in this study. In some cases, it is possible to make "by hand" tolerance analyses by taking this work into account when the stiffness of each part is bigger than the bearing's for example. So the domain used for tolerancing is the clearance allowed with a chosen "ur".

3.3. Solving and post-processing

A finite element solving makes the resolution of this analysis. External forces, every element and deviations are input. Then deviations from tolerancing must modify relative positions (i.e. translations and rotations) of corresponding nodes. Thus the computation of displacements and forces can be done. Nodal displacements if initial clearances and geometrical specifications are right or not. Nodal forces can give ur.

4. CONCLUSION

In this article, two methods to analyse the behaviour of an elastic assembly have been shown. The first one seems to be simple but leads some problems in finding contact configurations. The simultaneous method is interesting because few hypotheses remain. There are lots of possibilities with it, solutions can be thought according to a fixed "ur" for an assembly.

The use rate introduced in this article with tolerancing allows to calculate tolerances by knowing technological limitations. In some cases building and solving assemblies (with very few degrees of freedom) and outputing optimal tolerances in regard with clearances and "ur" can be performed "by hand".

Mechanisms are divided into few elements so there are few degrees of freedom. Computing for such problems is fast with a computer , so it is possible to test a lot of deviation combinations. Even if a lot is possible, we will not test all the combinations. First a subspace of input combinations will have to be built by seeking the worse ones.

REFERENCES

[Giordano et al., 1999] Giordano M., Pairel E., Samper S., "Mathematical representation of Tolerance Zones", In 6th CIRP Inter. seminar on Computer-Aided Tolerancing, Univ. of Twente, Enschede, The Netherlands, 22-24 March 1999.

[Samper & al., 2001] Serge Samper; Max Giordano, In: *Proceedings of the Qualita 2001 4th international congress on quality and reliability* Annecy 2001.

[Samper & al. 1998] « Taking into account elastic displacements in 3D tolerancing - Models and applications Journal of Material Processing Technology - Elsevier Science.

[Perotto 1999] Mémoire CNAM « Jeux, efforts et déformation dans les roulements à billes », Juin1999.

Towards robust kinematic synthesis of mechanical systems

E. Sacks[1], **L. Joskowicz**[2], **R. Schultheiss**[3], **M. Kyung**[1]

[1] *Computer Science Department, Purdue University, West Lafayette, IN, USA.*
[2] *School of Computer Science and Eng., The Hebrew University of Jerusalem, Israel.*
[3] *Ford Werke AG, Kln, Germany*
E-mail: eps@cs.purdue.edu, josko@cs.huji.ac.il

Abstract: We describe our research in kinematic synthesis of planar mechanical systems based on configuration space manipulation. We present a design scenario that illustrates our methodology and describe two algorithms that support robust parametric design. The first algorithm helps designers select nominal parameter values and identify failure modes. The second algorithm helps optimize tolerance allocation. These are the first general algorithms for these tasks, as prior work is limited to lower pairs and to a few custom higher pairs.

Keywords: parametric design, robust design, tolerancing, kinematics, configuration space.

1. INTRODUCTION

We describe our research in robust kinematic synthesis of mechanical systems. Kinematic synthesis is the task of devising a system of mechanical parts that implements specified motion transformations. The design must meet its specifications despite part variations due to manufacturing. An optimal design achieves this goal at minimal cost. Kinematic synthesis is central to mechanical design because kinematics largely determines mechanical function.

Kinematic synthesis is an iterative process in which the designer selects a design concept, constructs a parametric model, assigns parameter values, and allocates tolerances. At each step, the designer makes changes, assesses their impact, and decides whether to advance to the next step or to return to a prior step. When a design fails due to part variations, the designer can change the nominal design or tighten the tolerances. Changing the nominal design is often better, since cost increases rapidly as tolerances decrease, but can be much harder.

Kinematic synthesis is difficult and time consuming. In the conceptual design step, the designer needs to compare competing concepts based on incomplete, high-level characterizations. In the later steps, he has to adapt the chosen concept to comply with numerous, often competing design specifications. The adaptation requires extensive kinematic analysis of many design instances. The analysis is difficult because it involves multiple part contacts that impose nonlinear motion constraints. Some contacts are part of the nom-

P. Bourdet and L. Mathieu (eds.),
Geometric Product Specification and Verification: Integration of Functionality, 135-144.

inal function, while others arise due to part variation. Both types can introduce failure modes that coexist with or supersede the correct function. Finally, the designer needs to formulate a realistic, application-specific cost function for tolerance allocation.

Software support for kinematic synthesis is limited. There are very few tools for conceptual design. Powerful commercial packages, such as CATIA and IDEAS, support construction and visualization of parametric designs. Kinematic analysis software is limited to multi-body systems: assemblies of parts that interact via a fixed set of feature contacts [Schiehlen, 90]. Prior research in synthesis provides algorithms for linkages [Erdman, 93], [Ramaswamy, 93] and cams [Angeles and Lopez-Cajun, 91], [Gonzales-Palacios and Angeles, 93] but does not address systems with contact changes. Tolerance analysis software is available for individual, user-specified system configurations, but not over a continuous work cycle [Chase et al, 97], [Solomons et al, 97], [Ballot and Bourdet, 97].

A new methodology, called robust design, has been developed to increase reliability and reduce redesign costs. In robust design, nominal and tolerance changes are evaluated together. The nominal design is modified to reduce its sensitivity to part variations. Then tolerances are allocated to guarantee correct function and to minimize cost. Robust design differs from the traditional design paradigm in which failure due to part variations is fixed primarily by tightening tolerances. Robust design is especially relevant to kinematic synthesis because failures due to tolerances are hard to detect and costly to correct [Schultheiss and Hinze, 99].

In this paper, we describe two algorithms that support kinematic synthesis of planar mechanical systems and illustrate the algorithms on a robust design scenario. A system is comprised of kinematic pairs with multiple contacts that form one or more open chains or closed loops. Each part has one degree of freedom: translation along a fixed axis or rotation around a fixed point. The first algorithm helps designers select nominal parameter values and identify failure modes [Kyung and Sacks, 2001]. The second algorithm helps optimize tolerance allocation. These are the first general algorithms for these tasks; prior work in the field is limited to lower pairs and to a few custom higher pairs. The algorithms build upon our kinematic analysis algorithm for the specified class of mechanical systems [Sacks and Joskowicz, 95].

2. DESIGN SCENARIO

We illustrate our kinematic synthesis algorithms on an optical filter mechanism from Israel Aircraft Industries (Figure 1a). The mechanism consists of a lens, a cam, and three filters mounted on identical followers. The lens is attached to a fixed frame (not shown). The followers are stacked on a shaft and can rotate independently. The cam consists of three slices that rotate together on a common shaft. Each cam slice drives the corresponding follower. Figure 1b shows the top cam slice and its follower. The cam slice consists of a driving pin and a locking arc. When the cam shaft rotates counterclockwise, the

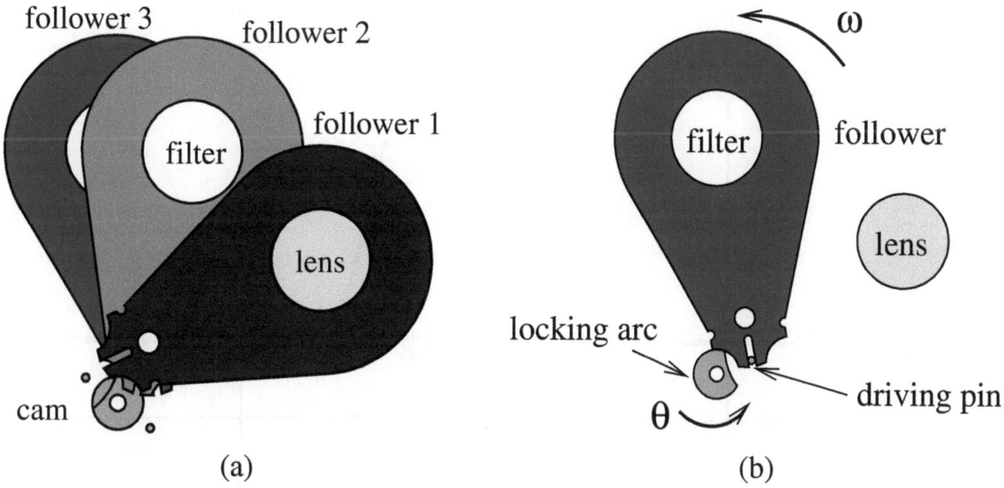

Figure 1: *Optical filter mechanism: (a) all filters, (b) top filter.*

pin engages the follower slot and rotates the follower until the filter covers the lens. The other two cam slices are identical, except that they are rotated by 90 and 180 degrees respectively. In the initial state, the filters are off the lens. When the cam shaft is rotated counterclockwise, the three followers are engaged in sequence. Rotating the cam clockwise resets the filters to the initial state.

The design task is to devise a mechanism to engage and reset the followers in the intended manner. The mechanism must be robust because it will be mounted on a vehicle and must be compact to fit in the alloted space. During conceptual design, the designer chooses a Geneva mechanism with one driver and one follower per filter. This concept dictates the functional geometric features: a pin/slot pair for the driving phase and a concentric concave/convex arc pair for the locking phase. The designer creates a parametric model of the pair with 25 functional parameters, including the centers of rotation, the pin and locking arc radii, and the slot dimensions.

The next step is to assign nominal parameter values that produce correct function. We perform this step via interactive manipulation of the cam/follower configuration spaces. Configuration space is a complete geometric representation of kinematics that reveals qualitative and quantitative function. We pick initial parameter values, compute the resulting configuration spaces, and evaluate them for correct function.

Figure 2a shows the configuration space of the top cam/follower pair. The coordinates are the part orientation angles. The configuration space wraps around at the top/bottom and left/right boundaries because the coordinates are angles. It is partitioned into free space where the parts do not touch (white area) and blocked space, where they overlap (gray area), separated by contact space where they touch (black curves). The dot marks the displayed configuration in Figure 1b. The horizontal contact curves correspond to the contact between the cam and the follower locking arcs. The slanted curves correspond to the contact between the cam pin and the driver slot. The gap between the curves represents

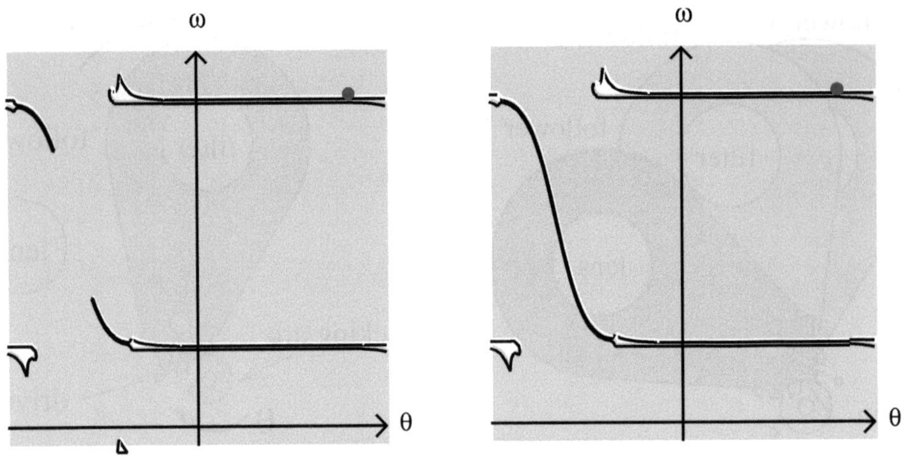

(a) (b)

Figure 2: *Configuration spaces for one filter: (a) blocked, (b) correct.*

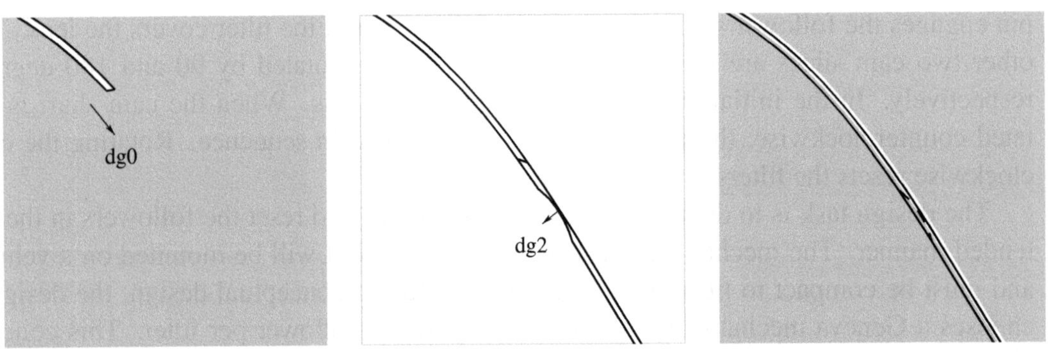

(a) (b) (c)

Figure 3: *Parametric modification of c-space: (a) detail of initial blocked space, (b) open channel, (c) wider midpoint.*

play. The configuration space shows that the cam blocks when the pin is partially engaged in the follower slot, since the slanted channel consists of two disjoint segments that end at these blocking configurations. Figure 2b shows a correct configuration space with a single slanted channel that connects the adjacent horizontal channels.

We must modify the initial parameter values to merge the two partial channels. We grab the channel bottom with the mouse and drag down (Figure 3a). The dragging causes the partial channels to meet (Figure 3b). The program implements dragging by computing parameter values that make the selected contact configuration track the mouse. Although now open, the channel is too narrow at its midpoint, so we widen it with a second dragging operation (Figure 3c).

Now that the nominal function is correct, we modify the parametric design by replacing the sharp corners of the follower slot with fillets (Figure 4). We assign values to the

(a) (b)

Figure 4: *Adding follower fillets: (a) before and (b) after.*

new parameters as before. The configuration space is only slightly different.

The final design step is to assign tolerances to the parameters and to assess their effects. We model kinematic variation by generalizing the configuration space representation to toleranced parts. The contact curves of a pair are parameterized by the touching features, which depend on the tolerance parameters. As the parameters vary around their nominal values, the contact curves vary in a band around the nominal contact space, which we call the contact zone. The contact zone defines the kinematic variation in each contact configuration: every pair that satisfies the part tolerances generates a contact space that lies in the contact zone. Kinematic variations do not occur in free configurations because the parts do not interact.

Figure 5a shows a detail of the cam/follower contact zone in the area where the cam unlocks the follower and the pin is about to enter the follower slot. The contact zone is bounded by the light grey curves. Its width varies with the sensitivity of the nominal contact configuration to the tolerance parameters. The upper and lower zones of the diagonal channel intersect, which implies that there are parameter values in the tolerance intervals that cause blocking.

In robust design, we prefer to remove the blocking by widening the channels, as described above. If this is impractical, we can tighten the tolerance intervals until the zones become disjoint, as shown in Figure 5b. We use our tolerance optimization algorithm to compute intervals that achieve this goal at minimum cost relative to an input cost function.

3. PARAMETRIC DESIGN ALGORITHM

We briefly describe the parametric design algorithm. For a detailed description, see [Kyung and Sacks, 2001]. The designer inputs a parametric model of a mechanical system and specifies initial parameter values. The synthesis program computes and displays configuration spaces for the kinematic pairs. These spaces encode the initial kinematics: feature contacts appear as contact curves and contact changes appear as curve adjacencies.

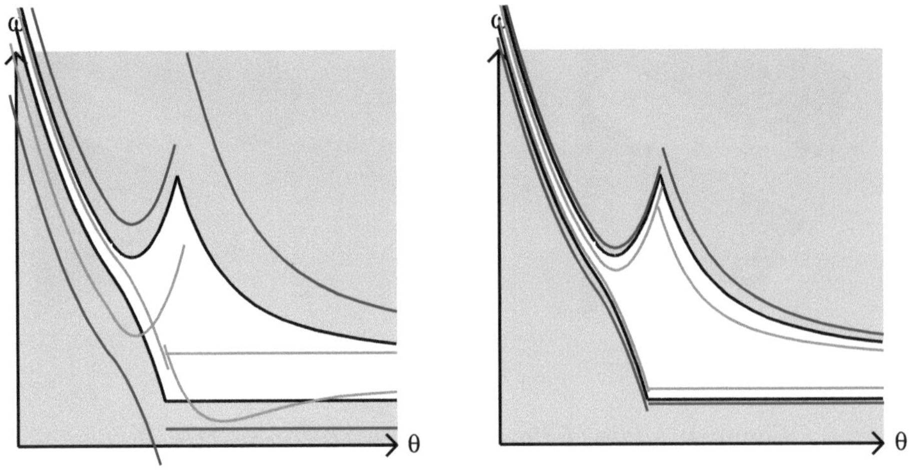

(a) (b)

Figure 5: *Detail of contact zones: (a) possible failure, (b) robust.*

Design objectives are expressed as changes in the contact curve geometry. The designer inputs the objectives with the mouse and the program achieves them by changing the design parameters.

The parameter update algorithm computes design parameter values that achieve specified kinematic changes. The input is a set of draggers (dgo and dg2 in Figure 3). Each dragger consists of a contact curve, a start point $\mathbf{q}_0$ on the curve, and a goal point $\mathbf{q}_0 + \delta\mathbf{q}$. Contact curves have the form $C(\mathbf{q}, \mathbf{p}) = 0$ where $\mathbf{q}$ is the two configuration space coordinates (θ, ω in our example) and $\mathbf{p}$ is the vector of design parameters. The start point, $\mathbf{q}_0$, satisfies $C(\mathbf{q}_0, \mathbf{p}_0) = 0$ with $\mathbf{p}_0$ the initial parameter values. The program computes a parameter update $\delta\mathbf{p}$ for which the curve goes through the goal point, $C(\mathbf{q}_0 + \delta\mathbf{q}, \mathbf{p}_0 + \delta\mathbf{p}) = 0$.

The contact equations are solved numerically because a closed-form solution is impractical. The program performs a sequence of small parameter updates governed by the linearized contact equation

$$\frac{\partial C}{\partial \mathbf{q}}(\mathbf{q}_0, \mathbf{p}_0)\delta\mathbf{q} + \frac{\partial C}{\partial \mathbf{p}}(\mathbf{q}_0, \mathbf{p}_0)\delta\mathbf{p} = 0.$$

There is one equation per dragger. The equations are normally under constrained because a typical number of draggers is less than five, while a typical number of design parameters is twenty. But they are over constrained when the draggers are inconsistent or when there are more draggers than parameters. We compute an exact, minimum-norm solution if possible and a least-squares solution otherwise, using singular value decomposition.

The program monitors updates for unintended kinematic changes. Contact curves that were not selected by the designer will often change shape because they share parameters with the selected curves. A sufficiently large change can cause a pair of disjoint curves to intersect or vice a versa. These events can cause structural changes in the system kinematics, such as jamming. The program detects these changes and modifies the parameter update to prevent them.

Our algorithm builds on prior work [Caine, 93], who designs planar part feeders via configuration space manipulation. The kinematic function is represented by a partial part/feeder configuration space. The designer requests a single change in the configuration space and the program changes the feeder geometry accordingly. The models are non-parametric and the modifications are heuristic. Structural changes are not addressed.

4. TOLERANCE ALLOCATION ALGORITHM

Our tolerance allocation algorithm uses constrained optimization to compute tolerance ranges for the design parameters that ensure correct kinematic function at minimum cost. The part shapes and motion axes are parameterized by a vector $\mathbf{p}$ whose nominal value is $\bar{\mathbf{p}}$. Parameter p_i is constrained to the interval $[\bar{p}_i - l_i, \bar{p}_i + u_i]$ with l_i and u_i non-negative. The l_i and u_i are the variables in the optimization problem. The constraints are bounds on the kinematic variation at a sequence of nominal system configurations. Minimum and maximum values for the variables can also be specified.

The kinematic constraints at a single configuration translate into linear inequalities among the variables. We formulate the constraint on a variable z in terms of the maximum increase in its value, δz^+, and the maximum decrease in its value, δz^-, due to variation in $\mathbf{p}$. Following standard tolerancing practice, we use the linear approximations

$$\delta z^+ = \sum_i \frac{\partial z}{\partial p_i} w_i \text{ with } w_i = \begin{cases} u_i & \text{if } \partial z/\partial p_i > 0 \\ -l_i & \text{otherwise.} \end{cases}$$

and

$$\delta z^- = -\sum_i \frac{\partial z}{\partial p_i} w_i \text{ with } w_i = \begin{cases} -l_i & \text{if } \partial z/\partial p_i > 0 \\ u_i & \text{otherwise.} \end{cases}$$

The signs are chosen so that both quantities are non-negative: each term is a non-negative number times a non-negative variable. The kinematic constraints are $\delta z^+ \leq s$ and $\delta z^- \leq t$ with s and t the input bounds on the kinematic variation.

In our design scenario, we bound the kinematic variation at the center of the diagonal channel. We pick configurations $c = (-1.47, 0.519)$ on the top curve and $d = (-1.47, 0.506)$ on the bottom curve. The nominal play is $0.519 - 0.506 = 0.013$, which means that the follower can rotate by 0.013 radians when the cam angle is 1.47 radians. The actual play equals $c^- - d^+$ with c^- the worst-case decrease in the top curve and d^+ the maximum increase in the lower curve. We specify the kinematic constraints $c^- \leq 0.05$ and $d^+ \leq 0.005$ to ensure a minimal play of 0.003 radians. The resulting linear equations are

$$\begin{aligned} c^- &= u_1 + l_2 + 0.91u_3 + 0.5u_4 + 0.87l_5 \leq 0.005 \\ d^+ &= l_1 + u_2 + 0.91u_3 + 0.5u_4 + 0.88l_5 \leq 0.005 \end{aligned}$$

with p_1 (and l_l, u_1) the pin radius, p_2 the slot width, p_3 the distance from the pin center to its center of rotation, p_4, and (p_4, p_5) the (x, y) coordinates of the follower center of rotation relative to the cam center of rotation.

We have developed an algorithm for computing the derivatives of the coordinates with respect to the parameters [Sacks and Joskowicz, 97] The first step is to formulate a parametric contact equation $y = f(x, \mathbf{p})$ for each pair of parts that is in contact. This is done by querying the configuration spaces of the pairs, as explained in our prior work. The second step is to differentiate the contact equations with respect to x and **p**. The final step is to compute the derivatives of the coordinates by the chain rule of calculus. The computation starts from the driving coordinates, which have zero kinematic variation by definition, and propagates from part to part via contact equations.

Choosing an objective function is a modeling decision based on specific product costs. Many options, linear and nonlinear, appear in the tolerancing literature. The research challenge is to find a function that reflects kinematic costs adequately for design optimization, yet yields a tractable optimization problem. The simplest option is to assign linear cost functions $w_i = a_i - b_i l_i - c_i u_i$ with $a_i, b_i, c_i \geq 0$ and to minimize the total cost $\sum_i w_i$. The signs are chosen so the cost of a parameter decreases as its lower and upper variations increase.

In our example, we set $a_i = 10$, $b_i = 1$, and $c_i = 1$ and obtained optimal tolerances of $l_2 = u_2 = 0.005$ and the other variables equal to zero. This is unrealistic because zero tolerances are unrealizable. The problem is that the linear cost function diverges from the true cost. We can improve the model by specifying a minimum value of 0.001 for every variable. The optimal tolerances are $l_2 = u_2 = 0.0017$ and the other variables equal to this minimum value. We see that the slot width dominates the cost of tolerancing for channel play.

5. CONCLUSIONS

We have presented two algorithms for robust kinematic synthesis of planar mechanical systems with one degree of freedom per part. The first algorithm supports parameter value selection for a nominal design. The second supports optimal tolerance allocation subject to kinematic constraints. Our next step is to apply these algorithms to large-scale problems in automotive transmission design and in other application areas. Our first research goal is to extend the algorithms to general planar systems with three degrees of freedom per part. The research challenge is structual change detection in three-dimensional configuration spaces. Developing an effective user interface is a significant technical challenge. Our second research goal is to handle spatial systems with one degree of freedom per part. The planar algorithms are applicable, but we need to derive parametric contact functions for all pairs of spatial features, such as planes, spheres, cylinders, and spatial curves.

ACKNOWLEDGMENTS

This research was supported by NSF grants CCR-9617600 and IIS-0082339, by the Purdue Center for Computational Image Analysis and Scientific Visualization, by a Ford University Research Grant, by the Ford ADAPT 2000 project, and by grant 98/536 from the Israeli Academy of Science. Dr. Avner Klunover from the Israeli Airforce Industry provided the optical filter mechanism.

REFERENCES

[Ballot and Bourdet, 97] Ballot, E. and Bourdet, P. A computation method for the consequences of geometric errors in mechanisms. in: *Proc. 5th CIRP Int. Seminar on Computer-Aided Tolerancing*, Toronto, 1997.

[Angeles and Lopez-Cajun, 91] Angeles, Jorge and Lopez-Cajun, Carlos. *Optimization of Cam Mechanisms*. Kluwer Academic Publishers, Dordrecht, Boston, London, 1991.

[Caine, 93] Caine, Michael E. The design of shape from motion contraints. AI-TR 1425, Massachusetts Institute of Technology, Artificial Intelligence Laboratory, 545 Technology Square, Cambridge, MA, 02139, 1993.

[Chase et al, 97] Chase, Kenneth, Magleby, Spencer, and Glancy, Charles. A comprehensive system for computer-aided tolerance analysis of 2d and 3d mechanical assemblies. In *Proc. of the 5th CIRP Int. Seminar on Computer-Aided Tolerancing*, Toronto, 1997.

[Erdman, 93] Erdman, G. Arthur. *Modern Kinematics: developments in the last forty years*. John Wiley and Sons, 1993.

[Gonzales-Palacios and Angeles, 93] Gonzales-Palacios, Max and Angeles, Jorge. *Cam Synthesis*. Kluwer Academic Publishers, Dordrecht, Boston, London, 1993.

[Kyung and Sacks, 2001] Kyung, Min-Ho and Sacks, Elisha. Computer-aided synthesis of higher pairs via configuration space manipulation. In *Proc. of the International Conference on Robotics and Automation*, Seoul, 2001.

[Ramaswamy, 93] Ramaswamy, Rajan. *Computer Tools for Preliminary Parametric Design*. PhD thesis, Massachussetts Institute of Technology, 1993.

[Sacks and Joskowicz, 95] Sacks, Elisha and Joskowicz, Leo. Computational kinematic analysis of higher pairs with multiple contacts. *Journal of Mechanical Design*, 117(2(A)):269–277, June 1995.

[Sacks and Joskowicz, 95] Sacks, Elisha and Joskowicz, Leo. Parametric kinematic tolerance analysis of planar mechanisms. *Computer-Aided Design*, 29(5):333–342, 1997.

[Schiehlen, 90] Schiehlen, W. *Multibody systems handbook*. Springer-Verlag, Berlin, 1990.

[Schultheiss and Hinze, 99] Schultheiss, Ralf and Hinze, Uwe. Detect the unexpected - how to find and avoid unexpected tolerance problems in mechanisms. In *Global consistency of tolerances, Proc. of the 6th CIRP Int. Seminar on Computer-Aided Tolerancing, F. van Houten and H. Kals eds., Kluwer*, 1999.

[Solomons et al, 97] Solomons, O.W., Van Houten, F., and Kals, H. Current status of CAT systems. In *Proc. of the 5th CIRP Int. Seminar on Computer-Aided Tolerancing*, Toronto, 1997.

Tolerance analysis and synthesis by means of clearance and deviation spaces

Max Giordano, Bassam Kataya, Eric Pairel
Laboratoire de Mécanique Appliquée
ESIA Université de savoie B.P. 806
74 016 ANNECY Cedex
giordano@esia.univ-savoie.fr

Abstract: A mathematical representation of tolerance zones is used to analyse toleranced mechanisms. This method permits to choose optimal geometric tolerances, in many cases. A mechanism is considered as constituted of solid bodies and joints. The set of displacements allowed by a joint is called clearance space. It takes into account the degrees of freedom of the joint but also the clearance between the two bodies linked by this joint. For each functional and toleranced surface, the set of displacements of the actual surface with regard to its nominal location and according to the tolerance zone, is called deviation space. For a simple closed loop mechanism, topological operations are made on the clearance spaces, and on the deviation spaces. In the case of a mechanism with a complex structure, the analysis is more complex but can be computed thanks to a few topological operations on the clearance and deviation spaces. This method allows to compare different tolerancings for each part. Consequently, it is possible to define optimal geometrical tolerances in many cases.
Key words : tolerancing, functional analysis, mechanisms, tolerance synthesis.

1. INTRODUCTION

For the generalisation of chains of dimensions in the 3D space, the concept of tolerancing equations is present in the literature. But those equations are not sufficient to analyse a given toleranced mechanism, it is necessary to generate the inequalities translated from the tolerances and from the contact conditions. Benis, Pino and Fortin [Benis et al., 1998] use the homogenous transformation matrices to compute the problem of 3D geometric tolerance transfer. Gaunet uses the small displacement torsors and build a system of equations and inequalities between the small displacement components due to the geometric deviations [Gaunet 1994]. But a great number of contact conditions are generated and the system is complex even for a mechanism with a limited number of parts and features.

The concept of configuration space has been used for the determination of assembling trajectories and applied to a kinematic method for the tolerance analysis of 2D mechanisms [Joskowicz et al., 1999]. The concept of feasibility space for automating tolerancing, introduced by Turner, leads to translate the tolerances into inequalities [Turner 1993]. In the case of small displacements, the deviation and

P. Bourdet and L. Mathieu (eds.),
Geometric Product Specification and Verification: Integration of Functionality, 145-154.

clearance spaces are introduced to the tolerance analysis in particular cases and in a 2D model [Giordano et al., 1992]. The set of inequalities is represented by polytopes by Teissandier and an appropriate data structure allows to build topological operations needed to the tolerance analysis, particularly the sum of Minkowski of the polytopes [Teissandier et al;, 1999].

In this paper, the equations and inequalities required to solve the problem of tolerance analysis are presented thanks to the concept of "deviation and clearance spaces", with the final objective to define a method for tolerance synthesis.

2. DEVIATION AND CLEARANCE TORSORS, TOLERANCING EQUATIONS

2.1. The hypotheses of the model

The mechanisms considered here are constituted of rigid bodies. The contact surfaces in the joints have no form defect, but only size, orientation and location deviations in regard to a nominal geometry assumed to be perfect.

Two types of small displacements are defined. First the deviation between the nominal surface and the actual surface attached to a part, and then the displacement due to the clearance between the two surfaces of the two parts in contact, which constitute the joint.

It is assumed that these displacements are small so that they can be represented in a given frame by a torsor constituted of two vectors, the rotation vector having 3 components (Rx, Ry, Rz) and the translation vector (Tx, Ty, Tz). In the 6D configuration space a torsor is then represented by a point. The given frame is defined in the 3D Euclidean space by a point and 3 orthogonal directions x, y, z.

2.2. The clearance torsor

For a given joint between two parts, the "clearance torsor" represents a small relative displacement between the two parts, when only this joint is taken into account. It is necessary to define a particular datum configuration between the two surfaces constituting the joint. In this particular configuration, the frames attached to each surface are coinciding and then, the clearance torsor is zero.

For example, for a perfect cylindrical joint, without any clearance between the two surfaces, only the rotation around and the translation along the axis of the cylinder are different of zero, and the four other components are zero. In a non perfect joint, the 6 components are non zero in the general configuration.

2.3. The deviation torsor

A deviation torsor is defined for a feature that belongs to a part. This torsor represent a small deviation between a datum frame attached to the part and the frame attached to the feature. For a perfect part, all the frames attached to each feature of the part are in nominal configuration between each other, so that all the deviation torsors are zero.

Any displacement keeps the geometry of a part unchanged, so that any set of datum frames is possible but all the datum frames are in an unique nominal position between each other. It is only their global position with respect to the part that is not determined. Therefore, if a part has only one feature, it is a nonsense to try to define a torsor deviation. If a manufactured part has two features A and B, for the datum frame (1) attached to the part, two deviation torsors E_{1A} and E_{1B} can be defined. For a different datum frame (1') attached to the part, the two deviation torsors will be $E_{1'A}$ and $E_{1'B}$. But the relative deviation torsor is $E_{AB} = E_{1B} - E_{1A} = E_{1'B} - E_{1'A}$ for any frame (1) or (1'). The only condition is that all the deviation displacements must be small.

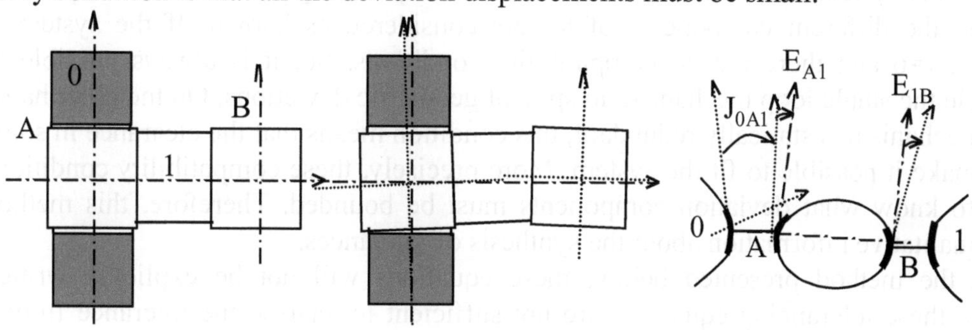

Figure1; Small displacements, nominal and actual frames

2.4. The tolerance equations

For a single closed loop mechanism, assuming that the nominal configuration and the different frames attached to each feature have been chosen, it is possible to obtain the geometric closed loop equation. In the general case, the exact equation can be obtained using the homogenous 4x4 matrix formalism. The closed loop is a product of these transformation matrices. But, in the assumption of small displacements between the real and nominal configurations of the different frames, the closed loop equation is translated into a sum of small displacement torsors.

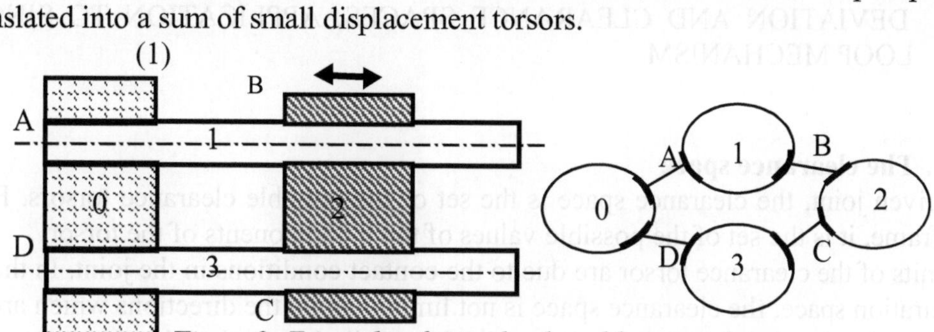

Figure2; Example of asingle closed loop mechanism.

For example, the mechanism shown fig.2, gives the following equation :

$$J_{0A1} + E_{A1B} + J_{1B2} + E_{B2C} + J_{2C3} + E_{C3D} + J_{3D0} + E_{D0A} = 0 \qquad (1)$$

Where the J terms are the clearance torsors for each joint and the E terms the deviation torsors for each part. This equation can be used for the determination of the deviation torsor components that have to be bounded. Therefore it gives qualitative information about tolerancing. The reasoning is as follows. In each clearance torsor, two kinds of components among the six of them are distinguished. On one hand there are those corresponding to the degrees of freedom of the joint considered as unknowns and are not bounded. And on the other hand, the other components are the ones that are allowed thanks to the clearance in the joint and are bounded and considered as known. These last components are equal to zero for a theoretical perfect joint or for an actual joint without any clearance.

The six components of each torsor of the equation (1) must be defined in a common frame. The six scalar equations (1) constitute a linear system. Setting r the rank of this system, there are 6-r conditions of compatibility that are linear equations between the different components of torsors considered as known. If the system is isostatic, r=6 and there are no compatibility conditions. So, it is always possible to assemble the single loop mechanism in spite of geometric deviations. On the other hand, if the mechanism is statically redundant, this condition means that the clearance in some joints make it possible to fit the system. More precisely, these compatibility conditions allow to know what deviation components must be bounded. Therefore, this method gives qualitative information about the synthesis of tolerances.

In the method presented below, these equations will not be explicitly written because these tolerancing equations are not sufficient to analyse the tolerance from a quantitative point of view. In order to do that, it is necessary to know the limits of the different terms that appear in the equation (1). When the limits of a component appear in the analysis, it means that this component has to be toleranced. The analysis of the boundaries of the clearance and deviation torsors gives more information than the simple analysis of the linear relations between their components, even if these relations are necessary. The method is developed in the next section but first, it is necessary to define the deviation and clearance spaces.

3. DEVIATION AND CLEARANCE SPACES, APPLICATION TO SINGLE LOOP MECHANISM

3.1. The clearance space

For a given joint, the clearance space is the set of the possible clearance torsors. For a given frame, it is the set of the possible values of the 6 components of the torsor.
The limits of the clearance torsor are due to the contact conditions in the joint. In the 6D configuration space, the clearance space is not limited along the directions which are the degrees of freedom of the joint. For example, for a cylindrical joint, the clearance space is bounded for the rotations around y and z (component called Ry and Rz) and for the translations along y and z (Ty and Tz). But it is considered as not bounded for rotation and translation of x axis (Rx and Tx).

For a clearance torsor noted J, its clearance space will be noted {J}.

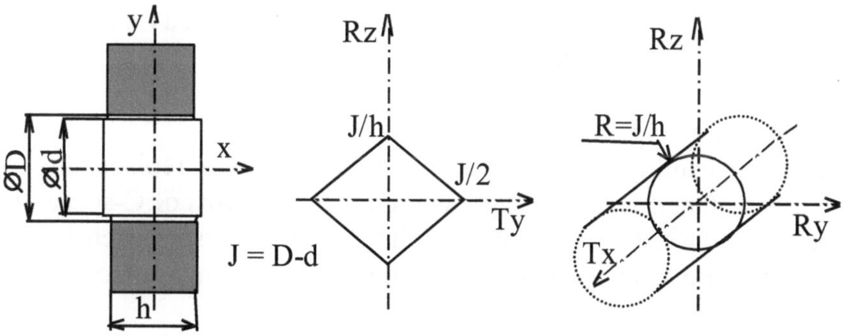

Figure 3; cylindrical joint, clearance space

3.2. The deviation space

For each part, the geometric tolerancing specifications consist in the definition of the limits of the deviation torsor. These limits can also be defined by inequalities between the components of the deviation torsor as shown in [Giordano et al., 1999]. The deviation space is the set of the components of the deviation torsor for given tolerance specifications relatively to a feature of a part. For a deviation torsor E, the deviation space is noted {E}. With the hypotheses of our model, the standard specifications can be translated into a deviation space, as we shown in the following examples.

Clearance or deviation spaces have similar properties from a mathematical point of view and they are here referred to as configuration space (or C-space for short). A C-space is always defined in a space having six dimensions (3 for rotations and 3 for translations). Generally, the graphic representation is done in a 2D or respectively 3D space (perspective), when the 4 or respectively 3 other components are given to particular values.

3.3. Serial composition of C-space

For two given torsors E_1 et E_2, each belonging to a C-space $\{E_1\}$ and $\{E_2\}$ respectively, the set of possible values of E_1+E_2, is a new C-space noted $\{E_1\}+\{E_2\}$ and called the sum of the two C-spaces. The sign + is used for this operation in the set of C-spaces.

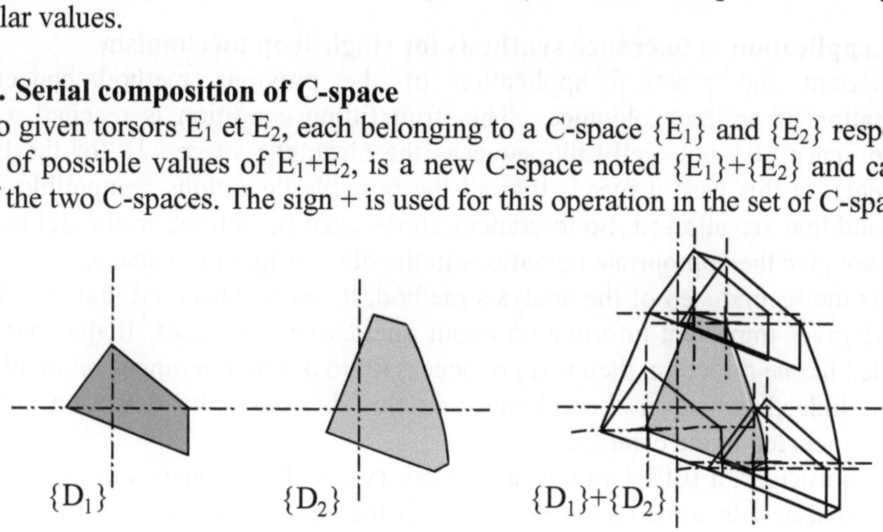

Figure 4; Minkowski sum of two C-spaces

For a geometric representation in 6D , the sum is built by the Minkowski sum of the C-space (fig.4). This operation is commutative and the neutral element {0} is constituted only of the torsor zero. This operation will be applied on deviation spaces and on clearance spaces or on both.

3.4. Analysis of a single loop mechanism

For the interchangeability, it must always be possible to replace any part of the mechanism by an other one respecting the tolerances. In terms of C-space, this property is translated as follows. For any value of the deviation torsors each belonging to its deviation space, there is at least one value for each clearance torsor belonging to its clearance space, so that the equation (1) is satisfied.

(1) is there written : $J_{0A1} + J_{1B2} + J_{2C3} + J_{3D0} = E_{A0D} + E_{D3C} + E_{C2B} + E_{B1A}$

setting $\{J\}=\{J_{0A1}\}+ \{J_{1B2}\}+ \{J_{2C3}\}+\{J_{3D0}\}$ and $\{E\}=\{E_{A0D}\}+\{E_{D3C}\}+\{E_{C2B}\}+\{E_{B1A}\}$ the fitting requirement is satisfied if and only if {E} is included in {J}or equal to {J} in the limit case.

$\{J\} \subseteq \{E\} \Leftrightarrow \forall E \in \{E\}, \exists J \in \{J\}$ so that J=E

The tolerance analysis for this condition can be decomposed in the following steps :

- building the clearance spaces for each joint having clearance,
- building the sum of these spaces {J}
- building the deviation spaces for given tolerances,
- building the sum of clearance spaces {E},
- checking the inclusion of {E} into {J}

This method can use graphic representations of the C-spaces but can also be computed thanks to an appropriate data structure in the software. From a numerical point of view, the method consists in checking the compatibility of a system composed of equations and inequalities.

3.5. Application to tolerance synthesis for single loop mechanism

An important and practical application of the previous method concerns the determination of optimal tolerances. The strict fitting condition is reached when the deviation space {E} is exactly the same as the clearance space {J} for the bounded components. In this case it means that all the possible deviations compatible with the fitting condition are allowed. So tolerances zones must be defined in the 3D real space so that they give the appropriate deviations in the 6D configuration space.

At the second step of the analysis method, it can be observed that the clearance space {J} gives important information about qualitative tolerances. If the space {J} is unbounded in one direction, then it is not necessary to define tolerances relatively to this direction. Indeed, whatever the tolerance in this direction, the deviation space will always be included in the clearance space.

In particular, a sufficient but not necessary condition consists in grouping some deviation spaces with some clearance spaces. If the equality is satisfied for each group, it will be satisfied for the sum.

Writing the equation (1) under the form :

$(J_{0A1} + E_{0A} - E_{1A})+(J_{1B2} + E_{1B} - E_{2B})+(J_{2C3} + E_{2C} - E_{3C})+ (J_{3D0} + E_{3D} - E_{0D}) = 0$

If the 4 conditions such as $\{J_{0A1}\} \subseteq \{E_{1A} - E_{0A}\}$ are satisfied then the global condition will be automatically checked.

For some joints there is no clearance. It is the case in particular for press fit cylindrical surfaces or for plane surfaces tightened by screws. The tolerance zone of those surfaces will then be equal to zero. Practically, they will be chosen as datum to define tolerance zones for other features. But this method may lead to have two or more features with a nil tolerance zone and this is not acceptable. Nevertheless, it is possible to generalise the previous method to avoid this problem. For each joint with clearance it is possible to define several tolerance zones so that the corresponding C-spaces $\{J\}$ and $\{E_A\}$, $\{E_B\}$...verify the condition $\{J\} \subseteq \{E_A\} + \{E_B\} + \ldots$

This method gives tolerance zones that are not close to the nominal feature. The concept of projected tolerance zone of the standard allows that, even if it is not commonly applied except for the few examples presented in the standard.

The example of the mechanism fig.2 illustrates the method. The features A and D are press fitted. The part (2) slides with a clearance at joint B and C. If Jm is the minimum difference between the diameters of the cylinders for the joint B and C, the optimal tolerances zones are defined fig. 5 with Jm = t0 + t1 + t3.

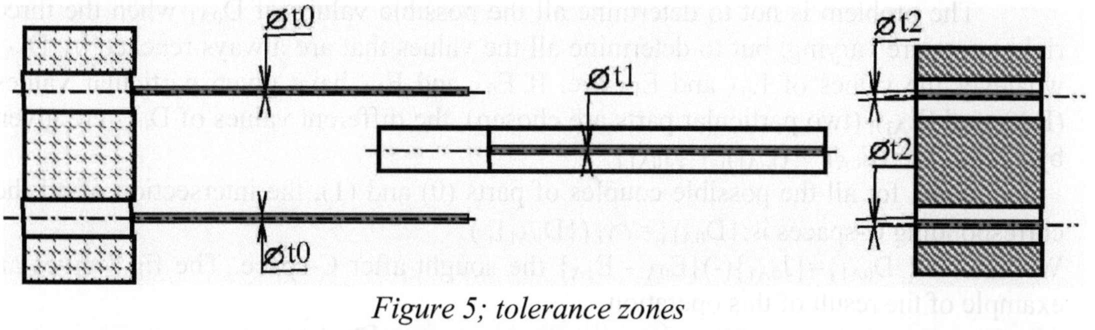

Figure 5; tolerance zones

When the cylinders are not in maximal material condition, the clearance is larger than its minimum Jm, so that the tolerance zone can be increased. The maximal material principle allowed by the standard has to be used here. The fig. 6 gives the corresponding specifications. The numerical values are given as an example.

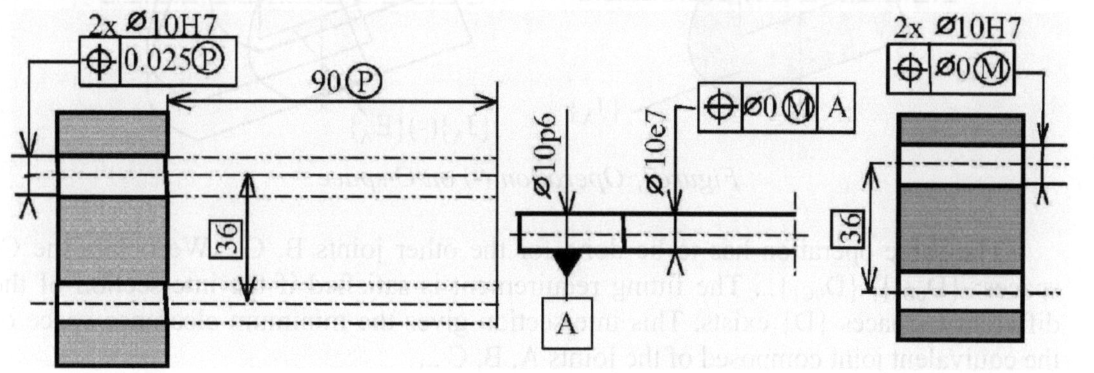

Figure 6; Example for specifications

Other solutions are possible, particularly if the manufacturing constraints are taken into account. The location tolerances of cylinders (1) and (3) can be replaced by an unique straightness tolerance of their axes, always with the maximum material condition. The diameters are swapped: H7p6 is replaced by P6h7 and H7e7 by E7h7. The other tolerances are unchanged.

3.6. The case of joints assembled in parallel

The fig.8 shows such a mechanism. One method consists in trying to generalise the previous method and to consider the different loops of the mechanism. For each loop, an equation between clearance torsors and deviations can be written. Nevertheless, in the C-space, it is more difficult to find an interpretation as simple as for a single loop. It is the reason why an other method is applied here. It allows a generalisation to complex mechanisms and will be more efficient for the optimal tolerance synthesis.

A frame is attached to each part (0) and (1). Each of them may be built from one or more feature of the two parts. When there is neither deviation nor clearance, the two frames are in nominal configuration. When there are deviations and clearance the focus is on the small relative displacement of the two frames. D_{0A1} is the torsor of relative displacements of the two frames due only to the joint A: $D_{0A1} = E_{0A} + J_{0A1} - E_{1A}$

The problem is not to determine all the possible values of D_{0A1} when the three right terms are varying, but to determine all the values that are always reached by D_{0A1} whatever the values of E_{0A} and E_{1A} are. If E_{0A} and E_{1A} have given particular values $(E_{0A})_i$ and $(E_{A1})_i$ (two particular parts are chosen), the different values of D_{0A1} are given by : $\{D_{0A1}\}_i = (E_{0A})_i + (E_{A1})_i + \{J_{0A1}\}$.

Now, for all the possible couples of parts (0) and (1), the intersection of all the corresponding C-spaces is: $\{D_{0A1}\} = \cap_i (\{D_{0A1}\}_i)$

We notice : $\{D_{0A1}\} = \{J_{0A1}\}(-)\{E_{0A} - E_{1A}\}$ the sought after C-space. The fig.7 gives an example of the result of this operation.

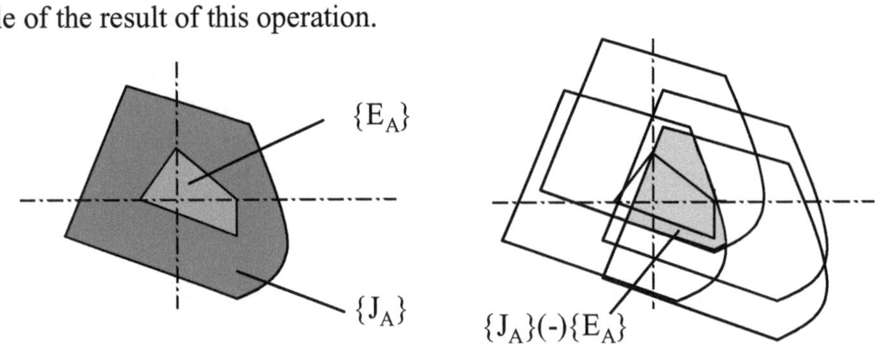

Figure7; Operation (-) on C-space

The same operation has to be done for the other joints B, C ...We obtain the C-spaces: $\{D_{0B1}\}$, $\{D_{0C1}\}$... The fitting requirement is satisfied if the intersection of the different C-spaces {D} exists. This intersection gives the minimum clearance space of the equivalent joint composed of the joints A, B, C ...

4. EXAMPLE

The mechanism fig.8 is simple but illustrate the method. There is no particular difficulties to built the different C-spaces. The qualitative tolerances are defined, but the values of the tolerance zones are not affected here.

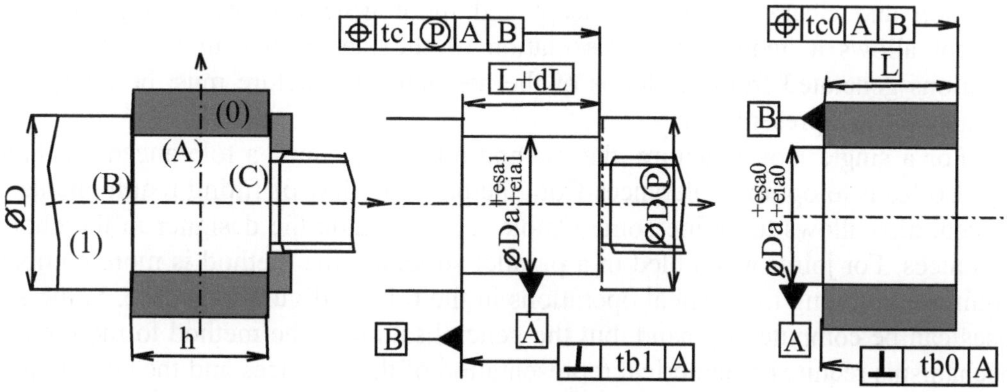

Figure8; example of a parallel structure

The 0x axis of the datum frame built on each part is the cylinder A axis. The plane Oyz is perpendicular to the x axis and is leaned against the actual plane B. In this frame the deviation spaces $\{E_{0A}\}$ and $\{E_{1A}\}$ are equal to zero, so $\{D_{0A1}\}= \{J_{0A1}\}$. The deviation spaces $\{E_{0B}\}$ and $\{E_{1B}\}$ are built from the perpendicularity tolerance. The C-space $\{D_{0B1}\}$ is then the same than $\{J_{0B1}\}$. The location tolerance of the plane C allows to build the C-space $\{D_{0C1}\}$. The C-spaces $\{D_{0B1}\}$and $\{D_{0C1}\}$ are unbounded for the components Ty, Tz and Rx corresponding to degrees of freedom for the joints B and C (fig.9). The strict fitting condition is obtained when the intersection of the 3 C-spaces is limited to the Rx axis. Jm = 0 and dL = (tc0 + tc1)/2.

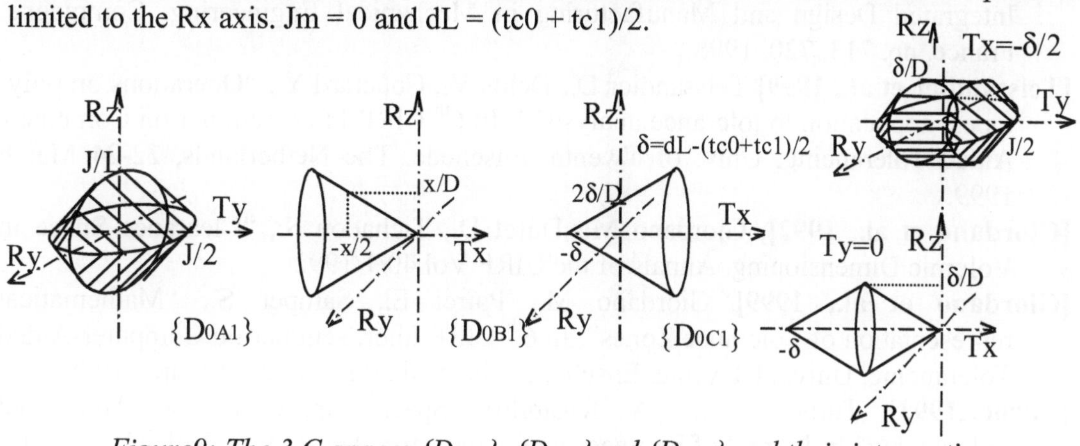

Figure9; The 3 C-spaces $\{D_{0A1}\}$, $\{D_{0B1}\}$and $\{D_{0C1}\}$ and their intersection

In the general case, the minimal clearance space is the intersection of the three C-spaces $\{D_{0A1}\}$, $\{D_{0B1}\}$and $\{D_{0C1}\}$. It is bounded for Ry, Rz, Tx, Ty, Tz but unbounded for the Rx axis that is the degree of freedom of the joint between (0) and (1).

CONCLUSION

Analysing a toleranced mechanism leads to verify system of equations and inequalities. The small displacement torsor formalism allows simpler expressions. The terms of the mathematical system generated are the components of small displacements. The deviation and clearance space concept and more generally the configuration space concept allows a geometric representation of the inequalities in a 6 D space. The equations generated from the loops of the mechanical structure must be interpreted in this geometric representation.

For a single loop structure, the method allows to check a toleranced mechanism, but also leads to optimal tolerances from the point of view of fitting requirements. This concept also allows to define some simple rules to guide the designer in the choice of tolerances. For joints assembled in a parallel structure, the method is more complex. It requires computing topological operations in the 6 D configuration space. Some simple cases can be computed by hand, but the generalisation of the method to more complex mechanisms requires a numerical representation of the C-spaces and the computation of the topological operations.

REFERENCES

[Joskowicz et al., 1999] Joskowicz L., and Sacks E. "The configuration space method for kinematic tolerance analysis"; In 6th CIRP Inter. seminar on Computer-Aided Tolerancing, Univ. of Twente, Enschede, The Netherlands, 22-24 March 1999.

[Ben et al., 1998] Bennis F., Pino L., Fortin C., "Geometric tolerance transfer for manufacturing by an algebraic method", In 2nd International Conference on Integrated Design and Manufacturing in Mechanical Engineering, Compiegne, France, pp. 713, 720, 1998.

[Teissandier et al., 1999] Teissandier D., Delos V., Couetard Y., "Operations on polytopes: application to tolerance analysis""; In 6th CIRP Inter. seminar on Computer-Aided Tolerancing, Univ. of Twente, Enschede, The Netherlands, 22-24 March 1999.

[Giordano et al., 1992], Giordano M., Duret D., Tichadou S., "Clearance Space in Volumic Dimensioning, Annals of the CIRP Vol 40/1/1992.

[Giordano et al., 1999] Giordano M., Pairel E., Samper S., "Mathematical representation of Tolerance Zones", In 6th CIRP Inter. seminar on Computer-Aided Tolerancing, Univ. of Twente, Enschede, The Netherlands, 22-24 March 1999.

[Turner,1993] Turner J.U., "A feasibility Space Approach for Automated Tolerancing",In Journal of Engineering for Industry, Agust 1993, Vol. 115, pp.341-346.

[Gaunet, 1994] Gaunet D., "Modèle formel de tolérancement de position. Contributions à l'aide au tolérancement des mécanismes en CFAO", PhD Thesis, Ecole Normale Supérieure de Cachan, feb. 1994.

Synthesis of tolerances starting from a fuzzy expression of the functional requirements

Bernard ANSELMETTI, Hédi MEJBRI, Kwamivi MAWUSSI
Laboratoire Universitaire de Recherche en Production Automatisée, ENS Cachan, 61, avenue du Président WILSON, 94235 Cachan Cedex
anselm@lurpa.ens-cachan.fr

Abstract: Functional tolerancing of mechanisms is a recognized industrial approach which comprises four main stages (definition of functional requirements of the mechanism, choice of the specifications of each part according to ISO standards, determination of the inequations corresponding to the chains of dimensions, assignment of a value to the tolerance of each specification).
The synthesis of tolerances distributes the tolerances of the specifications of all the parts in order to respect the functional requirements of the mechanism. A functional requirement is now defined in fuzzy form by fixing a limit of high quality and a limit of refusal. A CAD model gives approximate dimensions of the parts which guarantee the mechanical behavior of the product. 4 stages allow to solve the system of inequations: validation of the coherence of the functional requirements, maximization of the quality score of the mechanism by adjusting dimensions of the parts, control of the offsetting, maximization of the tolerances of the parts.
Keywords: tolerancing, synthesis of tolerances, fuzzy representation, optimization.

1. INTRODUCTION

1.1. Functional tolerancing of a mechanism

The functional tolerancing approach requires four main steps:

- The functional analysis gives functional requirements.
- The tolerancing consists in the choice of specifications using ISO standards to define available variations of geometrical characteristics of parts.
- The determination of the tolerance chains provides a system of inequations including dimensions and tolerances of parts.
- The synthesis of tolerances distributes tolerances on functional requirements among part tolerances with some criteria.

Figure 1 presents a gearbox in three positions and shows a lot of functional requirements. Those have only one limit and are often looped. A complete study would give about 700 requirements.

P. Bourdet and L. Mathieu (eds.),
Geometric Product Specification and Verification: Integration of Functionality, 155-164.

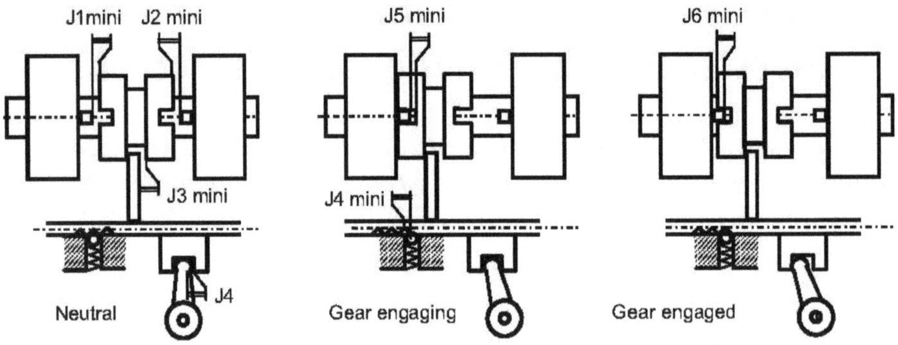

Figure 1; Functional requirements of a mechanism.

1.2. Synthesis of tolerances

In their state of the art of tolerancing, Zhang and Hug [Zhang, 1992] studied geometrical models, dimensional tolerances chain techniques, statistical and probabilistic approaches, analysis and synthesis methods and applications in manufacturing, assembly and design.

Taguchi [Taguchi, 1989], Kapur [Kapur, 1989], Pillet [Pillet, 1999] propose to define a target value for each functional requirement and consider that the variability of each dimension of parts must be the smallest possible in order to reduce the quality loss.
Speckhart [Speckhart, 1972], Spott [Spott, 1973] and Chase [Chase, 1990] suppose that each requirement can be defined by an interval of tolerances and determine part tolerances using minimization production cost with a simple function given cost relating to tolerances. Jeang [Jeang, 1997] associates manufacturing cost and quality loss in one model.
Schneider [Schneider, 1991] firstly decomposes tolerances of functional requirement in design tolerances and secondly decomposes design tolerances in manufacturing tolerances according to the real manufacturing process and the capabilities of machine-tools.
These approaches distribute functional tolerances but consider that nominal sizes of part are fixed by CAD model.
Anselmetti [Anselmetti, 1998] notes that functional requirements have often only one limit and proposes to adjust CAD model in order to increase part tolerances. The Δl model used requires all functional, mechanical and logical requirements before optimization. If this lot of requirements is in conflict, it is very difficult to extract the opposite requirements.
With real industrial applications, it is difficult to define the limit of each requirement with accuracy and it is necessary to adjust part sizes. Optimization must be possible with only the main requirements.
This paper proposes a fuzzy representation of requirements and an optimization in four stages aimed at controlling quality, size deviation and part tolerance ranges.

2. FORMALIZATION OF THE PROBLEM

2.1. Functional requirements

The simple mechanism presented in Figure 2 is composed on two parts fixed by 6 screws. The main function of this mechanism requires a good coaxiality noted C1 of the right extremity with the datum A.

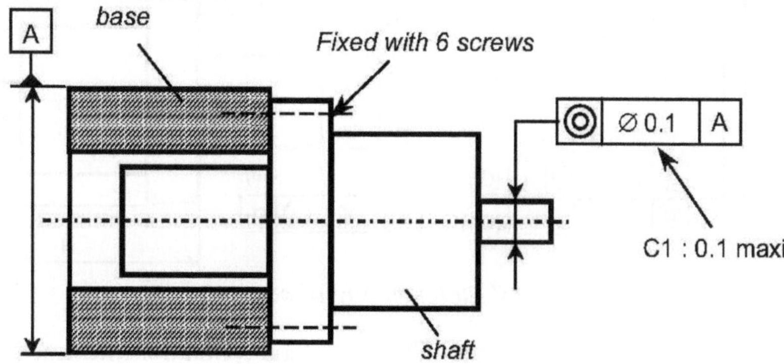

Figure 2; Basic mechanism.

The assembly of these parts requires two conditions. C2 is required during the introduction of the shaft in the hole of the base. C3 allows the contact between the planes of the two parts.

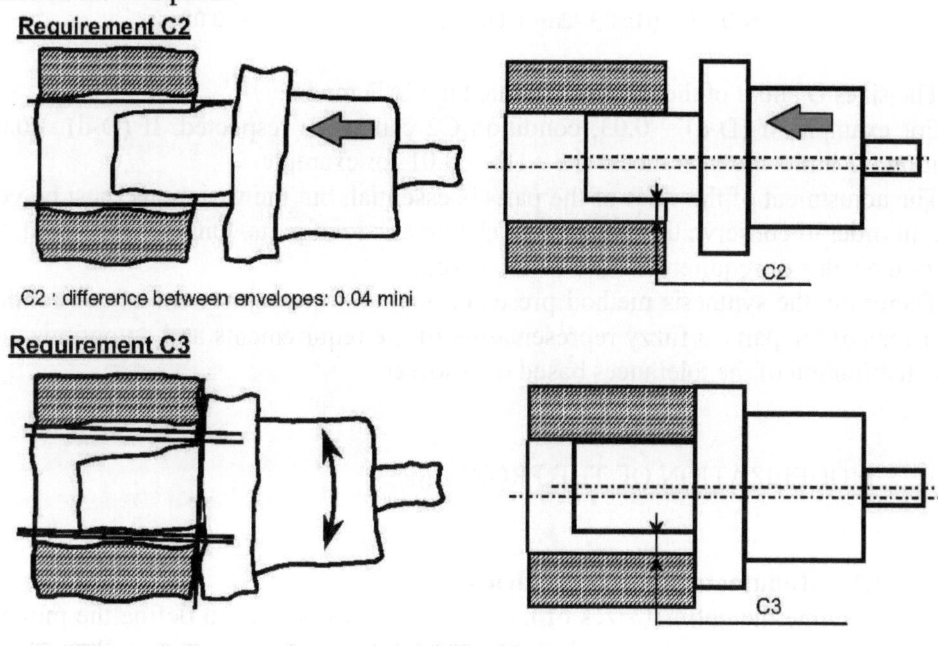

Figure 3; Assembly requirements.

The choice of the value of the requirements is very difficult because the value, which can assure a good quality, generally requires accuracy in manufacturing, hence a high cost.

2.2. ISO tolerancing and chains of tolerances

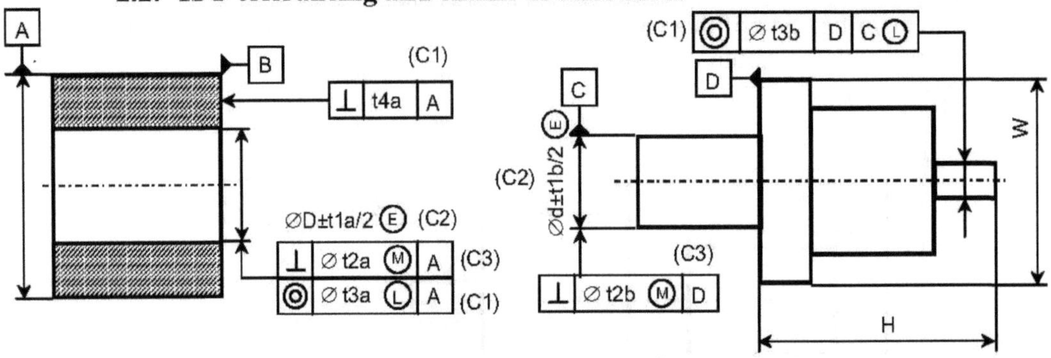

Figure 4; ISO tolerancing of parts.

Tolerancing has been obtained by the CLIC method (Cotation in Location with Influences of Contact) [Anselmetti, 2000-2]. The tolerance chains give 3 relations to respect:

$$C1 : (D - d) + (t1a/2 + t3a + t1b/2 + t3b + t4a \frac{2.H}{W}) \leq 0.1$$

$$C2 : (D - d) - (t1a + t1b)/2 \geq 0.04$$

$$C3 : (D - d) - (t1a/2 + t2a + t1b/2 + t2b) \geq 0.02$$

The sizes D and d of the parts are defined in CAD model.

For example, if (D-d) = 0.03, condition C2 cannot be respected. If (D-d) = 0.05, condition C2 limits the tolerances: t1a = t1b = 0.01 for example.

The adjustment of the sizes of the parts is essential, but the variations must be very small, in order to conserve the mechanical characteristic of parts. On the other hand, the limit of the value of requirement can be discussed.

Therefore, the synthesis method presented includes the adjustment of the nominal dimensions of the parts, a fuzzy representation of the requirements and an optimization of the distribution of the tolerances based on a solver.

3. MODELIZATION OF THE PROBLEM

3.1. Manufacturing requirements

To determine the optimal sizes of parts, manufacturers have to define the minimal values of part tolerances. For example, the inequations system changes as follows.

$$C1 : (D - d) + \underbrace{(t1a/2 + t3a + t1b/2 + t3b + t4a \frac{2.H}{W})}_{0.09 \text{ mini}} \leq 0.1 \Rightarrow D - d \leq 0.01$$

$$C2 : (D - d) - \underbrace{(t1a + t1b)/2}_{0.02 \text{ mini}} \geq 0.04 \Rightarrow D - d \geq 0.02$$

$$C3 : (D - d) - \underbrace{(t1a/2 + t2a + t1b/2 + t2b)}_{0.04 \text{ mini}} \geq 0.02 \Rightarrow D - d \geq 0.06$$

These constraints impose opposite conditions: (D - d $\geq$ 0.06 and D – d $\leq$ 0.01). There is no solution without a compromise.

3.2. Fuzzy representation of a requirement

The functional requirement is now defined by two limits: the Limit of Satisfaction (LS) and the Limit of Refusal (LR). A quality score shows the level of the designer's satisfaction.

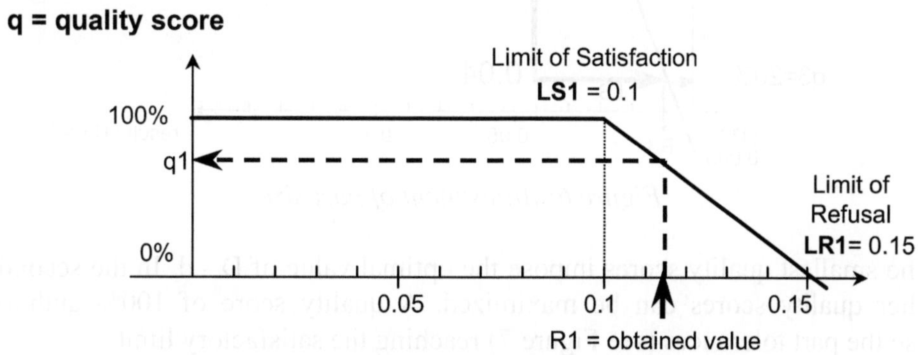

Figure 5; Fuzzy representation of C1 requirements.

For example, for requirement C1, if the coaxiality obtained is less than or equal to 0.1, the designer is very satisfied (quality score 100%). A coaxiality greater than 0.15 is not acceptable (quality score < 0%). Between these two limits, q1 describes the satisfaction level of the designer.
For requirement C2: LR2 = 0.01 and LS2 = 0.04. For C3: LR3 = 0.005 and LS3 = 0.02

3.3. Principle of adjustment of parts sizes

The optimization requires to choose the size D-d in order to maximize the smallest quality scores of all requirements taking manufacturing constraints into account.

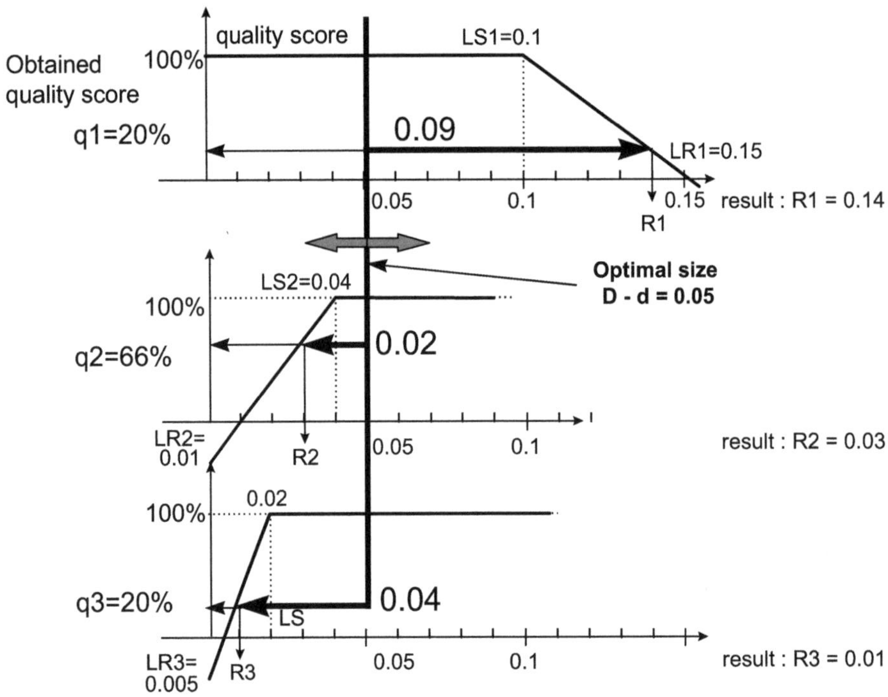

Figure 6; Adjustment of part size.

The smallest quality scores impose the optimal value of D - d. In the second stage, the other quality scores can be maximized. A quality score of 100% authorizes to increase the part tolerances (see Figure 7) reaching the satisfactory limit.

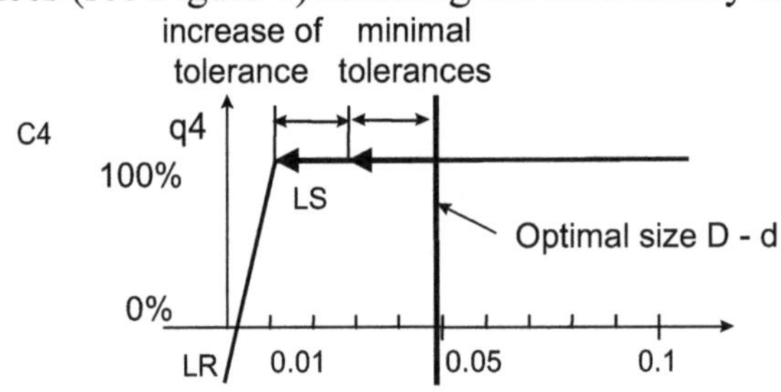

Figure 7; Maximization of tolerances.

4. GENERALIZATION IN CLIC SYSTEM

4.1. Principle of optimization

The optimization with EXCEL solver (Microsoft) requires 4 main stages:

- Checking of the feasibility.
- Optimization of the quality.
- Control of the offsetting.
- Maximization of the tolerances.

4.2. New system of inequations

CAD models give all the characteristics of part surfaces. To adjust the sizes of parts, an offset is offered to each functional surface and defined by a small variation noted vi.

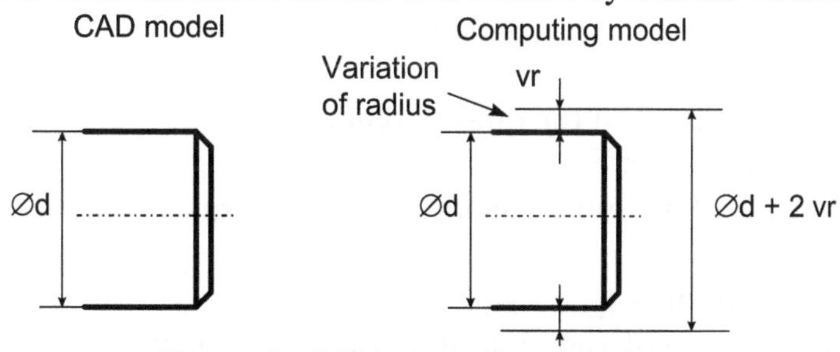

Figure 8; Offsetting of part surfaces.

For each part tolerance ti, the manufacturer imposes a minimal value ti_{mini} and a maximal value ti_{maxi}. But, when it is possible, an increased value ti* will be added to ti_{mini}.

In the inequations, ti will be replaced by $ti_{mini} + ti^*$.

With this more detailed definition each inequation will change. For example, inequation C2: $(D - d) - (t1a + t1b)/2 \geq 0.04$ becomes:

$$R2 = [(D-2.vR)-(d+2.vr)] - [(t1a_{mini} + t1a^*)+(t1b_{mini} + t1b^*)]/2.$$

The fuzzy relation of requirement C2 gives the quality score q2 function of vR, vr, t1a* and t1b*.

4.3. Analysis of feasibility

In this stage, each part tolerance is fixed at its minimal value (ti* = 0).

The parameters of the solver are:

- Maximization of the smallest value of qi noted q_{min}.
- Set of variables: vi
- No Constraint

The main result of the solver is the quality scores of critical requirements q_{min} which must be accepted by the designer. Figure 9 presents a simple part with conflict between requirements.

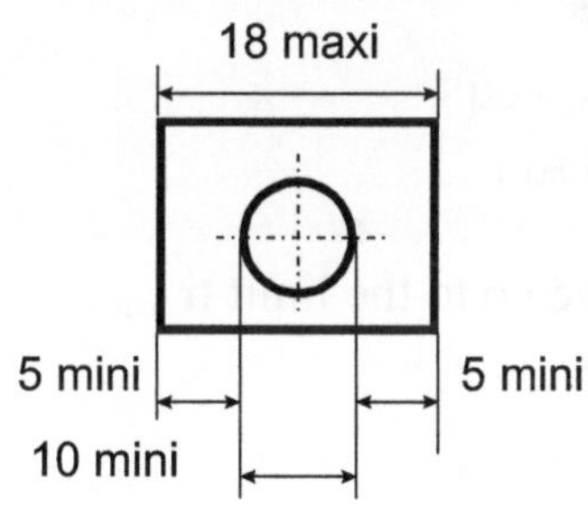

Figure 9; Requirements in conflict.

In this case, the maximization of the smallest quality scores gives negative values or too small values for each requirement in conflict. So, the manufacturer can chose the most

accurate process for the concerned dimensions of the parts, the designer can change the limits of the requirements or can modify the mechanism.
This fuzzy representation of requirements allows detecting requirements in conflict in a complex set of inequations (700 inequations).

4.4. Maximization of the quality score

The parameters of the solver are:

- Maximization of the sum of quality score $S = \Sigma\, qi$
- Set of variables: vi
- Constraint: $\min(qi) \geq q_{min} - \varepsilon q$

The margin εq lets a range of variation of all offset vi, which allows the maximization of all qi without decreasing the quality of the critical requirements.
The main result of this step is the optimal quality score of each requirement noted qi_{opt}.

4.5. Control of offsetting

The parameters of the solver are:

- Maximization of the sum of variation $S = \Sigma\, (vi)^2$
- Set of variables: vi
- Constraint: $(qi) \geq qi_{opt} - \varepsilon q$

This step limits the difference between the adjusted model and the CAD model. The main result of this step is the optimal variations of each surface noted vi opt.

After this step, the designer must define two limits vi_{mini} and vi_{maxi} around the optimal variation vi_{opt} for each surface. Sometimes, a new optimization of vi is required to control vi in these limits.

4.6. Maximization of tolerances

The cost of the part decreases when the interval of tolerances increases. The cost can be controlled with a simple function $S = \Sigma\, (1/ti)$.
The parameters of the solver are:

- Minimization of part cost: $S = \Sigma\, (1/ti)$.
- Set of variables: vi, ti*
- Constraint: $(qi) \geq qi_{opt} - \varepsilon q$
- $ti^* \geq 0$ and $ti + ti^* \leq ti_{maxi}$
- $vi_{mini} \leq vi \leq vi_{maxi}$

In this stage, tolerances increase up to the limit ti_{maxi}.

5. APPLICATION

An application with CATIA V4 CAD system (Dassault System) and EXCEL Solver (Microsoft) can automatically define tolerancing and synthesis of tolerances, with arithmetic or probabilistic approach.

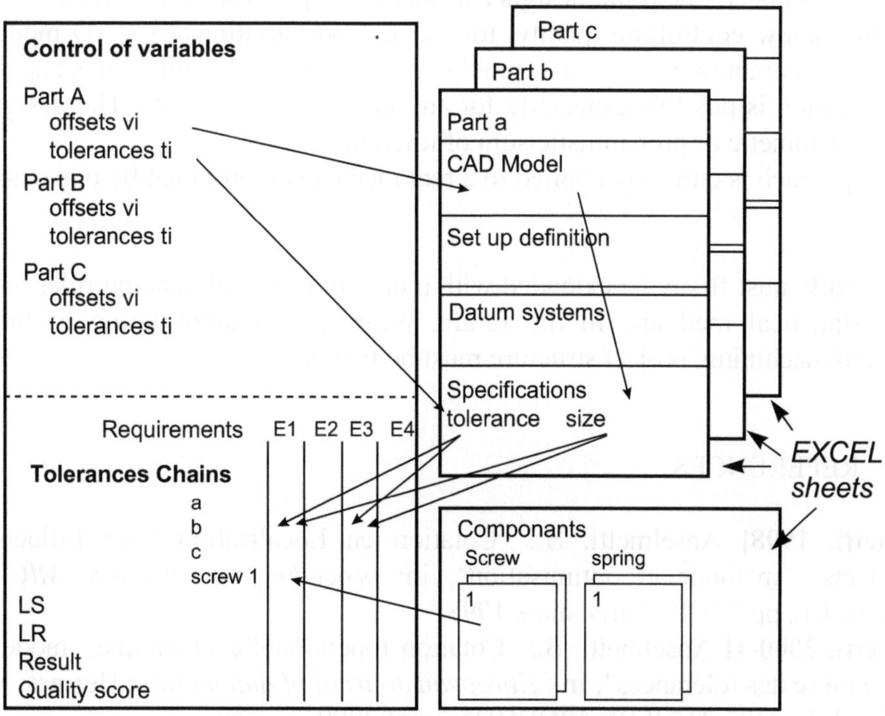

Figure 10; Architecture of CLIC system.

A specific offsetting is defined for each type of feature:

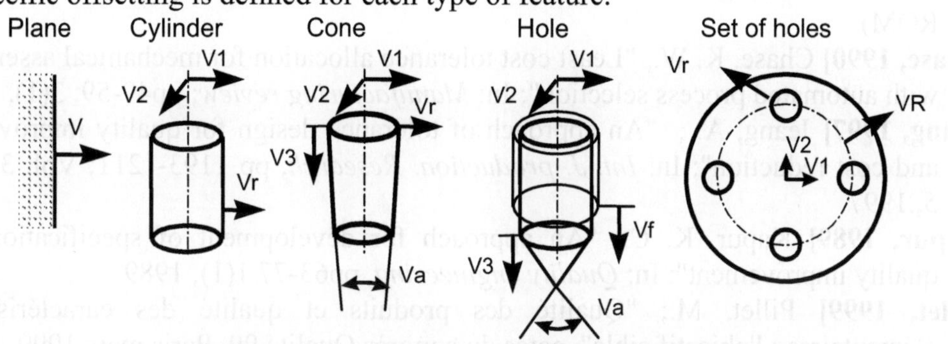

Figure 11; Variation of different features.

The probabilistic approach is based on the statistical addition of the variances of each dimension (ex: $Ri = \sum m_j + \sqrt{3} \cdot \sqrt{\sum (t_j)^2}$) to determine the quality score obtained [Anselmetti, 2000-1].

6. CONCLUSION

This method of synthesis of tolerances can seem to be complex. Simplified methods were tested but they cannot detect incompatible requirements. Fuzzy representation of these requirements can help the designer to make compromise. The different stages of optimization allow controlling quality, tolerances and variations of CAD models. For each stage, the computing time is acceptable (about 2 s for 35 requirements and 3 parts). The optimization is possible especially for any critical requirements. The designer can choose the arithmetic or probabilistic sum of tolerances.

This approach is currently applied to a real mechanism provided by one automobile industry.

This work must firstly be extended with more complete tolerancing rules and more accuracy statistical methods. In the future, wearing and deformation of the parts, welding and machining, project structure must be treated.

7. REFERENCES

[Anselmetti, 1998] Anselmetti, B.; "Cotation en Localisation avec Influence des Contacts : méthode et optimisation"; In: *procedings of colloque MICAD 98*; volume C6, pp 17-20 ; Paris, mars 1998.

[Anselmetti, 2000-1] Anselmetti, B.; "Cotation fonctionnelle statistique : modélisation et synthèse des tolérances"; In : *European Journal of automation*; Hermes, Volume 345 n°2-3 pp305-315 ISSN 1269-6935; avril 2000.

[Anselmetti, 2000-2] Anselmetti, B.; "Functional ISO tolerancing with 3D influences"; In: *actes du congrès international IDMME'2000*; Montréal 15-19mai 2000 (CD ROM).

[Chase, 1990] Chase, K. W.; "Least cost tolerance allocation for mechanical assemblies with automated process selection"; In: *Manufacturing review*; pp49-59; 3(1),

[Jeang, 1997] Jeang, A. ; "An approach of tolerance design for quality improvement and cost reduction"; In: *Int. J. production. Research*, pp 1193-1211; Vol. 35, N°. 5,.1997

[Kapur, 1989] Kapur, K. C.; "An approach for development of specifications for quality improvement"; In: *Quality engineering*, pp63-77 1(1), 1989

[Pillet, 1999] Pillet, M.; "Qualité des produits et qualité des caractéristiques élémentaires : l'objectif cible". actes du congrès Qualita 99, Paris mars 1999.

[Schneider, 1991] Schneider, F; Rémy-Vincent, J; "Définition des ensembles mécaniques : conditions d'obtention de cotations optimales"; In: *Proceeding of 23eme séminaire international sur les systèmes de production*; CIRP, juin 1991.

[Speckhart, 1972] Speckhart, F. H.; "Calculation of tolerance based on a minimum cost approach"; In: *Journal of engineering for industry*, pp 447-453. Vol 94(5) 1972,

[Spotts, 1973] Spotts, M. F.; "Allocation of tolerances to minimize cost of assembly"; In: *Journal of engineering for industry;* pp762-764; August 1973.

[Taguchi, 1989] Taguchi, G.; "Introduction to quality engineering, White plains"; *NY: Unipub*. pp. 21-22

[Zhang, 1992] Zhang H. C.; Hug, M. E.; "Tolerancing techniques: the state-of-the-art"; In: *Int Journal of Production Research;* pp2111-2135; Vol. 30, NOV. 9, 1992.

An Accurate Fixture Model for Precision Fixturing

Michael Y. Wang
Department of Automation and Computer-Aided Engineering
The Chinese University of Hong Kong
Shatin, N.T., Hong Kong
yuwang@acae.cuhk.edu.hk

Abstract: The conventional point-kinematic model of fixtures has only treated point geometry of the contacts between locators and workpiece. However, this model, which ignores the underlying surface properties of the locators-plus-workpiece system, is inherently incapable of capturing the effects of the geometric properties important to accurate positioning of the workpiece. In this paper, we present a fixture model based on the full kinematics of locator-workpiece contact. This model incorporates a "virtual" kinematic chain with meshing parameters of contact kinematics. It is shown that the conditions of deterministic localization are related to surface properties of both the workpiece and the locators, including the surface curvature, torsion and scale factors, as opposed to the conventional point-kinematic model. The full-kinematic model developed here has a strong implication for designing fixtures with high locating precision requirements.
Keywords: Fixture model, contact and grasp, fixturing, grasping

1. INTRODUCTION

Proper fixture design is crucial to product quality in terms of precision and accuracy in part fabrication and assembly. Fixturing systems, usually consisting of clamps and locators, must be capable of positioning, holding, and supporting a workpiece during machining, assembly, or inspection. Complex processes such as laser drilling of air cooling holes near the leading and trailing edges of a turbine airfoil require a high precision in locating and restraining the workpiece. Therefore, an accurate fixture model for analysis and design of a locating scheme and a fixture structure is essential for reducing dimensional variation in the final product.

While the subject of fixture analysis and design has been extensively studied in the fields of robotics and manufacturing over many years (Murray, Li and Sastry 1994), these studies are almost entirely based on a *simplified kinematic* model for the geometric or kinematic analysis of conditions of workpiece positioning and total constraint (or form

P. Bourdet and L. Mathieu (eds.),
Geometric Product Specification and Verification: Integration of Functionality, 165-174.

closure). In the model, each locator (or a robot finger) and workpiece contact is considered as a theoretical point without involving surface properties of the locators and the workpiece (Asada and By 1985, Chou, Chandru and Barash 1989). This model has been widely used for the studies of fixture synthesis (Chou et al. 1989, Wang 2000), fixture contact type and friction effects (DeMeter 1994), modular fixtures (Brost and Goldberg 1996), machining fixtures (DeMeter 1994), and traditional or computer-based fixture design and planning. The conventional model is referred to as *point-kinematic model* in the paper.

In a recent development of methods for fixture diagnosis and fixture tolerancing schemes (Chou et al. 1989, Wang 2000, Carlson 2000), it raised an issue of the accuracy of the conventional *linearized point-kinematic model* for high precision applications such as the airfoil manufacturing. A quadratic sensitivity analysis is recently presented in (Carlson 2000). It is also based on the same *simplifying assumptions* and is a second order Taylor approximation of the nonlinear geometric constraints defined by the point-kinematic assumptions. Aimed for a more accurate description of a locating scheme the quadratic theory deals with the effects of the workpiece curvature around the contact points and interaction effects of locator positioning errors. However, the effect of locator geometry is not included in the quadratic analysis and it is quite complicated for the use in fixture layout and setup design.

In either linear or nonlinear analysis, the point-kinematic model will underestimate the positioning errors of a *non-prismatic* workpiece in the presence of locator errors. In neglecting the geometric properties of the locator and workpiece surfaces, the insufficient prediction of the positioning errors could have a significant impact in industrial practice. A good example is a turbine airfoil which has highly complex geometry. In the process of laser drilling of air cooling holes, the positioning accuracy of the fixtured airfoil is critical for producing the cooling holes with high precision in their locations. Another case is in dealing with small parts in assembly, for instance, of electronics devices. The positioning errors of the locators may be relatively large compared to the dimension of the part. Therefore, it is important to fully take into account of the locator dimensional errors in fixture analysis and design, especially when the fixture structure may involve long tolerance chains.

This paper presents a fixture model considering the surface geometry of locators and the workpiece. The fixture model is based on a *linear* description of the full kinematics of two rigid objects and the motion of their contact point over the surfaces of their bodies (Montana 1988). The linear model yields a full consideration of the effects of curvature in the conditions of kinematic localization of the workpiece by the locators. Thus, this model is referred to as the *full-kinematic model*. The development of this model is aimed to provide the foundation not only for precision fixture design but also for tolerance budgeting and specification in fixture manufacturing and assembly. Furthermore, when including the curvature effects, the fixture model offers the potential for a more realistic analysis of the stiffness properties of a locator-workpiece fixture system based on a contact mechanics description.

2 THE CONTACT KINEMATICS

In this section we discuss the concept of contact kinematics concerning two rigid bodies that move while maintaining contact with each other. The contact kinematics were fully described in (Montana 1988). As the foundation for the fixture model to be presented, it is necessary to describe the relevant information here largely following that of (Montana 1988).

2.1. The Meshing Contact

Consider two rigid bodies O and i with smooth surfaces in contact at a point c 1. Choose body frames C_o and C_i fixed relative to body O and i respectively. At the contact point c, we define a contact frame C_{co} fixed on body O with z axis aligning with the outward normal of the surface of O. Similarly, another contact frame C_{ci} is fixed on body i. There exists an angle between the x axes of these two body frames, which is called contact angle ψ and can be defined such that a rotation of C_{co} through angle $-\psi$ around its z axis would align the x axes (Montana 1988).

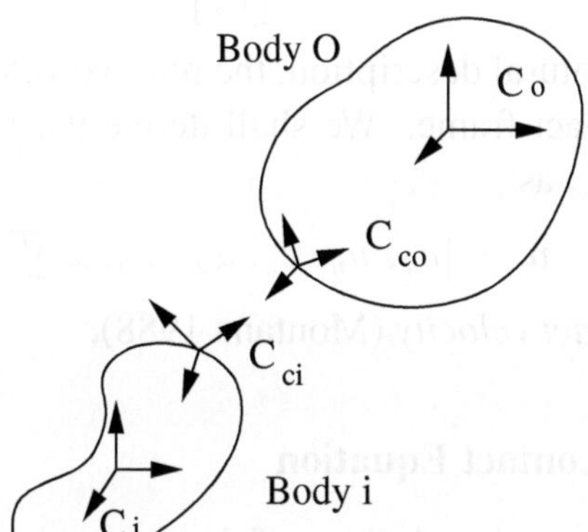

Figure 1: The coordinate frames for contact kinematics.

Furthermore, a surface is defined to be a continuous mapping from R^2 to R^3, $\boldsymbol{f}(\boldsymbol{\xi})$, with surface parameters $\boldsymbol{\xi} = (u, v)$ in a mathematical rigor. The unit normal vector of the surface is given as

$$\boldsymbol{n} = \frac{\boldsymbol{f}_u \times \boldsymbol{f}_v}{||\boldsymbol{f}_u \times \boldsymbol{f}_v||} \tag{1}$$

In the so called Gaussian frame, three properties of curvature $\boldsymbol{S}$, torsion $\boldsymbol{T}$, and scale $\boldsymbol{M}$

of the surface are defined as

$$S = \begin{bmatrix} \boldsymbol{n}_u \cdot \boldsymbol{f}_u & \boldsymbol{n}_u \cdot \boldsymbol{f}_v \\ \boldsymbol{n}_v \cdot \boldsymbol{f}_u & \boldsymbol{n}_v \cdot \boldsymbol{f}_v \end{bmatrix} \tag{2}$$

$$T = \begin{bmatrix} \frac{\boldsymbol{f}_v \cdot \boldsymbol{f}_{uu}}{\|\boldsymbol{f}_u\|^2 \cdot \|\boldsymbol{f}_v\|} & \frac{\boldsymbol{f}_v \cdot \boldsymbol{f}_{uv}}{\|\boldsymbol{f}_u\| \cdot \|\boldsymbol{f}_v\|^2} \end{bmatrix} \tag{3}$$

$$M = \begin{bmatrix} \|\boldsymbol{f}_u\| & 0 \\ 0 & \|\boldsymbol{f}_v\| \end{bmatrix} \tag{4}$$

As clearly described in (Montana 1988) the relative motion of the two bodies has five degrees of freedom, when they maintain a single point of contact between their surfaces with only sliding or rolling motions allowed. Among these five degrees of freedom, two are for the position of the point of the contact on the surface of body O and, similarly, another two degrees of freedom are for body i. The last degree of freedom is for the relative rotation ψ_i around the common surface normal. We shall refer to these *internal* motion parameters as *meshing parameters* and denote them in a vector form by

$$\boldsymbol{\varphi}_i = \begin{bmatrix} \boldsymbol{\xi}_i \\ \boldsymbol{\xi}_o \\ \psi_i \end{bmatrix} \tag{5}$$

For a more direct and natural description, the relative velocity between the two bodies can be expressed in a contact frame. We shall define the velocity of body i relative to body O in contact frame C_{ci} as

$$\boldsymbol{u}_i = [u_x, u_y, u_z, \omega_x, \omega_y, \omega_z]_i^T \tag{6}$$

This is also called the *contact velocity* (Montana 1988).

2.2. Montana's Contact Equation

With the internal and external descriptions of the contact motion using the meshing parameters $\boldsymbol{\varphi}$ (or the meshing velocity $\dot{\boldsymbol{\varphi}}$) and the *contact velocity* $\boldsymbol{u}_i$ respectively, the kinematics of contact are defined as the relationship between these two velocity descriptions. A linear relation has been derived by Montana (Montana 1988) as

$$\boldsymbol{u}_i = \boldsymbol{H}_i \dot{\boldsymbol{\varphi}}_i \tag{7}$$

where the matrix $\boldsymbol{H}$ is given by

$$\boldsymbol{H}_i = \begin{bmatrix} -\boldsymbol{M}_i & \boldsymbol{R}_{\psi_i} \boldsymbol{M}_o & 0 \\ 0 & 0 & 0 \\ \Delta \boldsymbol{S}_i \boldsymbol{M}_i & \Delta \boldsymbol{R}_{\psi_i} \boldsymbol{S}_o \boldsymbol{M}_o & 0 \\ -\boldsymbol{T}_i \boldsymbol{M}_i & -\boldsymbol{T}_o \boldsymbol{M}_o & 1 \end{bmatrix} \tag{8}$$

and

$$\boldsymbol{R}_\psi = \begin{bmatrix} \cos\psi & -sin\psi \\ -\sin\psi & -\cos\psi \end{bmatrix} \quad \Delta = \begin{bmatrix} 0 & 1 \\ -1 & 0 \end{bmatrix} \tag{9}$$

This linear relation is called the full-kinematics of contact in (Montana 1988).

3. THEFIXTUREMODEL

We now present the fixture model based on the contact kinematics of two bodies. We assume that the workpiece is sufficiently rigid with piecewise differentiable surfaces. Each locator is considered to be in contact with the workpiece without friction. At each locator contact, the surface properties of the locator and the workpiece are assumed to be well-defined. For a unique localization, six locators are used. In this paper we discuss the kinematics of fixture localization only.

3.1. The Kinematics of Locators-Plus-Workpiece System

In the contact equation (7), let body O be the workpiece and body i be a locator. Similar to the case of a multi-fingered robotic grasp (Montana 1995), the fixture system of the locators plus workpiece can be represented as a kinematic "chain" shown in Fig. 2. Each locator-workpiece contact represents a virtual kinematic subchain with "joints" defined by the meshing parameters φ_i. The contact equation (7) defines the forward kinematics of the subchain.

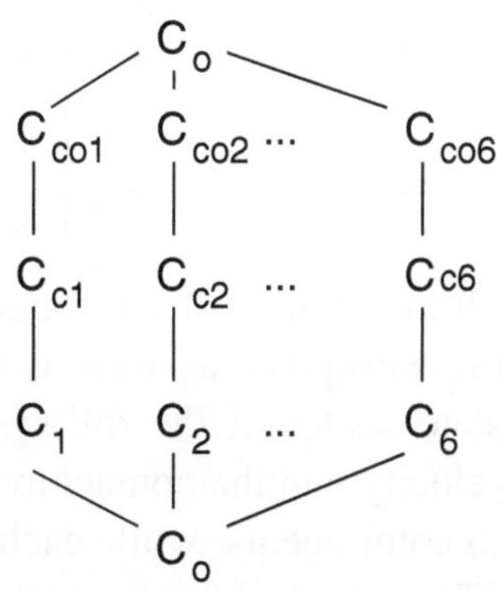

Figure 2: The kinematic "chain" of locators-plus-workpiece fixture system.

However, the contact velocity $\boldsymbol{u}_i$ in Eq. 7 is defined in the contact frame C_{ci}. In order to represent the full kinematics of the locators plus workpiece system, $\boldsymbol{u}_i$ should be transformed into the body frame C_i of its locator. For further convenience we shall let each C_i be coincidental to the body frame C_o of the workpiece. Therefore, the velocity $\boldsymbol{v}_i$ of locator i relative to that $\boldsymbol{v}_o$ of the workpiece O is expressed in body frame C_o as

$$\boldsymbol{v}_i - \boldsymbol{v}_o = \boldsymbol{A}_i \boldsymbol{u}_i \tag{10}$$

Here, $\boldsymbol{A}_i$ is the velocity transformation for the locator i. In the well known mathematics of rigid-body motions (Murray et al. 1994, Montana 1995), when a body frame transform from C_i to C_{ci} is given as

$$\boldsymbol{Q}_i = \begin{bmatrix} \boldsymbol{R} & \boldsymbol{r} \\ 0 & 1 \end{bmatrix}_i \tag{11}$$

then, the velocity transform $\boldsymbol{A}_i$ is given by

$$\boldsymbol{A}_i = \begin{bmatrix} \boldsymbol{R} & \widehat{\boldsymbol{r}}\boldsymbol{R} \\ 0 & \boldsymbol{R} \end{bmatrix}_i \tag{12}$$

where $\widehat{\boldsymbol{r}}$ is defined to be equivalent to $\boldsymbol{r}\times$.

Finally, applying Eq. 7 to Eq. 10 yields the following

$$\boldsymbol{v}_i - \boldsymbol{v}_o = \boldsymbol{A}_i \boldsymbol{H}_i \dot{\boldsymbol{\varphi}}_i \tag{13}$$

for each locator-workpiece contact $i = 1, \cdots, 6$. Therefore, the closure conditions of the six locator sub-chains can be succinctly expressed by

$$\boldsymbol{v}_o = -\boldsymbol{A}_1 \boldsymbol{H}_2 \dot{\boldsymbol{\varphi}}_1 + \boldsymbol{v}_1 = \cdots\cdots = \boldsymbol{A}_6 \boldsymbol{H}_6 \dot{\boldsymbol{\varphi}}_6 + \boldsymbol{v}_6 \tag{14}$$

Thus, the kinematics of the full locators-plus-workpiece fixture system are expressed as a single equation

$$\begin{bmatrix} \boldsymbol{I} & \boldsymbol{A}_1\boldsymbol{H}_1 & 0 & \cdots & 0 \\ \boldsymbol{I} & 0 & \boldsymbol{A}_2\boldsymbol{H}_2 & \cdots & 0 \\ \vdots & \vdots & \vdots & \vdots & \vdots \\ \boldsymbol{I} & 0 & 0 & \cdots & \boldsymbol{A}_6\boldsymbol{H}_6 \end{bmatrix} \begin{bmatrix} \boldsymbol{v}_o \\ \dot{\boldsymbol{\varphi}}_1 \\ \vdots \\ \dot{\boldsymbol{\varphi}}_6 \end{bmatrix} = \begin{bmatrix} \boldsymbol{v}_1 \\ \boldsymbol{v}_2 \\ \vdots \\ \boldsymbol{v}_6 \end{bmatrix} \tag{15}$$

This equation defines the full kinematics of the fixture system in terms of the body velocities of the locators and the workpiece as well as the internal meshing velocities of each locator contact. The state vector of the full system is an “extended” set of a combination of the workpiece velocity and the contact meshing velocities of all locators. The workpiece velocity $\boldsymbol{v}_o$ has 6 components, while each of the 6 locator contacts has 5 “internal” meshing velocities. Thus, the full fixture system has a total of 36 degrees of freedom.

3.2. Deterministic Localization

The velocity equation (15) of the full fixture system allows us to derive a *linear* model of the fixture with consideration of the surface properties of the locators as opposed to the point-kinematic fixture model of (Asada and By 1985). Small perturbations of location including both the orientation and the position can be approximated by $\delta\boldsymbol{q}_o \approx \boldsymbol{v}_o$ for the workpiece and by $\delta\boldsymbol{q}_i \approx \boldsymbol{v}_i$ for every locators ($i = 1, 2, \cdots, 6$). Thus, the following linear equation defines the relation between the workpiece location errors and the errors of the locator locations, with associated errors in the meshing parameters:

$$\boldsymbol{G}\,\delta\boldsymbol{q}_s = \delta\boldsymbol{q}_l \tag{16}$$

where

$$G = \begin{bmatrix} I & A_1H_1 & 0 & \cdots & 0 \\ I & 0 & A_2H_2 & \cdots & 0 \\ \vdots & \vdots & \vdots & \vdots & \vdots \\ I & 0 & 0 & \cdots & A_6H_6 \end{bmatrix}$$

$$\delta q_s = \begin{bmatrix} \delta q_o \\ \delta \varphi_1 \\ \vdots \\ \delta \varphi_6 \end{bmatrix} \quad \delta q_l = \begin{bmatrix} \delta q_1 \\ \delta q_2 \\ \vdots \\ \delta q_6 \end{bmatrix}$$

As discussed in (Asada and By 1985) a fundamental function of a fixture is *deterministic localization* which requires the location (including the position and orientation) of the workpiece to be uniquely determined by the 6 locators of the fixture. In other words, when $\delta q_l = 0$ in Eq. 16, there should exist only solution $\delta q_s = 0$ for the resulting equation

$$G\,\delta q_s = 0 \tag{17}$$

Therefore, if and only if rank$(G) = 36$, *then the location of the workpiece is kinematically determined, i.e., the fixture has deterministic localization.*

Further, it is convenient to rearrange Eq. 16 as

$$\begin{bmatrix} I & A_1H_1 & 0 & \cdots & 0 & 0 \\ 0 & A_1H_1 & -A_2H_2 & \cdots & 0 & 0 \\ \vdots & \vdots & \vdots & \vdots & \vdots & \vdots \\ 0 & 0 & 0 & \cdots & A_5H_5 & -A_6H_6 \end{bmatrix} \begin{bmatrix} \delta q_o \\ \delta \varphi_1 \\ \vdots \\ \delta \varphi_6 \end{bmatrix} = \begin{bmatrix} \delta q_1 \\ \delta q_1 - \delta q_2 \\ \vdots \\ \delta q_5 - \delta q_6 \end{bmatrix} \tag{18}$$

This would allow a partition of the system equations to separate the "internal" meshing parameters $\delta\varphi$ from the "external" variables δq_o such that

$$\delta q_o + B\delta\varphi_l = \delta q_1 \tag{19}$$

$$D\delta\varphi_l = \overline{\delta q_l} \tag{20}$$

with matrix partitions

$$B = \begin{bmatrix} A_1H_1 & 0 & \cdots & 0 & 0 \end{bmatrix}$$

$$D = \begin{bmatrix} A_1H_1 & -A_2H_2 & \cdots & 0 & 0 \\ \vdots & \vdots & \vdots & \vdots & \vdots \\ 0 & 0 & \cdots & A_5H_5 & -A_6H_6 \end{bmatrix}$$

and

$$\delta\boldsymbol{\varphi}_l = \begin{bmatrix} \delta\boldsymbol{\varphi}_1 \\ \vdots \\ \delta\boldsymbol{\varphi}_6 \end{bmatrix} \quad \overline{\delta\boldsymbol{q}_l} = \begin{bmatrix} (\delta\boldsymbol{q}_1 - \delta\boldsymbol{q}_2) \\ \vdots \\ (\delta\boldsymbol{q}_5 - \delta\boldsymbol{q}_6) \end{bmatrix}$$

Thus, the kinematic solution is given by

$$\delta\boldsymbol{\varphi}_l = \boldsymbol{D}^{-1}\overline{\delta\boldsymbol{q}_l} \tag{21}$$

$$\delta\boldsymbol{q}_o = \delta\boldsymbol{q}_1 - \boldsymbol{B}\delta\boldsymbol{\varphi}_l = \delta\boldsymbol{q}_1 - \boldsymbol{A}_1\boldsymbol{H}_1\delta\boldsymbol{\varphi}_1 \tag{22}$$

provided that $\boldsymbol{D}$ is not singular. In fact, this solution shows that the condition for deterministic localization can be reduced to $\text{rank}(\boldsymbol{D}) = 30$ under the assumption that all 6 locators maintain contact with the workpiece.

3.3. The Full-Kinematic Model vs. The Point-Kinematic Model

With the above developed fixture model considering the full surface geometry of the locators-plus-workpiece system, it is straightforward to show that the conventional point-kinematic fixture model is subsumed by this full-kinematic fixture model.

In the expression of the contact equation (7), there exists a special condition for the contact velocity

$$u_{zi} = 0 \; (i = 1, 2, \cdots, 6) \tag{23}$$

for each locator contact. This explicitly expresses the necessary kinematic constraint on the relative motion between the workpiece and each locator to maintain all contacts. When applying the velocity transform $\boldsymbol{A}_i$ to this equation, we would obtain

$$\text{the 3rd row of } \boldsymbol{A}_i^{-1}(\boldsymbol{v}_i - \boldsymbol{v}_o) = 0 \tag{24}$$

If the locator is applied on the workpiece at position $\boldsymbol{r}_i$ and the orientation of the contact frame C_{ci} at the point is represented by $\boldsymbol{R}_i = -[\boldsymbol{t}\,\boldsymbol{b}\,\boldsymbol{n}]_i$ in terms of the workpiece surface normal and bi-normals, then it is easy to derive the following relation

$$\delta y_i = -\left[\boldsymbol{n}_i^T \; (\boldsymbol{r}_i \times \boldsymbol{n}_i)^T\right]\delta\boldsymbol{q}_o = \boldsymbol{h}_i\delta\boldsymbol{q}_o \tag{25}$$

where δy_i represents the projection of the locator positioning errors along the normal direction of the workpiece surface (Wang 2000). Collecting the equations for all 6 locators would yield the conventional point-kinematic model of fixtures with the pure point geometry assumption (Asada and By 1985, Wang 2000):

$$\delta\boldsymbol{y} = \boldsymbol{J}\delta\boldsymbol{q}_o \tag{26}$$

where $\delta\boldsymbol{y} = \{\delta y_1\,\delta y_2 \cdots \delta y_6\}^T$ and $\boldsymbol{J}^T = [\boldsymbol{h}_1^T\,\boldsymbol{h}_2^T \cdots \boldsymbol{h}_6^T]$. Therefore, the conventional fixture model represents a *subset* of the full kinematics of the locators plus workpiece fixture system. For the point-kinematic model, it is well known that the deterministic localization is true if and only if the matrix $\boldsymbol{J}$ has full rank, i.e., $\text{rank}(\boldsymbol{J}) = 6$ (Asada and By 1985). Clearly, this condition is subsumed by the necessary and sufficient condition for deterministic localization of the full-kinematic model that $\text{rank}(\boldsymbol{G}) = 36$ as shown by Eq. 17.

There exist two special cases of the full-kinematic model worthy mentioning:

1. $\boldsymbol{S}_o = 0$. This is the case of a prismatic workpiece. The surface properties of the locators will have no effects on the fixture kinematics. Thus, the full-kinematic model would not yield any more information than the conventional model.

2. $\boldsymbol{S}_i = \infty$ *for* $i = 1, 2, \cdots, 6$. This is indeed the case of all point locators. It should be noted that the full-kinematic model would not degenerate into the conventional fixture model. The geometric effects of the workpiece surface remain to be captured by the full-kinematic model as opposed to the point-kinematic model which neglects all surface properties.

4. STATISTICAL ACCURACY OF LOCALIZATION

In general, the locator positioning errors $\delta\boldsymbol{q}_i$ depend on the dimensioning and tolerancing scheme assigned to the fixture assembly and its components. In the early stage of fixture design, only variations or tolerances of the error sources may be known. Therefore, the locator errors are usually considered as statistical variables and the resulting fixture localization errors $\delta\boldsymbol{q}_o$ can be characterized statistically.

If it is assumed that the locator errors $\delta\boldsymbol{q}_i$ $(i = 1, 2, \cdots, 6)$ follow an independent normal distribution $N(\boldsymbol{0}, \sigma^2\boldsymbol{I})$ with equal standard deviation σ for all locators, then it is easy to show that the variance matrix of the workpiece localization error $\delta\boldsymbol{q}_o$ is given as

$$\text{Var}(\delta\boldsymbol{q}_o) = \sigma^2\left(\boldsymbol{I} + 2(\boldsymbol{A}_1\boldsymbol{H}_1)(\boldsymbol{D}^T\boldsymbol{D})^{-1}(\boldsymbol{A}_1\boldsymbol{H}_1)^T\right) \tag{27}$$

For the point-contact model it is easy to determine the variance of the localization error as

$$\text{Var}(\delta\boldsymbol{q}_o) = \sigma^2(\boldsymbol{J}^T\boldsymbol{J})^{-1} \tag{28}$$

based on Eq. 26 (Wang 2000). Since the point-kinematic model consists of a subset of the full-kinematic model, this predicted localization error would also be a subset of that predicted by the full-kinematic model. Thus, the point-kinematic model would underestimate the positioning error of the fixture.

5 CONCLUSIONS

In this article we present a fixture model based on the full kinematics of locator-workpiece contact. This model incorporates the surface properties of both the workpiece and the locators, including the surface curvature, torsion and scale factors, as opposed to the conventional point-kinematic model. The necessary and sufficient condition for deterministic localization is also given with a statistical characterization of the positioning accuracy of the fixture.

This full-kinematic fixture model could have a strong implication in analysis of fixturing precision. In general, it is understood that the full model will provide a more accurate estimate of the localization error than the conventional model. It would be of practical relevance to further determine if a theoretic upper bond can be derived on the precision gain over the conventional model. Ongoing and future research also includes the development of fixture synthesis methods using the full fixture model to minimize the influence of fixture tolerances on fixturing precision.

REFERENCES

Asada, H. and By, A. B.: 1985, Kinematics analysis of workpart fixturing for flexible assembly with automatically reconfigurable fixtures, *IEEE Journal of Robotics and Automation* **RA-1**(2), 86–93.

Brost, R. C. and Goldberg, K. Y.: 1996, A complete algorithm for designing plannar fixtures using modular components, *IEEE Trans. on Robotics and Automation* **12**(1), 31 – 46.

Carlson, J. S.: 2000, Quadratic sensitivity analysis of fixturing and locating schemes for rigid parts, *ASME Journal of Manufactring Science and Engineering*. Accepted.

Chou, Y.-C., Chandru, V. and Barash, M. M.: 1989, A mathematical approach to automated configuration of machining fixtures: Analysis and synthesis, *Journal of Engineering for Industry* **111**, 299 – 306.

DeMeter, E. C.: 1994, Restraint analysis of fixtures which rely on surface contact, *Journal of Engineering for Industry* **116**(2), 207 – 215.

Montana, D. J.: 1988, The kinematics of contact and grasp, *International Journal of Robotics Research* **7**(3), 17 – 32.

Montana, D. J.: 1995, The kinematics of multi-fingered manipulation, *IEEE Trans. on Robotics and Automation* **11**(4), 491 – 503.

Murray, R. M., Li, Z. and Sastry, S. S.: 1994, *A Mathematical Introduction to Robotic Manipulation*, CRC Press, Boca Raton, FL.

Wang, M. Y.: 2000, An optimal design for 3-d fixture synthesis in a point set domain, *IEEE Trans. on Robotics and Automation* **16**(6), 839 – 846.

An analytical approach to machining deviation due to fixturing

Armillotta A., Carrino L., Moroni G., Polini W.*, Semeraro Q.*
**Dipartimento di Ingegneria Industriale, Università degli Studi di Cassino*
Via G. Di Biasio, 43 – 03043 Cassino, Italy
Ph. +39 0776 299679 Fax +39 0776 310812 E-mail: polini@unicas.it

Abstract
When designing fixtures, a relevant aspect for positioning accuracy is the configuration of locators, which establish a relationship between part datum references and machine axes. This work aims to investigate on how locators configuration affects precision of manufacturing operations. It considers how deviations on locators position used in manufacturing set-up propagate on toleranced features.
The paper focuses on calculating position errors on machined features due to fixture configuration. An analytical approach has been used: the position of each locator has been simulated by a density probability distribution and, consequently, the probability density function of the position of the machined feature has been calculated. 2D parts have been considered to validate this approach. In this way it is possible to model both real cases, such as drilled plates, and 3D cases when assuming a perfect accuracy on the support plane.
Keywords: tolerance control, fixture design, statistical positioning

1. INTRODUCTION

The fixture equipment is commonly used to locate and to keep steady the parts for machining, assembly, inspection and other operations. It should guarantee accuracy and repeatability in positioning of the part, no deformation, and accessibility of the tool to the workpiece. Other criterions can include easy load and unload of the parts, a reduced weight and a fast cleaning of chipping.

Two fixture equipments exist: a special purpose and a modular one. A special-purpose fixture is fabricated according to the shape of the workpiece. The cost and time spent in making the dedicated fixture can only be justified when the quantity of the production is high or the product sales can cover the cost of making the fixture.

A modular fixture conforms to parts of different machines and processes at a minimum reconfiguration cost. Modular fixturing systems are the best solution available to date for machining, where the forces on the parts are very high, and assembly or finishing or inspection, where the stresses are low. They are built by standard components that can be rapidly mounted on a holed or T-slotted baseplate, so that fixture configuration could be easily modified according to part geometry. These

P. Bourdet and L. Mathieu (eds.),
Geometric Product Specification and Verification: Integration of Functionality, 175-184.

standard components may be divided into locators (or fixels) and clamps. The locators are used to locate part inside machine volume, while clamps keep part into contact with locators. The modular fixture is cost and time efficient, particularly for small to medium batch production, because it can be assembled, disassembled and reassembled for a variety of workpieces.

The fixture design represents a critical step of the process planning for a manufacturing system. The cost and the time needed to design the fixture equipment can be easily justified for mass productions, but they are too high for small to medium batch production. In this field the researches have been oriented towards the automation of the fixture design, called Computer Aided Fixture Design, in order to strongly decrease the lead time and the cost connected with design. The first studies have involved robotic engineering [Asada et al., 1985]. However, there are important differences between the requirements of a mechanical grip that must hold a part and those of a fixture equipment keeping steady a part. The first concerns with the functional requirements: the positioning accuracy needed for machining is higher than that for manipulating. The second difference is connected with friction. The friction is exploited by robot to hold the part, while it is not taken into consideration for machining since the cutting forces are very high and dynamic.

The design of a modular fixture should ensure that the part is not free to move relative to the workspace (kinematics analysis); each part of the batch assume the same spatial configuration within tolerance; the fixture can withstand forces acting on the part during the task. When a part is located inside the fixture equipment, it should assume a stable posture. The fixture should provide a deterministic positioning of the part in order to guarantee the required accuracy of the machining. When the clamping elements are applied on the part, the fixture strengths should not modify the stable and accurate position previously assumed by the part.

The existing research provides essential steps towards completely automating and fully integrating fixture-design systems. In particular the deterministic positioning of the part has been studied by means of geometric reasoning on the possible movements of a rigid body in contact with other parts. Some approaches are limited by the complexity of calculation or the difficulty to model complex 3D parts. The more largely used formalism is based on the screw theory due to its compactness and its general applicability [Ohwovoriole et al., 1981]. The screw theory allows to identify both the causes of a wrong positioning of the part and the possible actions to intervene. The elements of the matrix of the constraints can be easily extracted by a solid modeller, whereas the CAD tools are commonly used to design the modular fixturing system. Moreover, it allows to integrate the kinematic analysis of the part with the automatic planning of modular fixture equipment beginning by the geometry of the parts. Bourdet and Clement used the displacement screw vector to mathematically describe the misalignment between part and machine [Bourdet et al., 1974]. They extended this work by developing a model to determine the nominal positions of locators minimising the magnitude of the screw displacement vector [Bourdet et al., 1988]. Weill connected the screw displacement vector to the geometric variation of critical part features and minimised this geometric function [Weill et al., 1991]. More recent studies deal with robust design of fixture configuration, by analysing the influence of workpiece surface errors and fixture set-up errors on the stability of part [Cai et al. 1997], develop algorithms for workpiece localization [Chu et al., 1998], or employ the screw parameters associated with TTRS to determine the position uncertainty of a part [Desrochers et al., 1997].

When designing fixtures, a relevant aspect for positioning accuracy is the configuration of locators, which establish a relationship between part datum references and machine axes. Once a workpiece has been located and clamped, the cutting tool is moved relative to the machine reference frame for the purpose of generating the machined surfaces of the workpiece. Assuming that the workpiece datum features and the locator positions are perfect and that no other sources of machining error exist, the machined surfaces will be true with respect to form and position in relation to the workpiece datum reference frame. Otherwise, if the positions of the locators are allowed to vary from their nominal values, there will be a misalignment between the part and the machine reference frames. This misalignment will result in position error of the part machined features. Very interesting is the work of Choudhuri that presents a method for modelling and analysing the impact of a locator tolerance scheme on the potential datum related, geometric errors of linear, machined features [Choudhuri et al., 1999]. This model considers profile and dimensional tolerances applied to spherical tip locators in contact with planar workpiece datum features; it is tested by means of simulation studies. The authors, however, do not consider the distribution of the machined features as a function of locator errors, but only the worst cases.

This work deals with the problem of kinematics analysis of modular fixtures. In particular, it proposes a method to calculate the position error of a machined feature due to the inaccuracy of the fixture configuration. It represents a design step that substitutes the deterministic positioning of the part due to the screw theory. In fact, the proposed approach allows to choose the positions of the locators able to minimise the machining deviation of a surface point. In this way it is possible to move the positions of locators in order to decrease the machining inaccuracy.

In a previous paper [Armillotta et al., 1996], a method for checking deterministic positioning from locator configuration has been proposed. Besides detecting positioning incorrectness, it derives an explanation of singularity reasons, in order to ease the redesign of a wrong fixture. A following study [Armillotta et al., 1999] extends the previous method in order to highlight quasi-singularity conditions, where part inaccuracies are likely to result in excessive geometric errors on machined features.

To integrate the approaches based on deterministic positioning, a statistical method has been proposed in this paper. The results of the study presented here show how an analytical approach, based on probability distributions, defines the distribution of positions of machined features as a function of the inaccuracy of the locators scheme. In this way it is possible not only to provide useful criteria for fixture design, but also to verify the manufacturing capability during the tolerance design stage. 2D parts have been considered in order to model some real cases, such as plates, and to simplify more complex problem. The 3-2-1 locating method has been taken into consideration, because it is widely used and it satisfies stability, accessibility and non-redundancy conditions.

2. PROPOSED APPROACH

The case study is shown in Figure 1: a plate to be drilled. The basic dimensions and a tolerance specify the position of the hole. The locating position of the workpiece is shown, with two fixels on the primary datum and one on the secondary, whose coordinates related to the machine reference frame *XOY* are represented by the three couples of values:

$$\text{fixel } 1(x_1, y_1);\ \text{fixel } 2(x_2, y_2);\ \text{fixel } 3(x_3, y_3); \qquad (1)$$

The proposed approach considers the uncertainty source in the positioning error of the machined hole due to the variance in the positioning of the fixels. It aims to minimise the machining uncertainty due to this source. It neglects the tool positioning error or the geometric deviations on datum elements.
In the following the capital letters represent the constants, while the small letters stand for the variables of the problem.
The six coordinates of the fixels (1) in the machine reference frame *XOY* are considered distributed according to a normal probability density function with mean equal to the nominal position of fixels and standard deviation equal 0.01 mm respectively:

$$\begin{array}{ll} x_1 \approx N(X_1 = 0,\ \sigma = 0.01) & y_1 \approx N(Y_1, \sigma = 0.01) \\ x_2 \approx N(X_2, \sigma = 0.01) & y_2 \approx N(Y_2 = 0,\ \sigma = 0.01) \\ x_3 \approx N(X_3, \sigma = 0.01) & y_1 \approx N(Y_3 = 0,\ \sigma = 0.01) \end{array} \qquad (2)$$

with Y_1, X_2 and X_3 constant and different by zero.

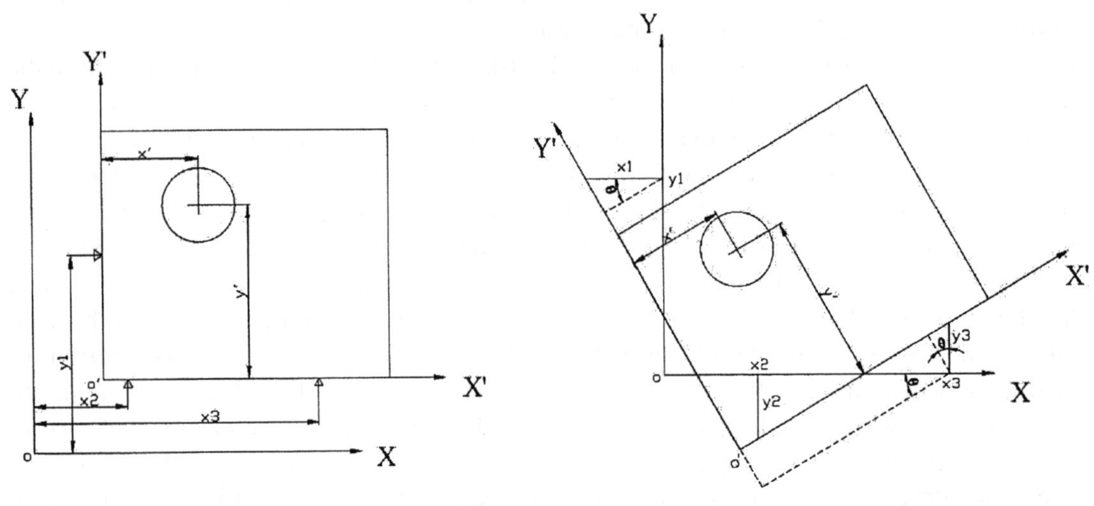

Figure 1: Case study

It has been chosen a normal distribution to simplify the problem to solve, while the variance value is due to the same tolerance conditions of the three fixels with respects the support plate.
The machine reference frame *XOY* (MRF) and the part reference frame *X'O'Y'* (PRF) have been considered. The first passes through the nominal value of fixel positions, while the second is related to the actual position of the fixels. In particular the axis X' is constituted by the straight lines passing through the actual positions of fixels 2 and 3, while the Y' axis is the straight line perpendicular to the X'-axis and passing through the actual position of fixel 1. The coordinates of the centre of the drilled hole are equal to the nominal values $P \equiv (X, Y)$ in the MRF, while they are variable as a function of fixel probability distributions $P' \equiv (x', y')$ in the PRF.
To determine the influence of the fixels position on the location of the drilled hole, the probability density function of the centre P' of the drilled hole in the PRF has been calculated as a function of the probability density function of the fixels (2). The coordinate of the hole centre x' and y' may be expressed as a function of P and fixel positions (1) by means of the following equations:

$$x' = (X - x_1) \cdot cos(\vartheta) + (Y - y_1) \cdot sin(\vartheta) \tag{3}$$

$$y' = (Y - y_3) \cdot cos(\vartheta) - (X - x_3) \cdot sin(\vartheta) \tag{4}$$

$$\vartheta = \text{arctg} \frac{y_3 - y_2}{x_3 - x_2} \tag{5}$$

The angle ϑ is very small in comparison with the part dimensions. Therefore, the ϑ functions may be appossimated as $sin(\vartheta) \approx \vartheta$, $cos(\vartheta) \approx 1$ and $\text{tg}(\vartheta) \approx \vartheta$ and the equations (3)-(5) become:

$$x' = (X - x_1) + (Y - y_1) \cdot \vartheta \tag{6}$$

$$y' = (Y - y_3) - (X - x_3) \cdot \vartheta \tag{7}$$

$$\vartheta = \frac{y_3 - y_2}{x_3 - x_2} \tag{8}$$

The equations (6)-(8) represent an indeterminate system of three equations in the six variables $x_1, x_2, x_3, y_1, y_2, y_3$. It has been calculated its Jacobian matrix:

$$J = \begin{bmatrix} \frac{\partial x'}{\partial(x_1)} & \frac{\partial x'}{\partial(x_2)} & \frac{\partial x'}{\partial(x_3)} & \frac{\partial x'}{\partial(y_1)} & \frac{\partial x'}{\partial(y_2)} & \frac{\partial x'}{\partial(y_3)} \\ \frac{\partial y'}{\partial(x_1)} & \frac{\partial y'}{\partial(x_2)} & \frac{\partial y'}{\partial(x_3)} & \frac{\partial y'}{\partial(y_1)} & \frac{\partial y'}{\partial(y_2)} & \frac{\partial y'}{\partial(y_3)} \\ \frac{\partial \vartheta}{\partial(x_1)} & \frac{\partial \vartheta}{\partial(x_2)} & \frac{\partial \vartheta}{\partial(x_3)} & \frac{\partial \vartheta}{\partial(y_1)} & \frac{\partial \vartheta}{\partial(y_2)} & \frac{\partial \vartheta}{\partial(y_3)} \end{bmatrix} = \tag{9}$$

$$= \begin{bmatrix} -1 & (Y - y_1) \cdot \frac{y_3 - y_2}{(x_3 - x_2)^2} & -(Y - y_1) \cdot \frac{y_3 - y_2}{(x_3 - x_2)^2} & -\frac{y_3 - y_2}{x_3 - x_2} & -\frac{Y - y_1}{x_3 - x_2} & \frac{Y - y_1}{x_3 - x_2} \\ 0 & -(X - x_3) \cdot \frac{y_3 - y_2}{(x_3 - x_2)^2} & (X - x_2) \cdot \frac{y_3 - y_2}{(x_3 - x_2)^2} & 0 & \frac{X - x_3}{x_3 - x_2} & -\frac{X - x_2}{x_3 - x_2} \\ 0 & \frac{y_3 - y_2}{(x_3 - x_2)^2} & -\frac{y_3 - y_2}{(x_3 - x_2)^2} & 0 & -\frac{1}{x_3 - x_2} & \frac{1}{x_3 - x_2} \end{bmatrix}$$

As can be seen in matrix (9), the change of P' position due to x_2, x_3 and y_1 are infinitesimal when compared with respect to x_1, y_2 and y_3. Therefore, x_2, x_3 and y_1 have been considered constant and equal to their nominal values X_2, X_3 and Y_1.
The system constituted by the three equations (6)-(8) is determined; the coordinate x', y' and ϑ of P' are function of only x_1, y_2 and y_3 fixel variables.
Therefore, the probability desity functions of the coordinates x' and y' depend only on the probability distribution of the coordinate x_1, y_2 and y_3 of the fixels.
The probability density function of the three independent variables x_1, y_2 and y_3 is equal to:

$$f_{x_1,y_2,y_3}(x_1,y_2,y_3)=f_{x_1}(x_1)\cdot f_{y_2}(y_2)\cdot f_{y_3}(y_3)=\frac{1}{\sqrt{2\cdot\pi}}\cdot e^{-\frac{1}{2}\cdot x_1^2}\cdot\frac{1}{\sqrt{2\cdot\pi}}\cdot e^{-\frac{1}{2}\cdot y_2^2}\cdot \frac{1}{\sqrt{2\cdot\pi}}\cdot e^{-\frac{1}{2}\cdot y_3^2}=\left(\frac{1}{\sqrt{2\cdot\pi}}\right)^3\cdot e^{-\frac{1}{2}\left[x_1^2+y_2^2+y_3^2\right]} \tag{10}$$

as a conseguence of hypotheses (same normal distribution and independence).
The probability density function of x', y' and ϑ is derived by (10) recalling the *statistical transformation method* of random variables [Mood et al, 1974] and the equations (6)-(8):

$$f_{x',y',z'}(x',y',\vartheta)=\left|det(J)\right|\cdot f_{x_1,y_2,y_3}\left(g_1^{-1}(x',y',\vartheta),g_2^{-1}(x',y',\vartheta),g_3^{-1}(x',y',\vartheta)\right)=$$
$$=\left|X_3-X_2\right|\cdot\left(\frac{1}{\sqrt{2\cdot\pi}}\right)^3\cdot e^{-\frac{1}{2}\left[(X-x'+(Y-Y_1)\cdot\vartheta)^2+(Y-y'-(X-X_2)\cdot\vartheta)^2+(Y-y'-(X-X_3)\cdot\vartheta)^2\right]} \tag{11}$$

Then the equation (11) has been integrated in ϑ in order to obtain a function in the plane (x', y'):

$$f_{x',y'}(x',y')=\int_{-\infty}^{+\infty}f_{x',y',\vartheta}(x',y',\vartheta)=\frac{\left|X_3-X_2\right|}{2\cdot\pi\cdot\sqrt{(Y-Y_1)^2+(X-X_2)^2+(X-X_3)^2}}\cdot$$
$$\cdot exp\left\{-\frac{1}{2}\cdot\left(\frac{(x'-X)^2\cdot\sigma_{y'}^2\cdot(X_3-X_2)^2}{(Y-Y_1)^2+(X-X_2)^2+(X-X_3)^2}+\right.\right.$$
$$\left.\left.\frac{2\cdot(x'-X)\cdot(y'-Y)\cdot(Y-Y_1)\cdot[(X-X_2)+(X-X_3)]}{(Y-Y_1)^2+(X-X_2)^2+(X-X_3)^2}+\frac{(y'-Y)^2\cdot\sigma_{x'}^2\cdot(X_3-X_2)^2}{(Y-Y_1)^2+(X-X_2)^2+(X-X_3)^2}\right)\right\} \tag{12}$$

The equation (12) is the probability density function we look for. It has been characterised by the following first and second order moments:

$$\mu_{x'}=X \tag{13}$$
$$\mu_{y'}=Y \tag{14}$$

$$\sigma_{x'} = \frac{\sqrt{2 \cdot (Y - Y_1)^2 + (X_3 - X_2)^2}}{|X_3 - X_2|} \tag{15}$$

$$\sigma_{y'} = \frac{\sqrt{(X - X_2)^2 + (X - X_3)^2}}{|X_3 - X_2|} \tag{16}$$

$$cov[x', y'] = \rho \cdot \sigma_{x'} \cdot \sigma_{y'} = \frac{(Y - Y_1) \cdot [(X - X_2) + (X - X_3)]}{(X_3 - X_2)^2} \tag{17}$$

As can be seen by equations (15)-(17), while the variables x_1, y_2 and y_3 determine the probability density function of the coordinates x' and y', the variables x_2, x_3 and y_1 contribute to calculate the standard deviation of the same probability density function. Therefore, all the coordinates of the locator positions should be used to compare different solutions in terms of part positioning.

3. SIMULATION STUDY

The locator tolerance analysis model described previously has been applied to the fixture-workpiece system shown in Figure 2. It is a plate to be drilled at a nominal position $P(X,Y) \equiv P(40,70)$.

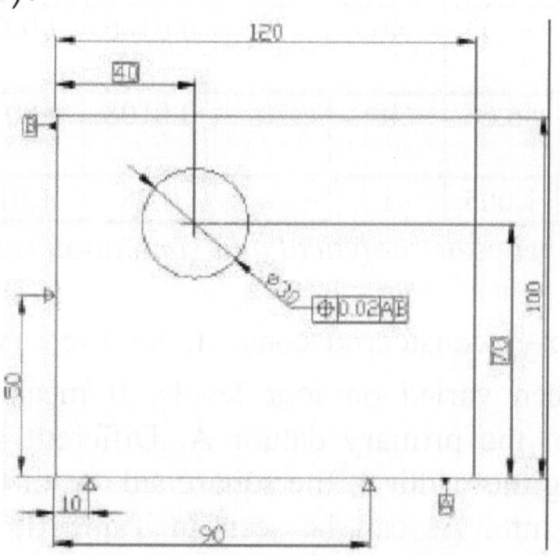

Figure 2: Application example.

The locating fixture is also shown, with two fixels on the primary datum and one fixel on the secondary, spaced according Y_1=50 mm, X_2=10 mm and X_3=90 mm respectively.

It is assumed that the distances of the locators from the X and Y axes are distributed according to a normal probability density function, with mean values equal to the nominal positions and standard deviations equal to 0.01 mm.

It has been calculated the probability of success that means the actual position of the hole, due to the fixturing error, is inside a shaped area centred around the nominal position P. Two shapes have been considered: a square with 0.02 mm side and a circle with 0.01 mm radius. This means to integrate the equation (12) with x' and y' variables ranging inside the considered square or circle area. To solve the integration problem it has been used Maple® (version 5.3), a mathematical software package.
The obtained results have been compared with the simulations carried out by means of a Monte Carlo approach. A set of values (10,000) has been randomly generated for each probability distribution of the coordinates of fixels. Then, the coordinates of the actual position of the hole have been calculated by means of the equations (6) and (7) for each generated instance and, finally, the percentage of success has been determined and compared with the value identified previously by the analytical approach. The Monte Carlo simulation has been carried out by means of Minitab®, a statistical software package.
The results, shown in Table I, underlines that the analytical approach seems to reproduce well the actual error of the machined feature simulated by means of the Monte Carlo method. The hypothesis of a different influence on the part orientation of the two coordinates of each fixel, on which is based the analytical approach seems to be verified.

	% square	% circle	$\mu_{x'}$ [mm]	$\mu_{y'}$ [mm]	$\sigma_{x'}$ [mm]	$\sigma_{y'}$ [mm]	$cov[x',y']$ [mm^2]	ρ
analytical approach	54.36	46.93	40	70	0.0106	0.0073	$0.06\ 10^{-2}$	$0.08\ 10^{-2}$
Monte Carlo simulation	53.96	46.66	40	70	0.0108	0.0072	$0.06\ 10^{-2}$	$0.08\ 10^{-2}$
anal./sim.	1.007	1.005	1	1	0.98	1.014	1	1

Table I – Comparison between analytical and simulation results due to fixel positions

Moreover, a variable (x_3) considered constant for the calculation of the probability density function has been varied on four levels. It means to increase the distance between the locators on the primary datum A. Different area of success have been considered by increasing the width of the square side or of the circle diameter to: 0.06 mm, 0.1 mm and 0.2 mm. As can be seen in Table II, the probability of success decreases by reducing the variable x_3 for the same shaped area, due to the increasing of the variances of the probability density function. When the area extension increases, the probability of success achieves 100%.
To summarise, all the coordinates of fixels seem to influence the machining deviation of a feature, any of them by modifying the shape of the probability density function of the coordinates of this feature in the part reference frame, the remaining by varying the variance of the same distribution. Both these factors are the variables of an otimizing model that aims to minimise the machining deviations.

[mm]	$\sigma_{x'}$ [mm]	$\sigma_{y'}$ [mm]	$cov(x', y')$ [mm^2]	% square (side)				% circle (Ø)			
				square side				circle Ø			
				0.02 [mm]	**0.06 [mm]**	**0.1 [mm]**	**0.2 [mm]**	**0.02 [mm]**	**0.06 [mm]**	**0.1 [mm]**	**0.2 [mm]**
X3=90	0.011	0.007	0.00062	54.36	99.53	100	100	46.93	99.30	100	100
X3=40	0.014	0.010	-0.0067	39.20	96.89	99.97	100	32.92	94.38	99.90	100
X3=20	0.030	0.036	-0.10	12.77	54.98	82.13	99.44	10.43	46.42	71.44	96.95
X3=15	0.057	0.078	-0.44	6.06	27.89	47.09	79.94	4.94	23.20	38.94	69.78

Table II – Analytical results by changing the variable x_3 and the shaped area extension

4. CONCLUSIONS

This work has dealt with the problems that locating fixture configurations and movement of the machine tool axes pose to the fulfilment of accuracy specifications on a part.

The presented analysis of the position error of a drilled hole resulting from inaccuracies on fixels position has shown that a statistical method based on the probability density distribution of the coordinate of locators position give results in accordance with reality. In particular, the two coordinates of each locator position seem to have a different influence on part orientation: one changes the form of the probability density function that describes the position of the machining feature in the fixture reference frame, while the other modifies the variance of the same distribution. Therefore, it is possible to opportunely choose the locators coordinate in order to minimise the machining deviation of a feature.

Results only applied to planar parts, which are a first approximation of three-dimensional cases when assuming a perfect accuracy on the support plate. Due to generality of the approach, extension to the general case should be possible. Further simplifications are the perfect form of datum elements and the perfect positioning accuracy of the machining tool.

Lastly, it has been assumed that tool working positions can be conveniently described in the fixture coordinate system: this is not true when machine-mounted probes are used to establish the co-ordinate transformation between machine and workpiece, thus reducing fixture effect but introducing calibration errors.

The release of all these assumptions is currently matter of further study.

ACKNOWLEDGMENTS

The work has funded partially by the italian M.U.R.S.T.(Ministry of University and Scientific and Technological Research) and CNR (National Research Council):

REFERENCES

[Armillotta et al., 1996] Armillotta, A.; Moroni, G.; Negrini, L.; Semeraro, Q.; "Analysis of deterministic positioning on workholding fixtures"; In: *Proceedings of International Conference on Flexible Automation and Intelligent Manufacturing*, pp. 274-284; Atlanta 1996; ISBN 1-5670-0067-3.

[Armillotta et al., 1999] Armillotta, A.; Bigioggero, G.F.; Moroni, G.; Negrini, L.; Semeraro, Q.; "Tolerance control in workpiece fixturing"; In: *Proceedings of the ASME 4th Design for Manufacturing Conference*, pp. 1-10; Las Vegas 1999; ISBN 0-7918-1967-1.

[Asada et al., 1985] Asada, H.; By, A.B.; "Kinematic analysis of workpart fixturing for flexible assembly with automatically reconfigurable fixtures"; In: *IEEE Journal of Robotics and Automation*, Vol. 1-2, pp. 86-94; 1985.

[Bourdet et al., 1974] Bourdet, P.; Clement, A.; "Optimalisation des Monrages d'Usinage"; In: *L'Ingenieur et le Techniciien de l'Enseignement Technique*, pp. 8-74; 1974.

[Bourdet et al., 1988] Bourdet, P.; Clement, A.; "A study of optimal-criteria identification based on the small displacement screw model"; In: *Annals of the CIRP*, Vol. 37/1, pp. 503-506; 1988.

[Cai et al. 1997] Cai, W.; Jack Hu, S.; Yuan, J.X.; "A variational method of robust fixture configuration design for 3-D workpieces"; In: *Journal of Manufacturing Science and Engineering*, Vol. 119, pp. 593-602; 1997.

[Choudhuri et al., 1999] Choudhuri, S.A.; De Meter, E.C.; "Tolerance analysis of machining fixture locators"; In: *Journal of Manufacturing Science and Engineering. Transactions of ASME*, Vol. 121, pp. 273-281; 1999.

[Chu et al., 1998] Chu, Y.X.; Gou, J.B.; Wu, H.; Li, Z.X.; "Localization algorithms: performance evaluation and reliability analysis"; In: *Proceedings of the IEEE International Conference on Robotics & Automation*, pp. 3652-3657; Leuven, 1998.

[Desrochers et al., 1997] Desrochers, A.; Delbart, O.; "Determination of part positioning uncertainty within mechanical assembly using screw parameters"; In: *Proceedings of 5th CIRP International Seminar on Computer Aided Tolerancing*, pp. 185-196; 1995; ISBN 0-412-72740-4.

[Grippo et al., 1988] Grippo, P.M.; Thompson, B.S.; Gandhi, M.V.; "A review of flexible fixturing system for computer integrated manufacturing"; In: *International Journal of Computer Integrated Manufacturing*, Vol. 1-2, pp. 124-135; 1988.

[Mood et al., 1974] Mood, A.M.; Graybill, F.A.; Boes, D.C.; "Introduction to the Theory of Statistics"; McGraw-Hill, Inc; 1974; ISBN 88-386-0614-5.

[Ohwovoriole et al., 1981] Ohwovoriole, M.S.; Roth, B.; "An extension of screw theory"; In: *Transaction of ASME: Journal of Mechanical Design*, Vol. 103, pp. 725-735; 1981.

[Weill et al., 1991] Weill, R.; Darel, I.; Laloum, M.; "The influence of fixture positioning errors on the geometric accuracy of mechanical parts"; In: *Proceedings of CIRP Conference on PE&MS*, pp. 215-225; 1991.

Three-dimensional geometrical tolerancing: quantification of machining defects

Olivier Legoff* - François Villeneuve**
** IRCCyN – Ecole Centrale de Nantes*
1, rue de la Noë 44321 Nantes cedex 3
Olivier.Legoff@irccyn.ec-nantes.fr
** *Laboratoire 3S, Domaine Universitaire, BP53*
38041 Grenoble cedex 9, France
Francois.Villeneuve@hmg.inpg.fr

Abstract: This paper is proposing to validate a three-dimensional model on manufacturing tolerancing for mechanical parts. The work presented relies on research conducted at the LURPA (Ballot and Bourdet) on the computation of three-dimensional tolerance chains for mechanisms. Models of the workpiece, the set-ups and the machining operations are provided. The concept of the Small Displacements Torsor (SDT) is used to model the process planning. The first part introduces the use of the concept of SDT in the case of manufacturing tolerancing. Then we propose, for a chosen workpiece, an experimental approach to measure and quantify the three-dimensional machining variations as torsors. At last, an analysis of the results is proposed.
Keywords : Tolerancing, Machining, Small Displacements Torsor, Process Planning

1. INTRODUCTION

In the context of integrated design and manufacturing it is essential to take into account the geometrical variations of parts of a mechanism. This should allow validating the correct functioning of the mechanism during the first steps of design. These analyses have to generate an optimal tolerancing, integrating two points of view, functional point of view and manufacturing (or machining) point of view. For the manufacturing (or machining) point of view, knowing the process planning, the usual method described in papers is the following:

- Generation of minimal Manufacturing Tolerance chains: each dimension and its tolerance are modelled as a vector.
- Machining simulation: for each Manufacturing Tolerance, the expert know-how gives a realistic minimal value. These minimal values are used to check if the part can be manufactured according to Manufacturing Tolerance chains and functional tolerances.
- Optimisation: when the manufacturability has been checked it is usually possible to enlarge some tolerances in order to facilitate manufacturing.

P. Bourdet and L. Mathieu (eds.),
Geometric Product Specification and Verification: Integration of Functionality, 185-196.

The papers take up the whole or a part of the above fields to model [Ballot et al., 1995] [Ji, 1999], to optimise [Ngoi et al., 1999] or to define cost approaches [He, 1996]. They accent on process planning assistance, set-up choice or product/process integration [Zhang, 1996]. They describe statistical models or worst-case models. But, in every quoted reference, the approach is unidirectional. It does not take the influence of rotation defects into account. Nevertheless, the three dimensional effects are not inconsiderable, especially when the lever arms are long.

The aim of our work is a three-dimensional approach in manufacturing tolerancing. Some works about 3D approaches of the CAT problem can be found in literature [Mathieu et al., 1991] [Desrochers et al., 1994] [Teissandier et al., 1997] [Laperriere et al., 1998] and, especially, in the field of manufacturing tolerancing [Kanaï et al., 1995] [Clément et al., 1996]. They differ in models used, just as well to describe the surfaces of the parts as to model the sum of the defects.

The method we propose relies on research on the computation of three-dimensional tolerance chains for mechanisms [Ballot et al., 1995]. Starting from these works, we proposed a formalisation of the problem within the more specific context of manufacturing tolerances. The main originality is to model the machining set-up as a mechanism [Legoff et al., 1999] [Villeneuve et al., 1999]. This work is based on the concept of Small Displacement Torsor (SDT) [Bourdet et al., 1996]. It opens up the way for the three-dimensional integration product/process because of the similarities between the concepts used in both points of view.

The works described in this paper concern the experimental studies carried out to valid the presented model. The aim is to answer to the following questions:

- Is it possible to identify the parameters of the model in a real machining job ?
- Which experimental method has to be carried out to identify these parameters ?
- Which experimental law is applicable to the identified parameters ?
- Is it possible to predict the three-dimensional behaviour of a mass production of parts ?

The first part recalls the principle of the modelling of surface variations with SDT as well as its application to the manufacturing tolerancing. The second part describes the experimental process and the results obtained on a machining job.

2. MODELING OF THE 3D MACHINING PROCESS DEFECTS

The works of E. Ballot and P. Bourdet [Ballot et al., 1995] model the interactions between the parts of a mechanism, so as to predict the position and orientation variations of these parts in a three dimensional space. The variations are supposed to be small enough to use the concept of the SDT excepted for the expected movements of parts. This principle is generally verified in all mechanisms as the presence of 'unwanted' degrees of freedom goes against the working of the mechanism considering the risks of jamming, of premature wear or of shocks (vibrations) that they generate.

2.1. SDT concept

The main idea consists in considering that the displacements of a rigid body or a surface, excepted for the expected movements, are supposed small as regards the other geometric dimensions (i.e. the nominal dimensions). The displacements can then be linearized in the first order. Knowing D_E, the small displacement vector of a point E of the feature considered and M, the small rotation matrix expressed in a coordinate system the small displacement D_{Pi} of any Pi point of the feature is obtained by:

$$\mathbf{D}_{Pi} = \mathbf{D}_E + \mathbf{M}\cdot\mathbf{EPi} = \mathbf{D}_E + \begin{pmatrix} 0 & -\gamma & \beta \\ \gamma & 0 & -\alpha \\ -\beta & \alpha & 0 \end{pmatrix}\cdot\mathbf{EPi} = \mathbf{D}_E + \mathbf{\Omega}\wedge\mathbf{EPi}$$

with $\mathbf{\Omega} = \alpha\mathbf{x} + \beta\mathbf{y} + \gamma\mathbf{z}$ where α, β and γ are the small rotations of the element, and $\mathbf{D}_E = u\mathbf{x} + v\mathbf{y} + w\mathbf{z}$ where u, v, w are the small translations of the point E and $\wedge$ is the cross product.

The SDT $\{\mathbf{T}_{feature}\}_{(E,\Re)}$ of the considered feauture expressed in a coodinate system $\langle\Re\rangle$ is:

$$\{\mathbf{T}_{feature}\}_{(E,\Re)} = \begin{Bmatrix} \mathbf{\Omega} \\ \mathbf{D}_E \end{Bmatrix} = \begin{Bmatrix} \alpha\mathbf{x} + \beta\mathbf{y} + \gamma\mathbf{z} \\ u\mathbf{x} + v\mathbf{y} + w\mathbf{z} \end{Bmatrix} = \begin{Bmatrix} \alpha & u \\ \beta & v \\ \gamma & w \end{Bmatrix}_{(E,\Re)}$$

Applying this concept to 'single' surfaces such as planes, spheres, cylinders, cones, torus, requires the introduction of undetermined components in the expression of the components of the associated torsor. These undetermined components are noted U; they reflect the components that leave the surface invariant in its local coordinate system. For instance, for a plane:

$$\{\mathbf{T}_{plan}\}_{OP} = \begin{Bmatrix} \alpha\mathbf{x} + \beta\mathbf{y} + U\mathbf{z} \\ U\mathbf{x} + U\mathbf{y} + w\mathbf{z} \end{Bmatrix}$$

where OP is a point belonging on to the plane and $\Re$ a local coordinate systemof reference such as z is the normal of the plane.

To compute operations on these torsors, the two following properties have been defined: Property 1: $\forall a \in R; a + U = U$ Property 2: $\forall a, b \in R^2; a\cdot U + b\cdot U = U$

2.2. Application to the machining of a part

The model developed for mechanisms has been enhanced to simulate in three dimensions the behaviour of a workpiece being machined.

The manufacturing of a part consists in creating new surfaces on a stock in a series of set-ups. Each set-up, noted Sj, is considered as a mechanism in itself made up of the following parts (figure 1):

- The part holder H, built upon the set-up surfaces Hi.
- The part in the state it is in at the end of the set-up (called workpiece P). It is built up, on the one hand, on the previously machined surfaces (or on the stock surfaces)

and, on the other hand, on machining features that result from the machining operations of the considered set-up. Each surface of the workpiece is noted Pi.

- The machining operations Mk built upon the surfaces Mki that result from the combination of the kinematics of the machine-tool used and the geometry of the tools.

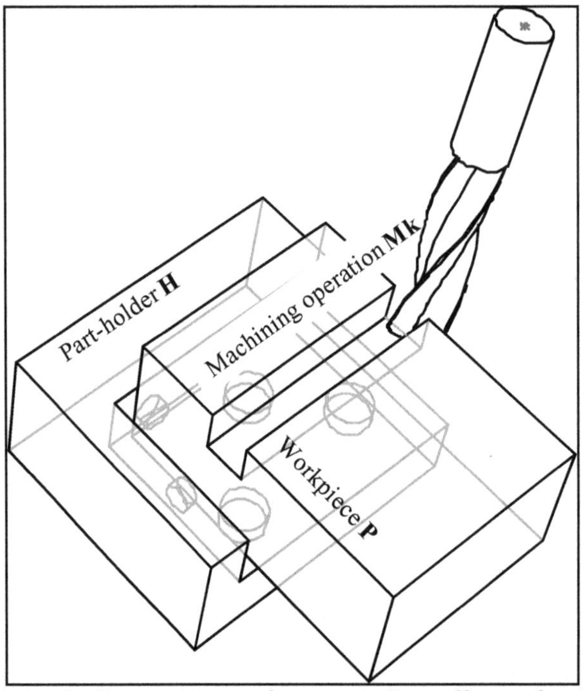

Figure 1; Components of a set-up Sj, milling of a slot.

Considering that the part cannot be deformed, then the surfaces created previously to the set-up considered are invariant. Hence Pi, once it is machined, remains invariant. To makes things clearer, the part holder and the machining operations are systematically named H and Mk though they differ from one set-up to the next. They will be distinguished by their set-up number Sj.

Each real surface of the part holder, the workpiece or the machining operations are modelled by a substitution surface. The substitution surface is a modelization of the real surface by a surface with a typology similar to the theoretical (or nominal) surface. It is considered that the variations of the substitution surfaces machined on the part as well as the variations of the part holder and the machining operations are of the second order relatively to the dimensions of the elements, which allows to model them thanks to a SDT.

The characteristic torsors and the 3D tolerance analysis and synthesis in machining have been described in previous papers [Villeneuve et al., 1999]. We resume hereafter some characteristic torsors and formula.

2.3. Definition of the characteristic torsors and formula

For each set-up the following SDT are defined:

$\mathbf{T}_{R,H(Sj)}$: global SDT of the part holder H relatively to its nominal position in set-up j.

$\mathbf{T}_{H,Hi(Sj)}$: deviation torsor of the surface Hi relatively to its nominal position on the part holder in set-up j.
$\mathbf{T}_{R,Mk(Sj)}$: SDT of a machining operation relatively to its nominal position in set-up j.
$\mathbf{T}_{Mk,Mki(Sj)}$: deviation torsor of the surface Mki relatively to its nominal position on the machining operation Mk in set-up j.
$\mathbf{T}_{Hi,Pi(Sj)}$: gap torsor that expresses the characteristics of the interface between the workpiece and the part holder at the level of the joint Hi/Pi.
For a workpiece the following SDT are defined:
$\mathbf{T}_{R,Pi(Sj)}$: deviation torsor of the machined surface Pi relatively to its nominal position in set-up j (variations linked to the machining). This torsor is the result of the removal of matter during the machining Mk on part P.
$\mathbf{T}_{P,Pi}$: deviation torsor of the surface Pi relatively to its nominal position on part P.
The deviation of a surface Pi relatively to its nominal position on the part depends upon the variations of the machining operation that generates Pi within the set-up Sj ($\mathbf{T}_{R,Pi(Sj)}$) and the positioning torsor of the part within the set-up ($\mathbf{T}_{R,P(Sj)}$).
$\mathbf{T}_{R,P(Sj)}$: SDT of the workpiece relatively to its nominal position in set-up j (positioning variations of the workpiece within the set-up).
The workpiece's SDT $\mathbf{T}_{R,P(Sj)}$ depends on the variations of the support surfaces of the workpiece and the part holder's surfaces opposite. It is obtained by the coupling of the torsors associated to each joint part/part holder. Thus, for any positioning surface in set-up Sj one gets:

$$\mathbf{T}_{R,P(Sj)} = \mathbf{T}_{R,H(Sj)} + \mathbf{T}_{H,Hi(Sj)} + \mathbf{T}_{Hi,Pi(Sj)} + \mathbf{T}_{Pi,P} = \mathbf{T}_{R,H(Sj)} + \mathbf{T}_{H,Hi(Sj)} + \mathbf{T}_{Hi,Pi(Sj)} - \mathbf{T}_{P,Pi}$$

Considering the position variations of the part holder within the set-up $\mathbf{T}_{R,H(Sj)}$ equals to a nil torsor, $\mathbf{T}_{R,P(Sj)}$ becomes:

$$\mathbf{T}_{R,P(Sj)} = \mathbf{T}_{R,Hi(Sj)} + \mathbf{T}_{Hi,Pi(Sj)} - \mathbf{T}_{P,Pi} \tag{1}$$

3. EXPERIMENTAL WORKPIECE, PART HOLDER AND MACHINING PROCESS

In order to carry out tests of machining and measurements to calculate the components of the deviation torsors, we have defined a test part, its machining process and the part holder. Only one set-up of the machining process is studied. We have chosen a prismatic part. The machine-tool on which were held the tests is a 4 axis milling center. A series of 45 parts was produced. The other criteria taken into account are evoked hereafter.

3.1. Part

We have chosen a massive part in order to be been free from possible problems of part deformation. The workpiece material is a plain carbon steel containing 0,48% carbon.

3.2. Part holder

The part holder (figure 2) realizes an isostatic positioning of each workpiece in order to ensure an optimal repeatability of workpiece positions. This led us to design a modular assembly made up of NORELEM elements. Three grinded positioning devices, whose surface of contact with the part is a 10 mm diameter disc, carried out the main joint. The secondary support is carried out by two supports of the same type. The tertiary support is obtained by a spherical surface. Finally the maintenance in position of the part (clamping) is ensured by three flanges. They are facing the main positioning devices in order to minimize the deformation of the part during clamping.

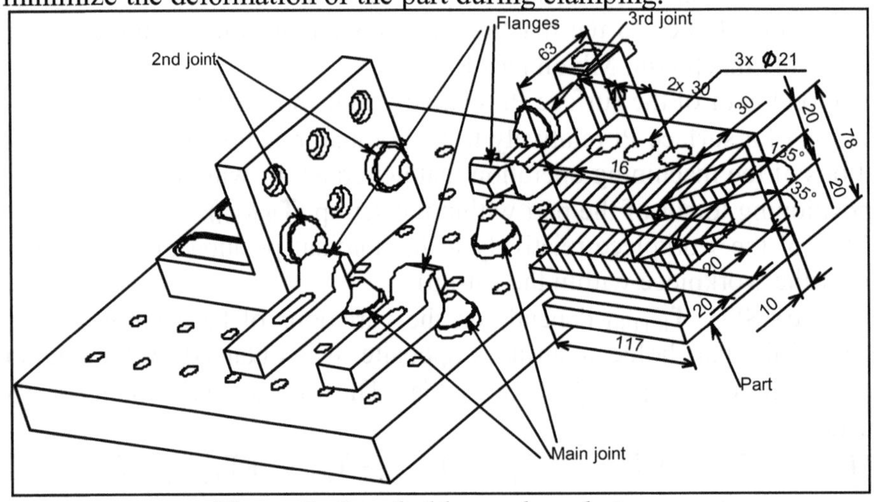

Figure 2; Part holder and workpiece.

3.3. Machining process

The machined surfaces in the considered set-up are the planes hatched on figure 2 as well as the three holes. The operations carried out before this set-up made it possible to obtain a semi-finished state compatible with the finishing operations on which we wish to make the identification.

The previous operations were:

a) Centering, center drill.
b) Drilling, drill ∅ 19.6.
c) Contourning, roughing end mill ∅ 30.
d) Boring semi-finishing, boring bar ∅ 20.5.
e) Contourning semi-finishing, HSS end mill ∅ 20.

So, the finishing operations in the studied set-up are:

a) Contourning finishing, HSS end mill ∅ 21.3 (axial depth of cut = 1 mm, radial depth of cut = 0.7 mm).
b) Boring finishing, boring bar ∅ 21 (radial depth of cut = 0.4 mm).

4. MEASUREMENTS

All the parts were measured with a Coordinates Measuring. Thus, taking into account the possibilities of the software, we adopted the process described hereafter.

4.1. Reference coordinate system

The results of measurements are expressed in the reference coordinate system $\Re$ It is defined by the three datum surfaces used during machining (figure 3):

- the lower plane SR1 whose normal gives the **z** axis,
- the plane SR2, perpendicular to SR1, whose intersection with SR1 gives straight line DR1 defining the **x** axis,
- the plane SR3, whose projection of a point (measured in the zone of the contact with the part holder) on DR1 enables us to define the origin of the reference coordinate system.

The third direction is automatically defined in order to build a Cartesian coordinate system. For the whole of these measurements, the points were measured in the support zones of the fixture in order to minimize the effects of the form defects and perpendicularity defects.

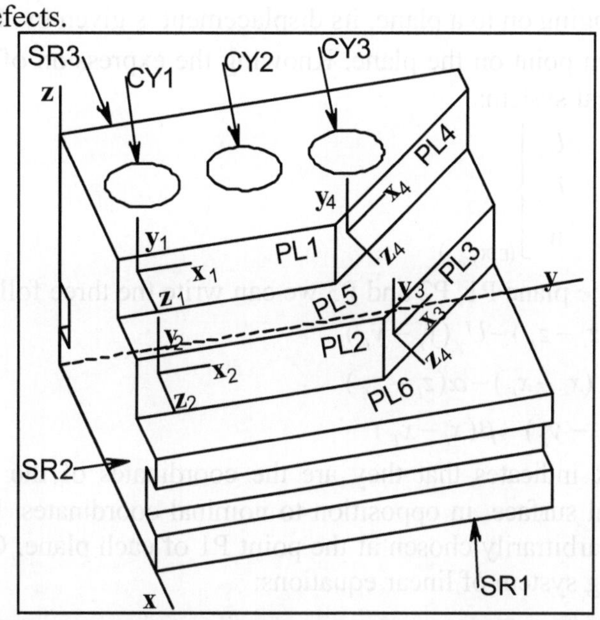

Figure 3; Measurements – local coordinate systems.

4.2. Measurements of the manufactured surfaces

The measured surfaces are: planes PL1 to PL6, cylinders CY1 to CY3. Each measured feature is automatically associated with optimized (least squares algorithm) perfect feature computed from the points measured on the surface. Taking into account the identification software, it is impossible to get the coordinates of these points. It was thus essential to carry out geometrical constructions in order to extract from measurements the features needed to calculate the deviation SDT. Thus for each plane, three points

P1, P2 and P3 belonging to the nominal plane are built in $\Re$ then projected onto the associated plane. Thus, we obtain a representation of the theoretical plane associated with real surface. For the cylinders, only the position of their axis is useful for calculations, so, we just need to build the intersection points between each axis with the higher plane and lower plane SR1.

5. CALCULATION OF THE DEVIATION TORSORS

5.1. Local coordinate systems

For each measured feature, it is necessary to define a local coordinate system in which calculations will be done. For the side milled planes (PL1, PL2, PL3 and PL4): the $\mathbf{z_i}$ axis is perpendicular to the plane, $\mathbf{x_i}$ axis follows the feed direction and $\mathbf{y_i}$ is parallel to the end mill axis. For the face machined planes and the holes, the reference axis **x**, **y** and **z** are used.

5.2. Deviation torsors of the planes

For any point belonging on to a plane, its displacement is given by: $\mathbf{D}_{\mathrm{Pi}} = \mathbf{D}_{\mathrm{E}} + \boldsymbol{\Omega} \wedge \mathbf{EPi}$, where E is a chosen point on the plane. Knowing the expression of the deviation torsor of a plane in its local system:

$$\{\mathbf{T}_{\mathrm{plane}}\}_{(\mathrm{E},\Re_{plane})} = \begin{Bmatrix} \alpha & U_x \\ \beta & U_y \\ U_\gamma & w \end{Bmatrix}_{(\mathrm{E},\Re_{plane})}$$

For each point of the plane P1, P2 and P3 we can write the three following equations:

$$\begin{cases} x_{Ri} - x_i = U_x + \beta(z_i - z_E) - U_\gamma(y_i - y_E) \\ y_{Ri} - y_i = U_y + U_\gamma(x_i - x_E) - \alpha(z_i - z_E) \\ z_{Ri} - z_i = w + \alpha(y_i - y_E) - \beta(x_i - x_E) \end{cases}$$

where the index R indicates that they are the coordinates of the points of the plane associated with real surface, in opposition to nominal coordinates. Let us consider now that the point E is arbitrarily chosen at the point P1 of each plane. One obtains for each plane, the following system of linear equations:

$$\begin{cases} z_{R1} - z_1 = w \\ z_{R2} - z_2 = w + \alpha(y_2 - y_1) - \beta(x_2 - x_1) \\ z_{R3} - z_3 = w + \alpha(y_3 - y_1) - \beta(x_3 - x_1) \end{cases}$$

So, the torsor components at point E in the local coordinate system associated to the plane are obtained with:

$$\begin{pmatrix} \alpha \\ \beta \\ w \end{pmatrix} = \begin{bmatrix} 0 & 0 & 1 \\ y_2 - y_1 & x_1 - x_2 & 1 \\ y_3 - y_1 & x_1 - x_3 & 1 \end{bmatrix}^{-1} \begin{pmatrix} z_{R1} - z_1 \\ z_{R2} - z_2 \\ z_{R3} - z_3 \end{pmatrix}$$

5.3. Deviation torsors of the holes

We are interested only in the position deviations of each cylinder defined by its axis. This last is defined by its two extreme points. Considering E a point chosen arbitrarily on the axis:

$$\{\mathrm{T}_{\text{cylinder}}\}_{(E,\Re_{cylinder})} = \begin{Bmatrix} \alpha & u \\ \beta & v \\ U_\gamma & U_w \end{Bmatrix}_{(E,\Re_{cylinder})}$$

For each point P1 and P2 ends of the axis, we have:

$$\begin{cases} x_{Ri} - x_i = u + \beta U_w - U_\gamma (y_i - y_E) \\ y_{Ri} - y_i = v + U_\gamma (x_i - x_E) - \alpha (z_i - z_E) \\ z_{Ri} - z_i = U_w + \alpha (y_i - y_E) - \beta (x_i - x_E) \end{cases}$$

All the points belong to the axis, so: $x_i = x_E$ and $y_i = y_E, \forall i$. Let us consider moreover that the point E is arbitrarily chosen at P1.

We obtain then, by considering only the equations without undetermined:

$$\begin{cases} x_{R1} - x_1 = u \\ y_{R1} - y_1 = v \\ x_{R2} = u + \beta (z_2 - z_1) \\ y_{R2} = v - \alpha (z_2 - z_1) \end{cases} \text{so:} \begin{pmatrix} \alpha \\ \beta \\ u \\ v \end{pmatrix} = \begin{bmatrix} 1 & 0 & 0 & z_2 - z_1 \\ 0 & 1 & z_1 - z_2 & 0 \\ 1 & 0 & 0 & 0 \\ 0 & 1 & 0 & 0 \end{bmatrix}^{-1} \begin{pmatrix} x_{R2} - x_2 \\ y_{R2} - y_2 \\ x_{R1} - x_1 \\ y_{R1} - y_1 \end{pmatrix}$$

6. RESULTS AND ANALYSES

The results shown on figures 4, and 5 are the components of the torsors resulting from calculations. The curves represent the results for the side milled planes (figure 4) and holes (figure 5). These data have been putted together considering the similarities between them.

Before analyzing of the values we have obtained, we will explain the significance of these various geometrical parameters when it is possible. For the side milled planes, α represents rotation around the feed direction, β the rotation around the milling spindle axis. Thus, excluding the deviations related to "adjustable" parameters (part holder orientation and position), we can do the following assumptions:

- α integrates the problems involved in the deformations of the machine-tool components (particularly the cutting tool),
- β integrates the defects of the machine-tool displacements during machining and the displacements from one workpiece to another one.

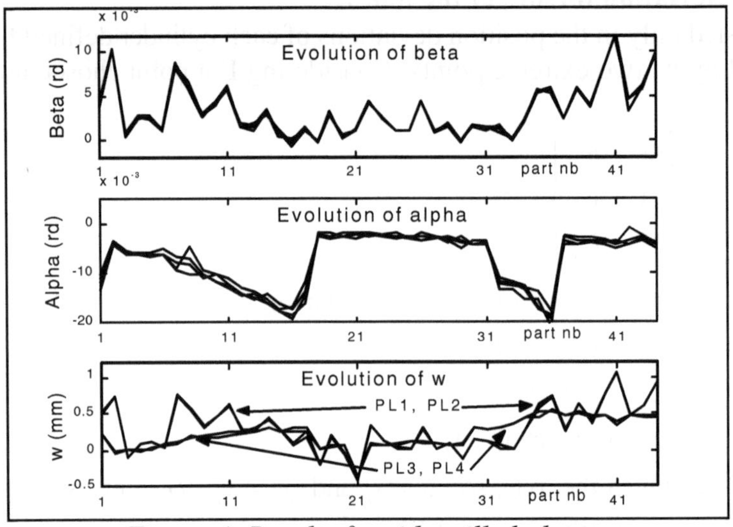

Figure 4; Results for side milled planes.

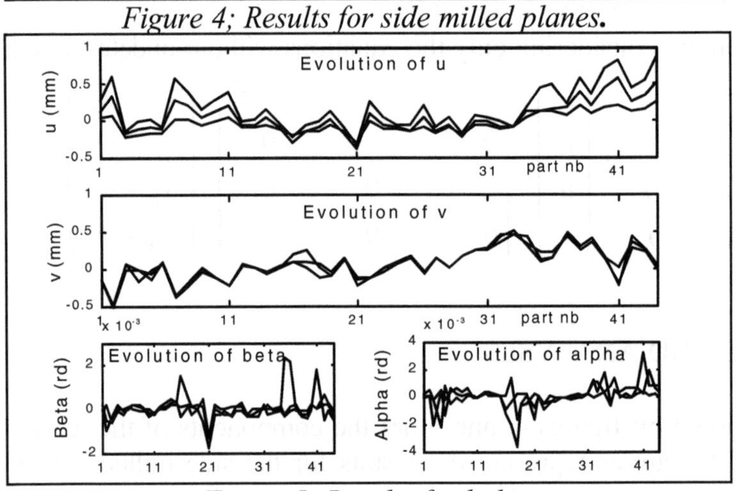

Figure 5; Results for holes.

6.1. Analyze of the side milled planes

We clearly distinguish four distinct areas for the rotation component α. These four areas correspond to the replacements of the end mill due to wear. The evolution of α follows then a slope characteristic of the tool wear. On the other hand, we see that the values of w for the side milled planes (figure 4) are problematic. Indeed the variations from one part to another can reach up to 0.7 mm on the series of the first 16 parts where there is no tool replacement. These values are a priori much too high considering the machining process we use. Let us analyze, for plan PL1 for example, what is really measured compared to the model described before. We identify $\mathbf{T}_{Pi,PL1}$, Pi being the positioning surfaces of the workpiece. However, $\mathbf{T}_{Pi,PL1} = \mathbf{T}_{Pi,P} + \mathbf{T}_{P,R} + \mathbf{T}_{R,PL1}$ so, taking into account equation 1:

$$\mathbf{T}_{Pi,PL1} = \mathbf{T}_{R,PL1} - \mathbf{T}_{R,Hi} - \mathbf{T}_{Hi,Pi}$$

If it is admitted that the part holder does not move from one part to another, $\mathbf{T}_{R,Hi}$ is the same for the whole series. So, the measurements we have done take into account the sum of the machining deviations ($\mathbf{T}_{R,PL1}$) and the gap SDT between the part holder and the workpiece ($\mathbf{T}_{Hi,Pi}$). Considering the evolution of β (figure 4), we can suppose that there is a rotation from a workpiece to another around **z** due to the gap SDT between the datum surface SR2 and the part-holder but there is always contact at the tertiary joint. This assertion is reinforced when we notice that the deviation u of the holes is closely related to the evolution of β for the side milled planes. So, the parameters β and w have been adjusted with a correction parameter calculated from the slope of the 3 holes in relation to SR2. Figure 6 shows the new expression of β and w after this correction: the four planes are now closely related.

So, we can express the standard deviation of w after correcting the slopes due to the tool wear and the intervals due to the tool changing.

	Plane 1	Plane 2	Plane 3	Plane 4
Standard deviation for w (mm)	**0.0178**	**0.0259**	**0.0260**	**0.0225**

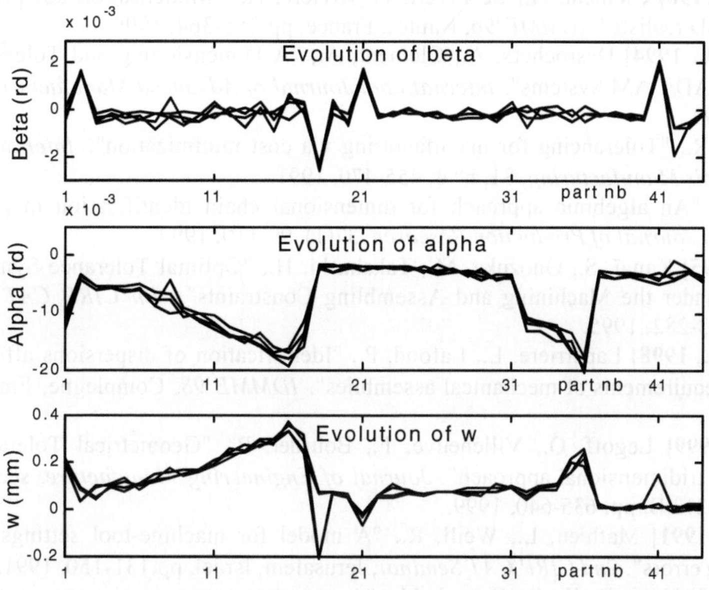

Figure 6; Results for side milled planes after correction.

7. CONCLUSION

An analysis process of a machining set-up in order to quantify the deviations due to the setting and the machining operations was proposed. This aims to validate the three-dimensional model on manufacturing tolerancing that we proposed. After a recall of the model employed, the experimental approach to measure and quantify the three-dimensional machining variations as torsors on a mass production of parts was described. The analysis of the results proves the importance of a three-dimensional consideration of the manufacturing deviations. It is shown, for example, the significant role played by the angular deviations. in the quality of the results. We also showed the difficulty of dissociating the effects due to the setting of those due to machining. The objective of works in progress is the improvement of the experimental protocol in order to more finely dissociate the identified components. The finality would be to specify a process of three-dimensional identification of machine-tool capabilities.

REFERENCES

[Ballot et al., 1995] Ballot, E., Bourdet, P., "Geometrical Behavior Laws for Computer Aided Tolerancing". *4th CIRP Seminar on Computer Aided Tolerancing*, Tokyo, Japan, pp. 143-154, 1995.

[Bourdet et al., 1996] Bourdet, P., Mathieu, L., Lartigue, C., Ballu, A., "The Concept of The Small Displacement Torsor in Metrology". In *Advanced Mathematical Tools in Metrology II*, Series on Advances in Mathematics for app. Sc., 40, pp. 110-122, 1996.

[Clément et al., 1996] Clément, A., Le Pivert, P., Rivière, A., "Modélisation des procédés d'usinage. Simulation 3D réaliste". *IDMME'96*, Nantes, France, pp.355-364, 1996.

[Desrochers et al., 1994] Desrochers, A., Clement, A., "A Dimensioning and Tolerancing Assistance Model for CAD/CAM Systems". *International Journal of Advanced Manufacturing Technology*, 9, 352-361, 1994.

[He, 1996] He, J.R., "Tolerancing for manufacturing via cost minimization". *International Journal of Machine Tools Manufacturing*, 31, n° 4, 455-470, 1991.

[Ji, 1999] Ji, P., "An algebraic approach for dimensional chain identification in process planning". *International Journal of Production Research*, 37(1), 99-110, 1999.

[Kanaï et al., 1995] Kanaï, S., Onozuka, M., Takahashi, H., "Optimal Tolerance Synthesis by Genetic Algorithm under the Machining and Assembling Constraints". *4th CIRP CAT Seminar*, Tokyo, Japan, pp.263-282, 1995.

[Laperriere et al., 1998] Laperriere, L., Lafond, P., "Identification of dispersions affecting pre-defined functional requirements of mechanical assemblies". *IDMME'98*, Compiegne, France, pp. 721-728, 1998.

[Legoff et al., 1999] Legoff, O., Villeneuve, F., Bourdet, P., "Geometrical Tolerancing in Process Planning : a tridimensional approach". *Journal of Engineering Manufacture*, section Manufacture and Design, 213 B, pp. 635-640, 1999.

[Mathieu et al., 1991] Mathieu, L., Weill, R., "A model for machine-tool settings as a function of positionning errors". *2nd CIRP CAT Seminar*, Jerusalem, Israel, pp.131-150, 1991.

[Ngoi et al., 1999] Ngoi, B. K. A., Ong, J. M., "A complete tolerance charting system in assembly". *International Journal of Production Research*, 37(11), 2477-2498, 1999.

[Teissandier et al., 1997] Teissandier, D., Couétard, Y., Gérard, A., "Three-dimensional Functional Tolerancing with Proportionned Assemblies Clearance Volume (U.P.E.L), application to setup planning". *5th CIRP CAT Seminar*, Toronto, Canada, pp.113-124, 1997.

[Villeneuve et al., 1999] Villeneuve, F., Legoff, O., Bourdet, P., "Three Dimensional geometrical tolerancing in process planning". *32nd International Seminar on Manufacturing Systems*, Leuven, Belgique, pp.469-478, 1999.

[Zhang, 1996] Zhang, G., "Simultaneous tolerancing for design and manufacturing". *International Journal of Production Research*, 34(12), 3361-3382, 1996.

Process plan validation including process deviations and machine-tool errors

Romulus BENEA, Guy CLOUTIER, Clément FORTIN
Department of Mechanical Engineering, École Polytechnique de Montréal
C.P. 6079, succ. Centre-Ville, Montréal, Québec, CANADA H3C 3A7
romulus@meca.polymtl.ca, guy.cloutier@polymtl.ca

Abstract: This paper proposes a tri-dimensional kinematics method for process plan validation confronting a design tree and a manufacturing tree. The main process deviations and machine-tool errors are taken into account. Starting from the CAD/CAM model, the elements involved in a process plan validation are emphasised considering a specific process plan and manufacturing process. A kinematical analogy is used, where the machined features are the elements of a mechanism. We propose an original enrichment process plan graph as an essential tool for the process plan validation, including the process deviations and the functional tolerances. A three stages model is built: Matrix computations, manufacturing simulations and virtual metrology. Different tolerance specifications are verified by simulation using virtual gauges. A further development in symbolical computation allowed performing a sensitivity analysis. This usefully isolates the influences of the dominant process deviations on different tolerances. The computational method is exemplified on an industrial part, and a process plan is validated considering some critical tolerances.
Keywords: Process plan, Graph, Process deviations, Tolerances, Virtual gauge.

1. INTRODUCTION.

Generating and interpreting tolerance specifications appropriately is important, but the respect of these specifications in manufacturing is essential for the success of engineering programs. Very few research works treated influences of process dispersions and machine tool errors on tolerance specifications, although a main issue in process plan validation. Bourdet questioned the validity of worst case analyses, based on the faulty assumption of statistically independent dimensions on machined parts [BOU 73], [BOU 75]. This assumption holds for assembly tolerance analysis: Dimensions of mating part manufactured on various machines could hardly be dependent. The same assumption generally does not hold within process plan analysis: Statistically *independent* surface variables yield statistically *dependent* dimensions between surfaces. It is thus wrong to forecast the variance of a resultant dimension by the sum of the variances of its constituents, as these share a covariance. Thus, worst cases (that really amount to a sum of standard deviations) must also account for this dependence. The ensuing process plan analysis method has successfully penetrated

P. Bourdet and L. Mathieu (eds.),
Geometric Product Specification and Verification: Integration of Functionality, 197-206.

some European manufacturers and Engineering schools. Rivest et al. [RIV 93] presented a kinematical approach for tolerance modelling using an assembly of 14 elementary links particularised for every tolerancing specification. Using this model, he was able to transfer tolerances as needed in some manufacturing cases. Recently, Legoff et al. [LEG 99] presented a 3D approach to analyse and to optimise manufacturing tolerances. Based on Small Displacement Torsors (SDT), the approach emanates from the works of Ballot and Bourdet [BAL 97], dealing with computational methods of geometric errors in mechanisms. The link between process planning and functional tolerances with torsor chains according to part set-ups is exposed.
This paper pursues the quest for statistically independent deviations further up the mechanical elements of the machine tool. We propose an original kinematics based model. A three dimensional six degrees of freedom (3D-6DOF) analysis is introduced to evaluate the influences of machine tool kinematics errors and process deviations on a machined part. Former works had only dealt with a uni dimensional analysis technique. In seeking for a such solution, the computing method developed is found to be applicable to design and assembly tolerance analyses also. Finally, an application of the model on an industrial part is presented, together with a sensitivity analysis.

2. MANUFACTURING TOLERANCING

In the design department, knowledge of functional dimensioning and tolerancing must ensure all functional requirements, taking into account realistic manufacturing constraints. The manufacturing department proposes the optimum machining and assembly processes. Ideally, the optimality is sought, based on the geometrical and technological data, considering available resources and weighing out the costs. In order to estimate the conformity of to be machined parts, a process plan validation based on a simulated manufacturing process should be performed. From a defined process plan, the validation incorporates machine tools kinematics and geometric errors, and process deviations. The validation is prone to the introduction of quasi-static manufacturing deviations. Dominant effects are sought, simplifying the equations stating the deviations between 'real' (simulated) features, and nominal features. The manufacturing deviations associated to a target feature must be verified against the imposed functional tolerances.

2.1. Links among part features.

A part manufacturing can be thought as a three dimensional complex multi-loop (Figure 1), where we distinguish two trees: The first, generated in design (D••), contains surfaces related by dimensions and tolerances. The second one, associated to the process plan, includes machining operations (M••) based on surfaces (S•), starting from the rough material (S_0). A more realistic process description by a process graph is planned in future works. Particular features serve as nodes for obtaining other machined surfaces (S_2, S_4 or S_i). Some pairs of surfaces are simultaneously related by superimposed D and M branches: (S_4, S_6) or (S_4, S_i). Other surfaces are solely related by one of theses types of branches: (S_1, S_3) or (S_2, S_3). Using the manufacturing tree, we obtain:

$$(S_1, S_3) \Rightarrow D_{13} \supseteq M_{13} = M_{01} \oplus M_{02} \oplus M_{23} \quad (1)$$
$$(S_1, S_6) \Rightarrow D_{16} \supseteq M_{16} = M_{01} \oplus M_{02} \oplus M_{24} \oplus M_{46} \quad (2)$$

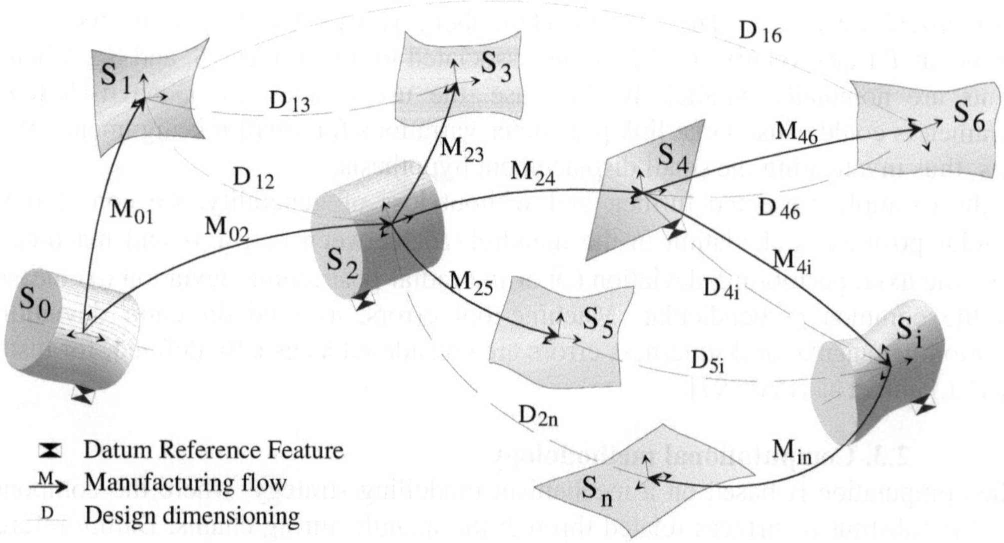

Figure 1: Kinematics chains containing part features.

The central tendency of the physical machining process can always be tuned using the available adjustments. Only the machine and process deviations (Dev), which represent the variations of dimension, position, orientation or form from the nominal value induced by the manufacturing process, are first taken into account.
The closure elements for a chain are the functional tolerances. In this case, between the surfaces S_5 and S_i for example, the tolerance is $T(D_{5i})$ and the constraint will be:

$$Dev(M_{25}) \oplus Dev(M_{24}) \oplus Dev(M_{4i}) \leq T(D_{5i}) \quad (3)$$

2.2. Process modelling

The manufacturing tolerance analysis is based on a process plan simulation including the main process deviations. We used substitute surfaces for real surfaces, associated to the nominal surfaces. Our approach allows for instance to model a nominal cylindrical surface by a substitute conical one. The roughness, tool wear, vibrations and heating phenomena, as well as the influences due to the tool setting or the chip geometry were neglected. The following main simplifying hypotheses were used: (i) the parts, tools and fixtures are considered rigid, (ii) the fixture/part clamping and positioning main deviations and the machine-tool/part manufacturing main deviations are considered, (iii) deviations are assumed to be in the micro-geometrical domain, allowing the linearisation of the small rotations, (iv) the machine tools kinematics errors (positional, straightness, angular and squarness errors) are considered.
One may note that form errors as defined by the standard [ASME 94] can be partially taken into account, and the model could be extended to the elastic domain using stiffness matrices.
The mathematical model uses SDT and transport matrices. The substitution surfaces are

provided by Minimum Geometric Reference Sets (MGRS), allowing to define their relative position. The transport matrices including process deviations and machine tool kinematics errors are used and the model allows including form errors and treating geometrical tolerances. The Denavit-Hartenberg Hayati (DHH) parameters allow to define an *'i'* frame relative to *'i-1'* frame associated to the surfaces S_i and S_{i-1} when two joints are nominally parallel. In this case, the use of classical Denavit-Hartenberg parameters could cause large link parameter variations for small misalignments of joint axis, thus invalidating the small displacement hypothesis.
In the example presented further, and without loss of generality, we considered the angular positioning deviation in the mandrel (α), between part axis and machine-tool axis, the axial positioning deviation (δ) or the radial positioning deviation (a) measured on the common perpendicular. Machine-tool errors, as feed direction deviation or positioning, angular and squarness errors are considered as usually defined, for instance by H.J.Pahk et al. [PAH 97].

2.3. Computational methodology

The computation is based on a mechanism modelling strategy, where the components are the substituted surfaces related through the manufacturing chains. Datum reference frames are associated to datum reference features, and R_{ij} are the reference frames associated to the surface *'j'* generated in the phase *'i'* of the process plan. A database for process deviations and machine-tool kinematics errors is created from the research and experimental data. Closed loops (Figure 1) composed by several manufacturing links and a design requirement closure link are used. The process plan graph including the machine-tool reference frames is essential for the process plan validation. The surfaces displacement is computed using the transport matrices, which include the process deviations and machine-tool errors. Finally, virtual gauges are built in order to verify the tolerance specifications. The computation is reduced to geometrical constraints between MGRS belonging to the related surfaces and chosen in the spirit and the interpretation of the ASME standard concerning tolerance specifications.

3. THE PROCESS PLAN GRAPH

We proposed the process plan graph, an original tool that includes the machine tools, the DRF with positioning deviations, the machining operations with process deviations, and the functional constraints. The machine tool kinematics errors are inputs, which are transported from within the machine tool to the machined surface generating process deviations associated to each surface. A simple example is presented in the Figure 2, where a part is turned in two phases, starting from the rough surfaces S_0 and S_0'. Using the design dimensioning, a design tree is built. The same surfaces $S_1 \ldots S_6$ defined in design are corresponding to the 'real' part features obtained in manufacturing, 11...23, which can be identified in a manufacturing tree. The process plan graph is the result of merging these two trees (Figure 2) by feature correspondence. The machine tool symbol (Mk) added to the Bourdet proposal allows considering machine tool kinematics with the dispersion inputs and introduces the possibility for 3D analysis using deviation torsors. Machining new features corresponds to the passing on an upper level in the graph, which is always done through the machine-tool reference frame (M_k) and includes the positioning deviations. The functional requirements (break lines) serve as closure link for the chains used to build the equations. In this example, we are studying the dimensions and tolerances on the axial direction. We denoted ε the positioning deviations on the DRF, and δ the manufacturing deviations. Thus, the deviations are computed following the graph and must respect the design constraints:

$$Dev(M_{13}) = \delta_{13} + \delta_{12} + \varepsilon_{12} + \delta_{21} \leq T(D_{13}) \quad (4)$$

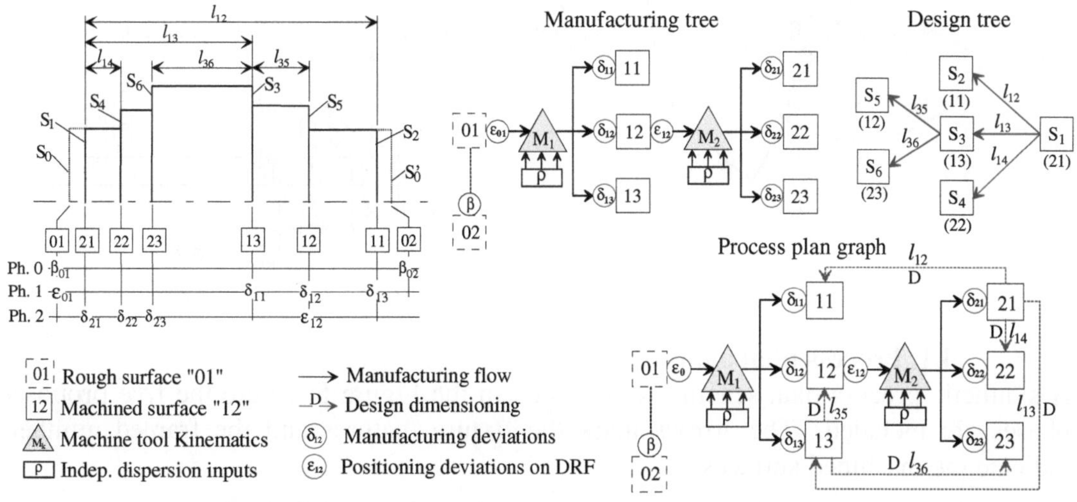

Figure 2: Process deviations in a process plan graph.

4. APPLICATION FOR AN INDUSTRIAL PART

This method was applied to validate a process plan for an industrial part and to evaluate the effects of the process deviations. Starting from a CAD model created in CATIA®, an isometric view of the part (Figure 3) and associated design draft (Figure 4) are presented, including several dimensional, orientation and positional tolerances. In order to integrate into the model the process deviations, a data base was created, including experimental data for clamping and positioning deviations extracted from [KOS 85] and machine-tool errors (positional, straightness, angular and squarness error) extracted from [PAH 97]. The modelling needs also geometrical data

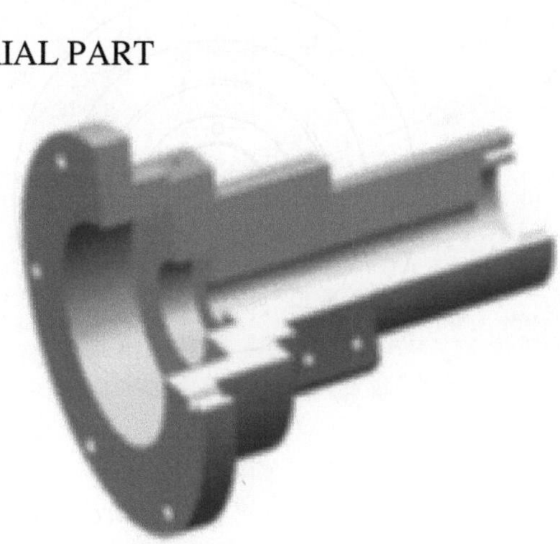

Figure 3: Isometric view.

from the CAD model (lengths and diameters) and technological data concerning the process plan (machining thickness, jaw dimensions).

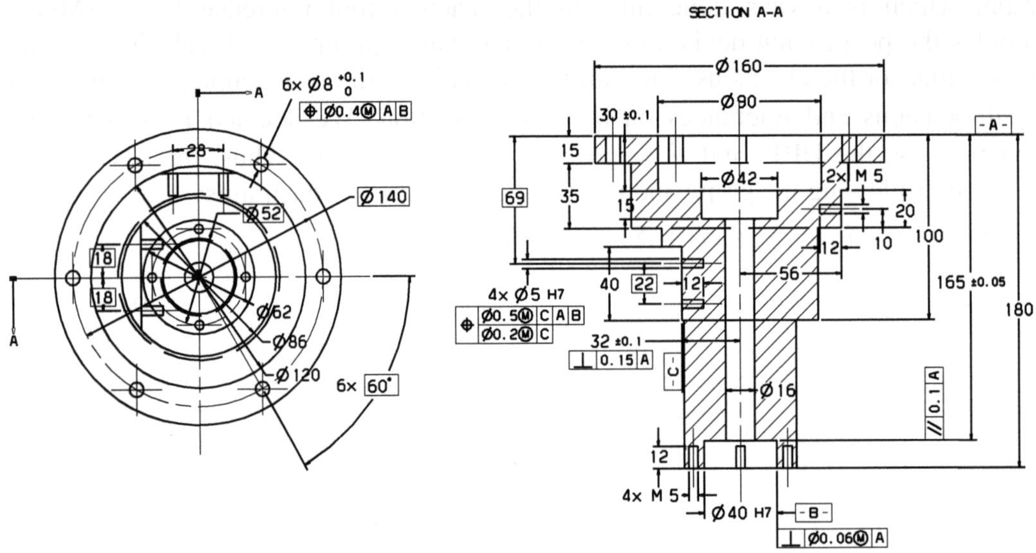

Figure 4: The part draft.

4.1. Process plan analysis.

A synthetic process plan schema is presented in the Figure 5, where the five proposed phases are included.. The arrows mark the fixture features and the framed numbers indicate the machined surfaces.

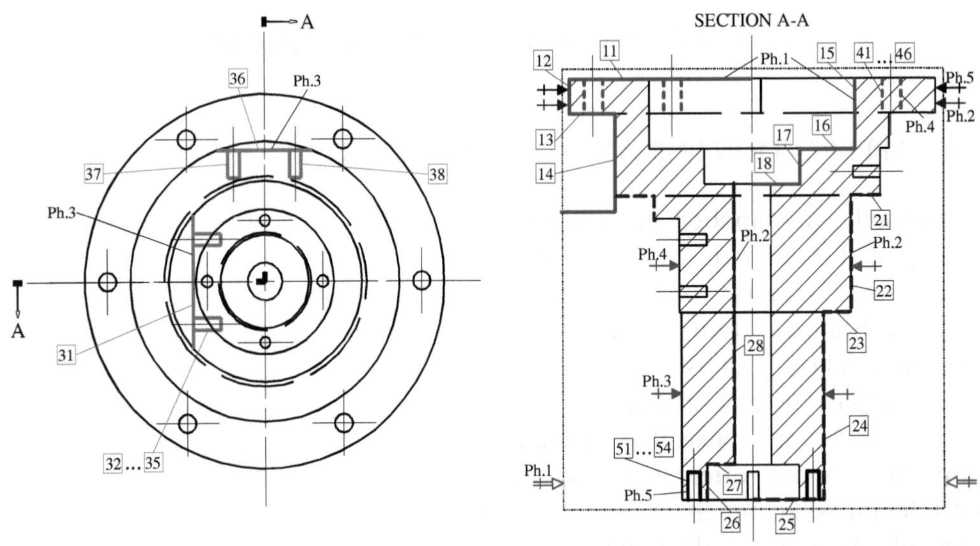

Figure 5: Process plan schema.

For the first two phases, a conventional lathe is used. The third phase is realised on a three axes horizontal boring and milling machine provided by a rotary table. The process plan is achieved with the last two phases on a boring and drilling machine. In order to evaluate the influence of the process deviations on the part geometry, a process plan graph (Figure 6) is used. We focused our attention on three types of tolerances :

- dimensioning tolerances: t1: 165±0.05 .
- orientation tolerances: t2: | // | 0.1 | A | and t3: | ⊥ | ϕ 0.06 Ⓜ | A | .
- positioning tolerances: t4: 6 x $\phi\ 8^{+0.1}_{\ 0}$ | ⊕ | ϕ 0.4 Ⓜ | A | B | and

  t5: 4 x $\phi\ 10^{+0.02}_{\ 0}$ | ⊕ | ϕ 0.5 Ⓜ | C | A | B | / | ϕ 0.2 Ⓜ | C |

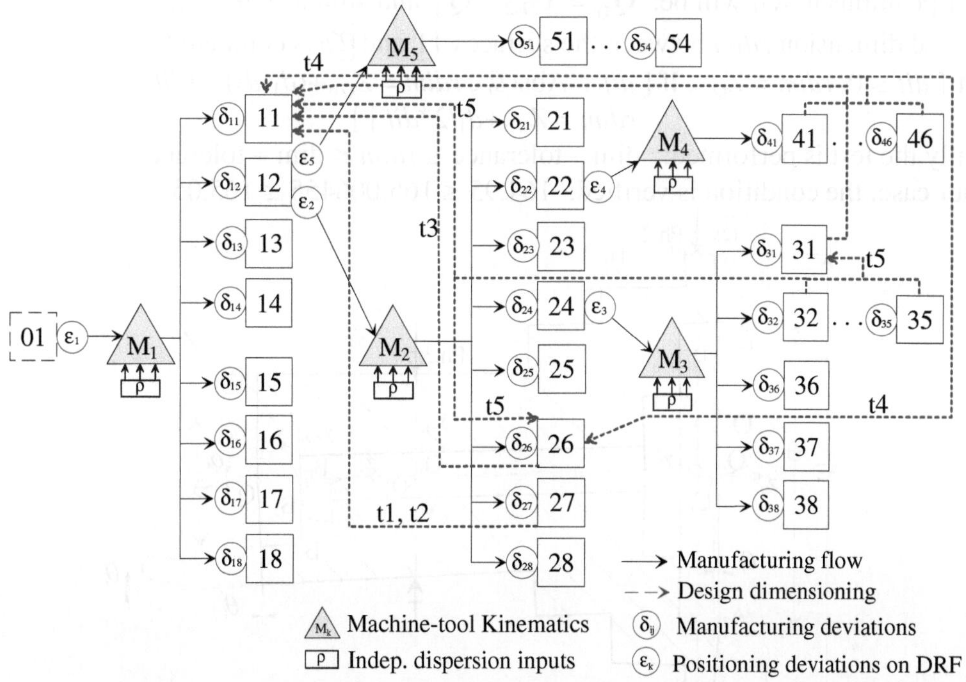

Figure 6: Process plan graph, including only critical specifications.

The dotted lines are the closure elements for the loops used in tolerance analysis. Several reference frames have been chosen corresponding to different sets-up. So, R_0 corresponds to the first set-up and $\mathbf{z_0}$ coincides with the rough part axis, R_1 to the second set-up and $\mathbf{z_1}$ is the lathe axis, and so on. The geometry of the features machined in the phase *'i'* is associated to the frame R_i, 'frozen' afterwards and serving in computation. The geometry description in the R_i frames considers the machine tool kinematics errors. The relative position of the frames takes into account the process deviations and the machine tool errors.

4.2. Development of closure equations.

A tolerance analysis is carried out starting from the loops identified in the process plan graph. In the following, some details are presented for the t1 tolerance.

t1: 165±0.05 This dimension refers to the relative position of the surfaces [11] and [27], and its tolerance analysis (Figure 7) requires a virtual gage. An adjustment movement is simulated around the axis $\mathbf{ax_a}$ with an adjustment angle (an_a): $\mathbf{ax_a} = \mathbf{unitv}(\mathbf{R_{21,1}.k_1} \times \mathbf{k_{21}})$, where **unitv** signify the unit vector, and **i, j, k** are the unit vectors of **x**, **y** and **z** axis.

$an_a = \text{Min}\,[\text{Abs}(\alpha_1), \text{Abs}(\theta_t)]$ and we denote $dif = \text{Abs}(\alpha_1) - \text{Abs}(\theta_t)$

$\mathbf{R_{21,22}}$ define this movement, a rotation around $\mathbf{ax_a}$ of an_a angle, which transports R_{21} into R_{22} . In the R_1 frame : $Z_{Q1} = Z_{C1} + \theta_t\,(d_1\text{-}d_7)/2$ and Z_{C1} is computed from geometrical and technological data.

$\mathbf{Q_1^{21}} = \mathbf{T_{21,1}} \,.\, [0, 0, Z_{Q1}, 1]^T$ and similarly for $\mathbf{C_1^{21}}$; $\mathbf{Q_1^{21}}$ is a vector that holds the coordinates of Q_1 in the R_{21} reference frame. $\mathbf{R_{21,22}}$ transports Q_1 into Q_{1r} and C_1 into C_{1r}.
Their positions in R_{21} will be: $\mathbf{Q_{1r}^{21}} = \mathbf{T_{21,22}} \,.\, \mathbf{Q_1^{21}}$ and similarly for $\mathbf{C_{1r}^{21}}$.
The real dimension *rdim* between the surfaces 11 and 27 is computed:

If [$dif \leq 0$, $rdim = Z_{Q1r}$, If [$dif \leq \text{Abs}(\theta_t)$, $rdim = Z_{Q1r} + (d_1\text{-}d_7)/2 * dif$,
$rdim = Z_{C1r} + d_1/2 * dif$]]

Finally the test is performed: dim – tolerance ≤ *rdim* ≤ dim + tolerance
In our case, the condition is verified: 164.95 ≤ 165.006455 ≤ 165.05

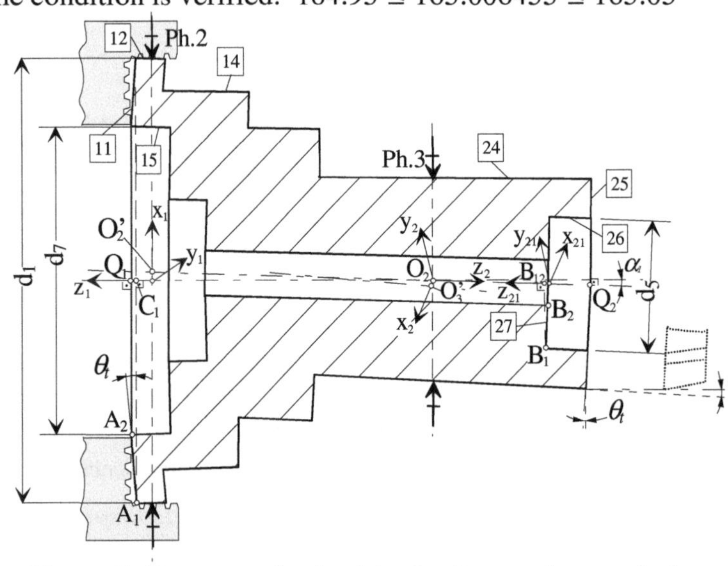

Figure 7: Features obtained in the first and second phase.

The orientation tolerances are verified in conformity with ANSI interpretation ([ASME 94], section 6.6). After computation, we obtained :
t2: Parallelism deviation, | // | 0.1 | A | : 0.016 ≤ 0.1 mm

t3: Perpendicularity deviation, | ⊥ | ϕ 0.06Ⓜ | A | : 0.005999 ≤ 0.06 mm

The position tolerance analysis considered the angular and squarness errors for the boring and drilling machine, as well as a positional error for the tool. The virtual gages required multiparametric optimisation procedures. The definition of references, the kinematics of gages and the degrees of freedom were established according to the ANSI interpretations [ASME 94], section 5.3.2 for (t4) and section 5.4.1 for (t5). The results are synthesised below, together with the corresponding graphics (Figures 8 and 9).

t4: Positional tolerance: 6 x $\phi\ 8^{+0.1}_{0}$ | ⊕ | ϕ 0.4Ⓜ | A | B | : 0.137598 ≤ 0.2 , which means that the tolerance is verified.

t5: Positional tolerance: 4 x $\phi\ 10^{+0.02}_{0}$

⊕	ϕ 0.5Ⓜ	C	A	B
	ϕ 0.2Ⓜ	C		

The maximum distances obtained were 0.112216 < 0.25 for PLTZF and 0.046769 < 0.1 for FRTZF which means that the composite positional tolerance is verified.

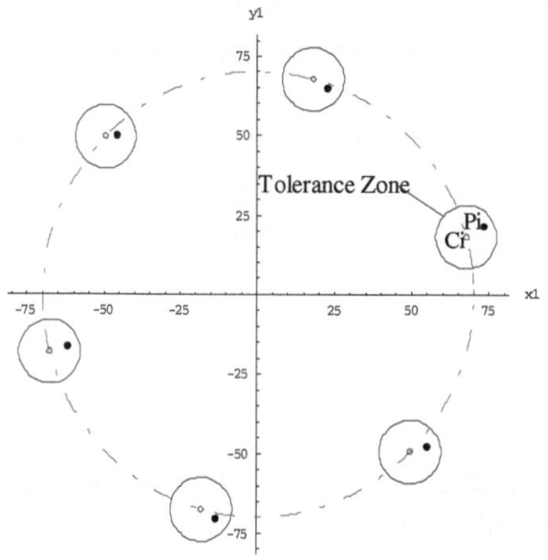

Figure 8: Positional tolerance (t4).

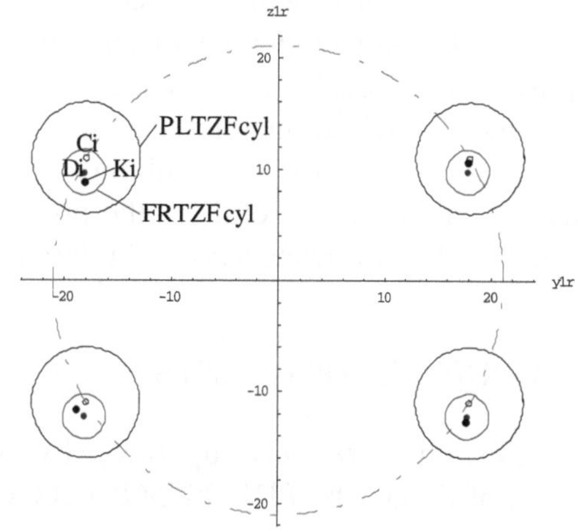

Figure 9: Composite positional tolerance (t5).

The software development was performed in Mathematica®, an environment providing symbolical computation and powerful graphical features.

4.3. Sensitivity analysis.

An interesting challenge is finding the input deviations that have a dominant influence on different tolerance specifications, allowing to the manufacturing engineer to react efficiently and improve the result on that specification. The symbolical computation tools of Mathematica® allowed us to use the symbolical input parameters in the tolerance analysis. A set of specific simplification rules for symbolic computation was build in order to eliminate the small combined influence of the input parameters through the transport matrices. Thus, a sensitivity analysis was performed, where we filtered the influences of the input deviations, being able to obtain an expression of the variation for a considered target tolerance in term of the main input process deviations.

The result for the dimensional tolerance specification (t1) is presented, where after a first filtering level we obtained: Dev(d) = f(a_1, α_1, α_1^2, δ_2, $\alpha_1.\delta_2$), where a_1 is the positional radial deviation, δ_2 is the axial positional deviation and α_1, α_2 the positional angular deviation for the phases 1 and 2 respectively. Eliminating the non-significant terms, we obtained finally :

$$\text{Dev(d)} = c_0 + c_1\alpha_1^2 - \delta_2 \quad (5)$$

In the Figure 10, we present graphically the contribution of these two parameters on the dimensional deviation. One can observe that δ_2 has the main influence.

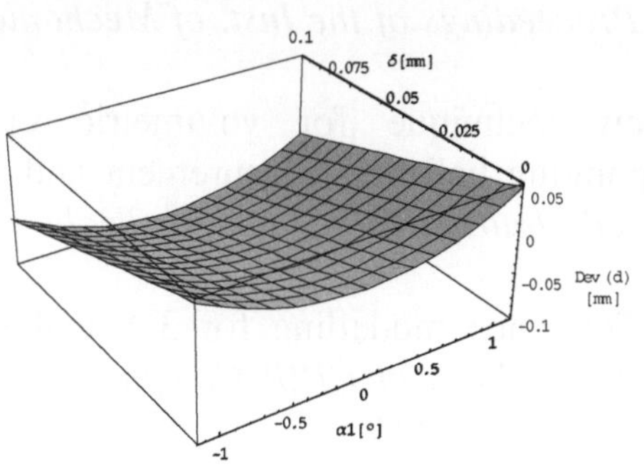

Figure 10: Dimens. tolerance sensitivity.

5. CONCLUSION.

The process plan validation is based on a manufacturing simulation including process deviations and machine tool errors. A process plan graph obtained by merging a design tree and a manufacturing tree is proposed to build the computational loops for tolerance analysis. A similarity with the mechanisms was observed, where the machined features are links of a kinematics chain and functional tolerances are equations closure. The proposed model deals with substitute surfaces and uses mathematical tools borrowed from robotics. Different tolerance specifications can be verified by simulation using virtual gages. The use of symbolical computation allowed performing a sensitivity analysis. This is very useful for the process planner, allowing to identify the dominant parameters that influence a critical tolerance.

ACKNOWLEDGEMENTS

This research work was supported by Natural Sciences and Engineering Research Council of Canada by PDF-207568 and OGP-0138478 grants.

REFERENCES

[ASME 94] American Society of Mechanical Engineers, ASME Y14.5M-1994, Dimensioning and Tolerancing, An ASME National Standard, 1994.

[BAL 97] Ballot E., Bourdet P., "A computational method for the consequences of geometrical errors in mechanisms; In: *Proceedings of CIRP Computer Aided Tolerancing 5th Seminar*, pp. 137-148, Toronto, Canada, 1997.

[BOU 73] Bourdet P., "Chaîne de cotes de fabrication - Première partie : Modèle", *L'ingénieur et le technicien de l'enseignement technique*, pp. 23-45, 1973.

[BOU 75] Bourdet P., "Chaîne de cotes de fabrication - Deuxième partie : Mode opératoire", *L'ingénieur et le technicien de l'enseignement technique*, pp. 15-23, 1975.

[KOS 85] Kosilova A., Mesheriakov R. et al., "The technologist handbook", Mashinostroenie Publishing House, Moscow, 4th edition, Vol. 1, Chapter 1: Machining precision for mechanical parts, pp. 6-88, 1985.

[LEG 99] Legoff O., Villeneuve F., Bourdet P., "Geometrical tolerancing in process planning: A tridimensional approach"; In: *Proceedings of the Inst. of Mechanical Engineers Part B*, **233**, pp. 635-640, 1999.

[PAH 97] Pahk H., Kim Y., Moon J., "A new technique for volumetric error assessement of CNC machine-tools incorporating ball bar measurement and 3D volumetric error model"; In: *International Journal of Machine Tools and Manufacture*, **37**, pp. 1583-1596, 1997.

[RIV 93] Rivest L., Fortin C., Desrochers A., "Tolerance modelling for 3D analysis: Presenting a kinematic formulation"; In: *Proceedings of CIRP Computer Aided Tolerancing 3rd Seminar*, Cachan, France, pp. 51-73, 1993.

Framework proposal for a modular approach to tolerancing

Prof. Alain Desrochers*, Prof. Luc Laperrière[1]

**Department of Mechanical Engineering*
University of Sherbrooke
Sherbrooke, J1K 2R1, Canada
alain.desrochers@gme.usherb.ca

Abstract: In current engineering practices, tolerancing activities are most often being done at the detail design phase, once drawings have been generated. The effects of tolerances are then propagated to other views of the product, such as process planning or quality control, where they will dictate choices of means and methods. Such an approach goes against current trend in product design processes and concurrent engineering.

In this perspective, the proposed framework aims at taking these dimensions into account through the development of a CAD tolerancing structure that will allow the gradual introduction of tolerancing information along the design process. The structure will also incorporate tools allowing the assessment and conciliation of tolerances types and values with respect to the various views of the product.

More specifically, the tolerancing structure is to take the form of a "Design Process Tolerancing Array" where the abscissa or columns will depict the stages of the design process in a temporal perspective and the ordinate or lines of the table will represent the various views of the products from an application point of view. Each element of the table or array will therefore contain information pertaining to a given view of the product at a particular step of the design process. Tolerancing requirements will characterize the elements of the table and proper mathematical representations (deterministic, statistic, screw, etc) will also be associated to them.

In addition to the proposed structure, tools or "bridges" are also to be provided to allow the transfer of information across elements of the table, therefore allowing the conversion and exchange of heterogeneous tolerancing information. These tools are to be further integrated into a conciliation procedure which is to formalize the concurrent engineering approach in the context of tolerancing. The paper will outline the various requirements for these tools without offering details on their mathematical formulations.

The proposed structure is generic which means that it can (and often has) to be adapted to the particular context of a given industry or type of product.
Keywords: tolerances, CAD, design process, concurrent engineering, integration

[1] Department of Mechanical Engineering, Université du Québec à Trois-Rivières, C.P. 500, Trois-Rivières, G9A 5H7, Canada

P. Bourdet and L. Mathieu (eds.),
Geometric Product Specification and Verification: Integration of Functionality, 207-216.

1. INTRODUCTION

In the Computer Aided Tolerancing research community, efforts have so far been devoted mostly to the modelling and tools for tolerancing. Indeed, little efforts have been dedicated in addressing *what* tolerancing information should be specified in *what context* and *when*.

This paper is concerned with this aspect of tolerancing and proposes a new approach for the assessment of this particular issue. Globally, the methodology implies the following steps :

1- splitting the tolerancing process into phases (section 3.1.);
2- associating these phases to corresponding phases of the design process (section 3.2.);
3- extracting specific tolerancing information pertaining to other views of the product (manufacturing, metrology, ...) in a concurrent engineering perspective (section 4.1.);
4- establishing bridges between segments of tolerancing information across views of the product (section 4.2.);
5- identifying the "drivers" for tolerance specification (section 5.1.);
6- extracting the proper tolerancing process in relation to a given product (section 5.2.).

As can be seen, each step will be presented in greater details in the sections outlined in parenthesis.

2. LITTERATURE OVERVIEW

From what has been investigated in the field, it appears that there is very little pertaining to the subject at hand. However, regarding tolerancing, there are important issues that would need to be looked upon, in terms of design processes and concurrent engineering. In particular, we feel that the various approaches for design such as Design For Assembly (DFA), Design For Manufacturing (DFM) and more generally, Design For X (DFX), where X represents a specific view of the product, would require careful examination.

There exist many different DFA methods. Certainly one of the most famous is that of Boothroyd and Dewhurst [Boothroyd et al., 1983]. The method addresses the problems of determining appropriate assembly methods, reducing the number of parts to be assembled and ensuring that the remaining parts are easy to assemble. Once the assembly method has been carefully determined, each product part goes through a very detailed assessment of how easy it is to handle and how easy it is to insert, using information such as its symmetry or its global envelope's nominal dimensions. This information is then systematically processed using tables from which a design efficiency score results.

Using this tabular approach, problematic parts can be easily identified (table entries with poor values) and proper design modifications or redesign can therefore be efficiently guided. The redesigned product can further be processed through the tables again and the new design efficiency score can be compared with that of the previous design, giving a feel of its overall improvement.

It is interesting to note that the DFA method makes use of tolerancing information only implicitly rather than explicitly. For example, in assessing a part's ease of insertion, the designer will have to select among tabulated situations such as "easy to align or position during assembly with resistance to insertion". Part's tolerancing is clearly underlying such a statement but is not explored in more details. Other DFA methods include Hitachi's Assemblability Evaluation Method or AEM [Miyakawa et al., 1986] and the Lucas DFA Evaluation Method [Redford et al., 1994].

Tolerances are not much more explicitly taken into account in DFM methods, i.e. there is no global understanding of how tolerances should be assigned for a particular product. Most DFM methods on the market simply aim at generating cost estimates for parts and tooling at the conceptual stages of design, sometimes even before tolerances have been selected.

Looking again at Boothroyd and Dewhurst DFM tool [Boothroyd et al., 1994], the software isolates the major cost drivers associated with a wide range of processes for part manufacture and finishing (design for machining, design for bending, etc...). Clearly, tolerancing should be considered as one of the major such cost driver, i.e. different tolerance choices may drastically affect manufacturing or tooling costs. But even so, the functional motivation behind proper tolerancing choices seems totally absent in these tools. Other DFX methods on the market, less important in the context of tolerancing, include Design For Services (DFS) and Design For Environment (DFE).

The discussion above exemplifies what could be termed the "tolerancing syndrome": tolerances are known to play a fundamental role in almost every aspect of product design, manufacturing, assembly and inspection, but the lack of formalisation behind its proper assessment and the lack of understanding of its sometimes subtle interactions in different engineering activities are such that tolerancing is never as predominant or explicit in these tools as it should really be. Thus each tool uses its own view of tolerancing without much concern on possible other interpretations for other domains. This reinforces the thesis of this paper which stipulates that a methodology is required that would allow the formulation of a tolerancing process in relation to a given design context or approach.

The "tolerancing syndrome" can also be verified by looking at state of the art softwares related to different aspects of design: the last decade has seen the emergence of very powerful 3-D modellers with texture rendering for aesthetics analysis; the same level of technological improvement is achieved using 3-D finite element modellers and analysers for structural analysis. However, very little has been done for tolerance analysis and synthesis, and CAD modellers today still have no mathematically rigorous representation of nominal dimension's variations.

To this end, Computer Aided Tolerancing (CAT) has itself become a research field of its own and today researches in the field are numerous. Still, CAT research tends to be very focussed on particular tolerancing aspects. For example, many researches deal with the mathematical foundations underlying proper tolerancing models, such as:

linearized models [Chase et al., 1995], screw and torsor models [Desrochers, 1999] [Adams et al., 1999], Jacobian models [Lafond et al., 1999], CAD with Design Of Experiments (DOE) models [Bisgaard et al., 2000], and so on.

Of course the mathematical tool for representing tolerances in future CAT systems is of prime importance. But there is certainly a need for developing some means of modularization and transformation among the models to take into account the various "flavours" of tolerancing in various engineering activities.

3. TOLERANCING AND THE DESIGN PROCESS

This general topic in fact covers the first two step of the approach outlined in the introduction. Section 3.1. presents a detailed account on how the tolerancing process can be split into distinct steps while section 3.2. exposes the classic decomposition of the design process itself.

3.1. The tolerancing stream

Globally, tolerancing should be incrementally specified, starting from comprehensive requirements at the assembly level to the definition of detailed GD&T specifications at the feature level.

The tolerancing stream, as we have decided to call it, describes a general procedure aiming at gradually but systematically translating general requirements into standard tolerance specifications.

1 - *At the assembly level*
  - a) Specification of general dimensions on parts and assemblies;
  - b) Specification of functional fits and clearances on assemblies;

2 - *At the part level*
  - a) Adjustment of dimensions on individual parts to meet fits and clearances;
  - b) Distribution of fits and clearances into tolerance zones on individual parts;

3- *At the feature level*
  - a) Decomposition of tolerance zones into GD&T specifications according to additional functional requirements;
  - b) Choice of acceptable values for the GD&T specifications selected.

3.2. The design process

Much has been said and written about the design process. In the literature however, it is generally recognised that there are four phases to this process: product (or project) definition, study of operating principles (or conceptual design), preliminary design, and finally detailed design (or drafting). In turn, each phase is characterised by a very large number of data processed using a large number of different activities. In this paper, in each phase we will focus only on the data and activities pertaining to the above mentioned tolerancing stream in section 3.1.

Naturally, there should be a match between the steps outlined in the tolerancing stream and the various phases of the design process. This is presented in the next sub section.

3.3. Integration of tolerancing and design process

Obviously, the chronology of both processes shall be respected. The issue here is more to establish a proper correspondence between the steps of these processes.

Table 1 presents such a correspondence:

Design phases	**Project definition**	**Conceptual Design**	**Preliminary Design**	**Detailed Design**
Tolerance requirements	Cost target; Quality level; Safety; Function; Aesthetics.	*Assembly level*: Dimensions; Functional fit & clearance.	*Part level*: Dimensions; Tolerance zones	*Feature level*: GD&T spec. with values.

Table 1 : Incremental tolerancing in the design process

4. TOLERANCING IN A CONCURRENT ENGINEERING CONTEXT

The next two steps of the global approach are presented in the following sub sections. The goal here is to put tolerancing in a concurrent engineering context, or, in other words, to take into account the various views of the product throughout its life-cycle.

4.1. Context dependent tolerancing information

Tolerances are being specified as part of the design process, as seen previously, but these have strong implications to other views of the product (manufacturing, metrology).

Example of tolerances seen from a manufacturing perspective includes machining tolerances (deterministic) according to a given process plan. In this instance, it is worth noting that these are usually different from those of design. Also, statistical tolerancing, with the specification of C_p and C_{pk} on the drawing, directly refer to process capabilities and hence provide a distinct framework for tolerancing.

In the field of metrology, tolerances are rather converted into corresponding tolerance zones in which the measurement point should be constrained.

Table 2 presents an extended view of tolerancing which includes constraints arising from manufacturing. Metrology and other views of the product should also be taken into account to built a more comprehensive table.

Phases ⇒ *Views* ⇓	**Project definition**	**Conceptual Design**	**Preliminary Design**	**Detailed Design**
Design	Cost target; Quality level; Safety; Function; Aesthetics.	*Assembly level*: Dimensions; Functional fit & clearance.	*Part level*: Dimensions; Tolerance zones	*Feature level*: GD&T spec. with values.
Manufacturing	Type of process (moulding, machining,..).	Choice of machine; Machine capabilities; Statistical or deterministic.	Mach. set-up; Tolerancing chain or Statistical evaluation.	Machining tolerances; Validation of process plan

Table 2 : Tolerancing in a multi-view perspective

1.2. Requirements for conversion mechanisms

Concurrent engineering can be a reality only if algorithms are available to convert tolerancing information in terms of the various views of the product, as seen in the previous sub section. Some of these conversion algorithm are well known and widely used while others still need to be investigated.

The most obvious examples of such algorithms are those pertaining to manufacturing. For instance, tolerance transfer are used to convert design tolerances into machining tolerances according to given machining settings and process plan. Tolerance transfers techniques have been traditionally used for solving unidirectional problems but can now be applied to more complex three dimensional cases.

However, conversion of statistical tolerancing information such as those pertaining to the machine capabilities can hardly be converted into deterministic, design tolerances. This may signal the need for developing a unified tolerancing model that could account for such situations.

In the field of metrology, conversion mechanisms take the form of fitting algorithms that allow point placement inside tolerance zones. Various optimisation techniques are available to achieve such goal and include root mean square, mini-max, largest and smallest enclosed surface, etc. The choice of a proper algorithm in a given context or for a particular GD&T specification is still a topic for debate.

Table 3 graphically presents an example of such tools or algorithms. On an horizontal axis, these tools support the evolution in time of tolerancing data throughout the phases of the design and manufacturing processes. Vertically, the proposed tools help support concurrent engineering by providing the required two-directional conversions between various form of tolerancing information.

Phases ⇒ Views ⇓	Project definition	Tools	Conceptual Design	Tools	Preliminary Design	Tools	Detailed Design
Design	Cost target; Quality level; Safety; Function; Aesthetics.	⇨ *Handbooks: DFA*	*Assembly level*: Dimensions; Functional fit & clearance.	⇨ *Tolerance synthesis*	*Part level*: Dimensions; Tolerance zones.	⇨ *Experience; Expertise.*	*Feature level*: GD & T spec. with values.
Tools	⇧ ⇩ *Experience; Expertise*		⇧ ⇩ *DFM*		⇧ ⇩ *Tol. transfert Tol. charting*		⇧ ⇩ *Conversion deterministic /statistic**
Manufac-turing	Type of process (moulding, machining,.)	⇨ *Production engineering*	Choice of machine; Machine capabilities; Statistical or deterministic	⇨ *Process planning*	Mach. set-up; Tolerancing chain or Statistical evaluation.	⇨ *Experience; Expertise.*	Machining tolerances; Validation of process plan

* applicable only if statistical tolerancing (with C_p and C_{pk}) has been used for manufacturing

Table 3 : Tolerancing in a multi-view perspective

5. CREATION OF DEDICATED TOLERANCING PROCESS

The previous sections have presented tolerancing in view of the design process and in a concurrent engineering perspective. However, these views are very general and need to be adapted according to the context of a specific product. Indeed, the tolerancing process itself is being conditioned by the type of product being dealt with and the corresponding manufacturing process.

This can be best illustrated by a simple, yet convincing comparison between the aerospace and the consumer good industries. In the aerospace industry, tolerances are most often dictated by design and are chosen to meet functional requirements. On the other hand, in the consumer good industry, tolerances, and more globally design, may sometimes be centred on the capabilities of a particular type of manufacturing equipment such as an injection moulding machine for instance.

Of course, more critical cases will often fall in between these two extreme examples. Consequently, developing dedicated tolerancing strategies seems to be an interesting, yet promising way of guarantying a final appropriate tolerancing for a given type of product or industry.

The development of such tolerancing processes is developed in these last two sub sections.

5.1. Identification of "key characteristics" for tolerancing

These are closely related and follow from the product initial requirements in terms of function, safety, cost, aesthetics, etc. In many ways, tolerances translate these requirements into geometrical constraints which, in turn, may dictate manufacturing and assembly processes. The analysis of customer requirements will therefore determine the "drivers" in terms of tolerancing at each stages of the design process. These "drivers" or tolerancing features may originate from any views of the product such as design, manufacturing or else and embrace the DFX (Design For X) philosophy.

5.2. Extraction of dedicated tolerancing process

The sequential fulfilment of these key characteristics constraints will form tolerancing processes adapted to a given product context. Such processes will remain valid for a specific family of parts or type of products and more generally for a particular industry.

Specifically, the tolerancing process will guide the designer towards the appropriate type of tolerances at the right moment in the design process. Therefore, deterministic, statistic, manufacturing or design tolerance information may be selected in relation to the customer requirement at each step of the design process.

The tolerance specification process will consequently insure incremental introduction of tolerances in CAD systems according to a set of initial objectives or customer requirements.

5.3. Application example

The concepts presented in the previous sections are best illustrated on a typical application. For the sake of the demonstration, the case of gear pump assembly for a car power steering has been conceptually selected.

Phases ⇒ ***Views*** ⇓	**Project definition**	*Tools*	**Conceptual Design**	*Tools*	**Preliminary Design**	*Tools*	**Detailed Design**
Design	50 $; 5 % leakage in operation; 1000 hours life.	⇨ *Handbooks; DFA*	*Assembly level*: Clearance to achieve leak spec..	⇨ *Tolerance synthesis*	*Part level*: Dimensions & tolerance zones on gears and shafts	⇨ *Handbooks and previous design.*	*Feature level:* GD & T spec. on gears & shafts.
Tools	⇩ *Handbooks (price and leakage)*		⇧ ⇩ *DFM*		⇧ *Tol. charting*		⇧ ⇩ *Conversion deterministic /statistic*
Manufac-turing	Casting of body; Machining of gears, shafts & cavity	⇨ *Production engineering*	Milling machine; Machine capabilities C_p and C_{pk}	⇨ *Process planning*	Mach. set-up; Evaluation of % of rejects	⇨ *Experience; Expertise.*	Machining tolerances for gears & shafts.

Table 4 : Tolerancing process for a gear pump assembly

6. CONCLUSION

The general objective of the project is to propose, within CAD systems, a framework for a tolerancing data architecture in a design process perspective. Such structure will establish links between GD&T specifications and customer or functional requirements.

Also of importance is the fact that the proposed approach introduces the notion of progressive or incremental specification of tolerances in CAD systems. Tolerancing therefore incorporates a temporal dimension that has not explicitly been taken into account previously. More importantly, such an approach will allow "tracability" of information from the final drawing with GD&T specifications back or upstream to the original *designer intent* and its initial formulation as customer requirements constraints.

Currently, the proposed work is still conceptual and awaits implementation in a CAD system, along with a validation phase in various industrial contexts. However, the concepts, as they were presented, are sound and translate into the tolerancing domain, a trend that has been emerging and growing in many areas of design.

REFERENCES

[Adams et al., 1999] Adams J. D. and Whitney D. "Application of Screw Theory to Motion Analysis of Assemblies of Rigid Parts", *Proc. of IEEE Int. Symp. on Assembly and Task Planning*, Porto, Portugal, pp. 75-80, 1999.

[Bisgaard et al., 2000] Bisgaard S., Graves S. and Shin G. "Tolerancing Mechanical Assemblies With CAD and DOE", *Journal of Quality Technology*, vol. 32, no 3, pp. 231-240, July 2000.

[Boothroyd et al., 1983] Boothroyd, G. and Dewhurst P.: Design for Assembly Handbook, Univ. of Mass., Amherst, MA, 1983.

[Boothroyd et al., 1994] Boothroyd, G., Dewhurst, P., and Knight, W.: Product Design for Manufacturing and Assembly, *Marcel Dekker inc.*, 1994, ISBN: 0-8247-9176-2.

[Chase et al., 1995] Chase K.W., Gao J. and Magleby S.P. "General 2-D Tolerance Analysis of Mechanical Assemblies With Small Kinematic Adjustments", *Journal of Design and Manufacturing*, vol.5, pp. 263-274, 1995.

[Desrochers, 1999] Desrochers A., "Modelling Three-dimensional Tolerance Zones Using Screw Parameters", *CD-ROM Proc. of 25th ASME Design Automation Conference,* Las Vegas, USA, September 1999.

[Lafond et al., 1999] Lafond P. and Laperrière L. "Modeling Tolerances and Dispersions of Mechanical Assemblies Using Virtual Joints", *CD-ROM Proc. of 25th ASME Design Automation Conference,* Las Vegas, USA, September 1999.

[Miyakawa et al., 1986] Miyakawa, S. and Ohashi, T.: The Hitachi Assemblability Method, *Proc. of 1st Int. Conf. on Product Design For Assembly*, Newport, RI, 1986.

[Redford et al., 1994] Redford, A. and Chal, J.: Design For Assembly Principles and Practices, *McGraw Hill*, 1994, ISBN 0-07-707838-1.

Method for Geometric Variation Management from Key Characteristics to Specification

Benoit MARGUET
Research engineer
Centre Commun de Recherche EADS, Suresnes, France
benoit.marguet@eads.net
Luc MATHIEU
Associate professor
Laboratoire Universitaire de Recherche en Production Automatisée (LURPA), ENS de Cachan, Cachan - France,
mathieu@lurpa.ens-cachan.fr

Abstract: For complex product like aircraft or car body, producibility improvement has to go through a better product assemblability. In order to reach this goal, effects of manufacturing variations have to be reduced. We propose to describe in this paper a pragmatic method analysing the impact of geometrical variations on product Key Characteristics (KC). The main idea of this method is that the flow of geometrical variations belongs to the assembly process. To identify the most promising assembly sequence allows then to minimise impact of geometrical variations. Based on this principle, the first step of the method is to identify where are the Key Characteristics for the product. This identification uses a top-down process going from functional product requirements to geometrical characteristics. Relative to this KC identification, our method performs a qualitative product analysis in order to eliminate the worst assembly sequences. This analysis is based on oriented graphs. At last a quantitative analysis allows to select the most promising assembly sequence. This integrated method has been performed with success on aircraft assemblies and has improved aircraft producibility.
Keywords: Management of tolerances, Product Function, Key Characteristics, Specifications.

1. INTRODUCTION

1.1. Producibility and assemblability

Car or aircraft industries are characterized by large and complex organizational structures and by an extremely aggressive business environment where customers are asking for a better product at a lower price. Product producibility as "a measure of the

P. Bourdet and L. Mathieu (eds.),
Geometric Product Specification and Verification: Integration of Functionality, 217-226.

relative ease of manufacturing a product" is a philosophy of designing a product so that it can be produced in an extremely efficient manner with the highest levels of quality. The influence on design concept to define a product easier to make is a vital relationship between manufacturing constraints and design rules in order to realize a complex product at a low cost.

As a complex product is made up from a high number of parts (a typical civil aircraft contains over 100,000 individual engineered parts) organized by various product levels, assemblability as a "measure of the relative ease of product assembly", plays a prominent role for product producibility. By focusing on designing a product easy to assemble, we will concentrate the product producibility benefits. Therefore efforts to improve product producibility have to be first put on assembly operation improvement This study of the ease of product assembly during the preliminary design steps is referred to "Design For Assembly" (DFA) [Boothroyd, Alting, 92] but is generally dedicated to small mechanical product.

1.2. Geometric variations and product assemblability

Most of the works done previously on DFA are generally focused on part reduction, modularity improvement, part standardization and minimization of the assembly failure.

This last item is directly related to part variations. Because there is no way to produce parts that fully conform to their nominal definition, geometrical and dimensional variations will always appear on parts. Impact of these variations can be more or less important for the product's assemblability. A level of geometrical variation out of tolerance on one part can lead to turn down the product quality or the product may fail to be assembled. In this case, part has to be reworked or repaired, adding extra cost for the product and increasing time required for the production process. Therefore reducing and managing geometric variations on all product's parts are a key step for improving product assemblability and consequently its producibility.

1.3. Questions about variation management method

When a company is decided to launch a variation management plan, two questions are generally asked: *where to manage product variations* and *how to control effect of variations on functional requirements*?

Unfortunately, no global answers are today given to resolve these two key questions. The goal of this paper is then to present a new integrated design method taking into account geometrical variation in the product development cycle as soon as possible. By managing variation reduction on product key characteristics and by analyzing assembly sequences, this method will improve the product producibility.

This paper will begin by describing the basic concepts used in our method. These concepts will allow us to propose an integrated design method for assemblability improvement. Following the method proposal, different analysis tools will be described in order to facilitate its deployment. At last we will illustrate the results of this method on a real complex industrial application.

2. INTEGRATED METHOD FOR GEOMETRIC VARIATIONS MANAGEMENT

2.1 Functional analysis and Key Characteristics

Designing products easy to assemble is feasible only with a concurrent product definition through Integrated Product Teams (IPTs). These teams gather together production, design, stress, facility equipment, manufacturing and tool engineering representatives. The first task dedicated to the IPTs is to fully understand how the product is supposed to work and to be used. By using functional analysis methods, IPTs define major customer requirements, technical requirements and product constraints to fulfill on the product. These major functional requirements and their cascades through all the component details allow focusing on what is really necessary about operation and performance for the product [Marguet, Mathieu, 97]. Moreover the functional analysis will be used in order to understand how the product design is related to the customer requirements. The design process can also evolve with a mapping of the functional product specifications into the realization of these specifications using appropriate geometric concepts. Amongst all geometric conditions related to functional specifications, assembly conditions will be related directly to the following assembly function: "parts have to fit perfectly together".

Unfortunately, for complex product like civil aircraft, millions of dimensions and geometric characteristics must be used to define all functional requirements. Each of these geometric characteristics will have variability introduced by the environment, the manufacturing processes and assembly processes. However only the variation in a small subset of these millions of features will significantly affect the functional requirements on the product. In order to be efficient, IPTs need to manage by hierarchy which characteristics are the most important. We call these characteristics "Key Characteristics" (KCs).

The first concept of our method is then to first focus variation reduction on these product's Key Characteristics.

2.2 Role of the Assembly Sequence

From a design point of view, assembly requirements between two parts will be generally viewed as geometric alignment or minimal gap between these parts. These assembly requirements can be modeled through assembly links between two assembly surfaces. Depending of their types, each assembly links will limit one or more degrees of freedom between the two related parts. D. Whitney goes further and defines two types of assembly link: the **mate** and the **contact** [Mantripragada, Whitney, 98]. A mate is an assembly link that establishes constraints and dimensional relationships between parts. A contact is an assembly link that provides partial constraint or just supports and fastens the part once it is located. The type of assembly link, mate or contact will be related to the sequence of assembly link realization. Mates are realized first until part is fully located on the product and only then are realized all contact links [Marguet *et al.*, 00]. Hence, for a hole to hole "panel-stringer and cleat" assembly, twelve assembly links between the four parts allows to locate and to fix together the parts (see figure 1). From these twelve assembly links, two on each cleat will be "mate" as they locate the cleats versus the panel as well as two in the two ends of the stringer as they locate the stringer versus the panel. All other assembly links between the parts are "contact" as they just fix the parts together but don't have any effect on the part location.

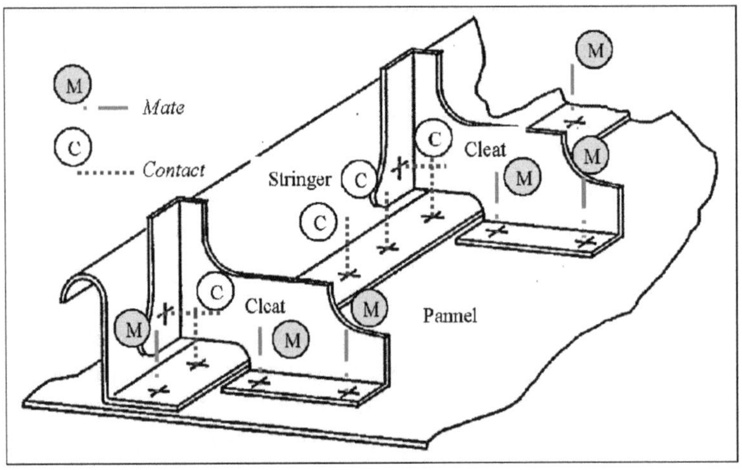

Figure 1. Mate and Contact

The principal cause of assembly failure will result from impact of geometrical variations on geometrical conditions. Impact of geometric variation will be directly related to assembly sequence and assembly link types [Marguet, Mathieu, 99]. Indeed, as mate establishes positional constraints between two components, any geometrical variation on this link has a direct effect on the position of the two components. But, if we consider only rigid parts, as contacts just supporting and fastening the part, geometrical variation has no impact on the relative position of the two parts. However these contacts have to be performed during the assembly operations. Contacts are then considered as assembly conditions.

Based on these observations, we can propose the following rule: geometrical variations on product flow from part to part *through mates only*. Selection of which assembly links will be mate and which will be contact allows the IPTs to lead variation flows through the product and hence to minimize geometrical variation impact on geometrical conditions.

2.3 Integrated Method

From the previous concepts, we can propose the following framework for a variation management method:

- Identify the major functional requirements to fulfill on the product with their admissible levels.
- Translate these major functional requirements into geometrical conditions with admissible tolerances.
- Select the Key Characteristics for the product.
- Select one admissible assembly sequence for the product.

- Identify where are the mates and the contacts based on the selected assembly sequence.
- Analyze the selected assembly sequence based on the impact of variation on geometrical conditions and variations on Key Characteristics.

The analysis of the assembly sequence may lead up to two solutions: the given assembly sequence and the variations on Key Characteristics allow to respect the geometrical conditions or not. If not, impact of geometrical variations have to be revalued based on a new admissible assembly sequence or new manufacturing process capabilities have to be selected for the product or functional analysis have to be modified based on new product design. This first solution is by far the most interesting one for the IPTs as it can be realized without extra cost for the product. This method is summarized on figure 2.

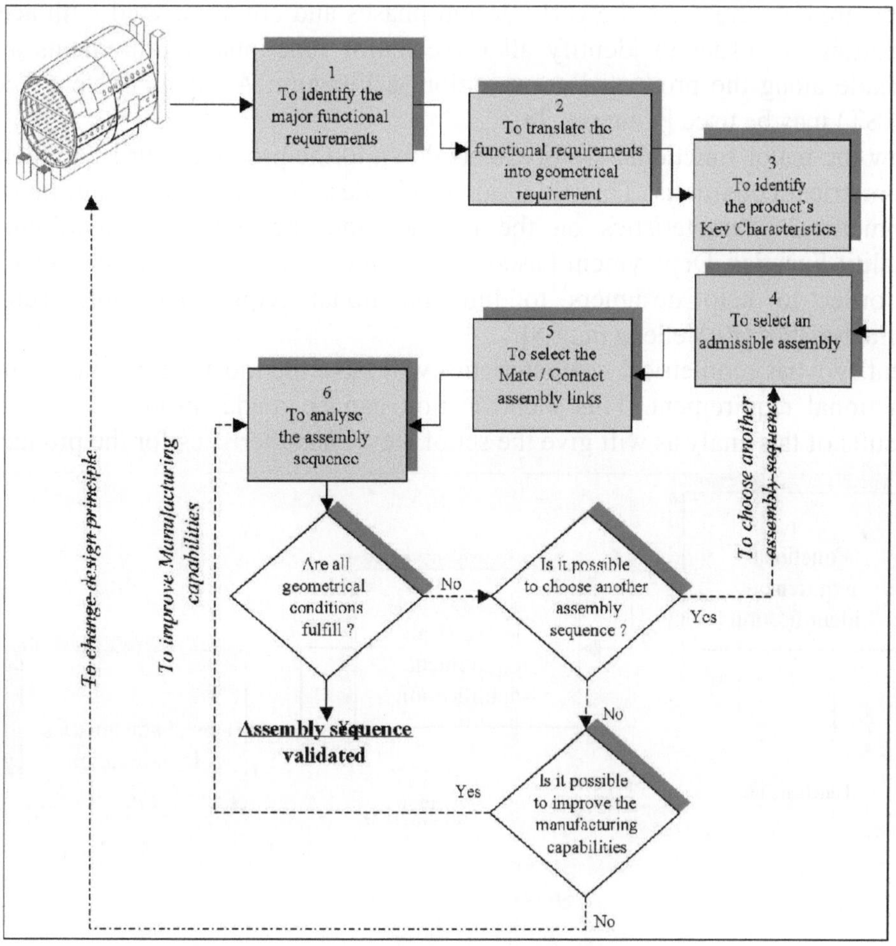

Figure 2. Integrated method

This pragmatic method will lead the IPT to first identify where variation reduction has to be implemented and secondly select which assembly sequence is the most interesting in order to minimise impact of variation on major functional requirements.

3. KEY CHARACTERISTICS IDENTIFICATION

As we have seen it before, Key Characteristics identification allows to focus variation reduction only where it is really necessary for the product. These Key Characteristics can be defined by the geometrical characteristics on the product for which a minute deviation from nominal specifications has a significant impact on the product's functions.

One of the principal difficulties from the Key Characteristic concept lies in the KC identification particularly for a new product design. The following steps describe briefly a global approach useful for the IPT in order to select what are the product's KCs.

1. Identify the major functional requirements that might be affected by variations. These requirements will be coming from the performance, safety or fit constraints. They are determined in the early design phases and are associated with acceptable variations. In order to identify all these major functional requirements and their cascade along the product decomposition, a Function Analysis System Technique (FAST) may be used [Yannou, 98].
2. Flow the major functional requirements down to the product in order to identify the geometrical conditions. This flow down will describe the relationship between the geometrical characteristics on the product and the functional requirements. A Quality Function Deployment based on the House of Quality concept may be used in order to help designers to link functional requirement into geometrical characteristics [ReVelle *et al.*, 98].
3. Identify what geometrical characteristics will have the most significant influence on functional requirement. This identification can be made through a risk analysis. Results of this analysis will give the set of Key Characteristics for the product.

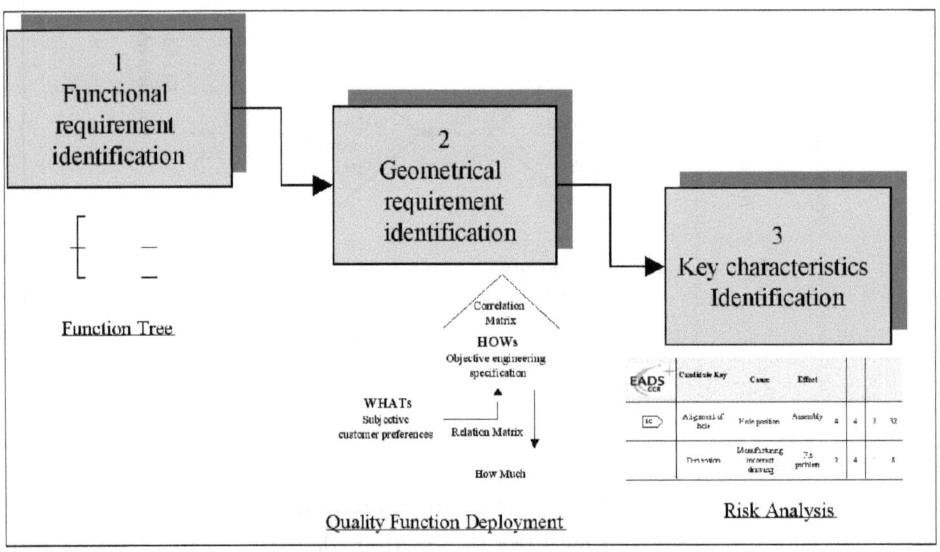

Figure 3. Key Characteristics Identification

4. ASSEMBLY SEQUENCE ANALYSIS

In order to analyze rapidly different admissible sequences and to select the most promising ones, we have developed a set of graphical tools. These tools will be able to analyze the product first from a qualitative aspect and secondly from a quantitative aspect.

4.1 Qualitative Assembly Analysis

In order to decide rapidly between the good and the bad assembly sequences, we propose to perform first a qualitative assembly analysis. This analysis will be based on an Assembly Oriented Graph. Assembly Oriented Graph (AOG) is an extension of the Datum Flow Chain [Mantriparagada, Whitney, 98] and of the liaison graph [Ballu, Mathieu, 99].

The AOG is a directed acyclic graphical representation of an assembly given a picture of the location dependencies of parts and surfaces. This graph is composed of nodes and oriented arcs. Each node represents assembly surface brought together by components. Oriented arcs represent mate between two assembly surfaces. The direction of the arcs specifies which component locates the others. The arrow points on the positioned component. Oriented dotted arcs represent contact between two assembly surfaces and dotted line represents geometrical conditions. Figure 4 depicts an AOG for a product composed from three parts A, B and C. Part A is the base part, this part gives the position of parts B and D and the position of part C is given by B.

Each AOG represents a different assembly sequence and allows the IPT to easily understand how parts are positioned, where are the geometrical conditions and how variations flow through the product.

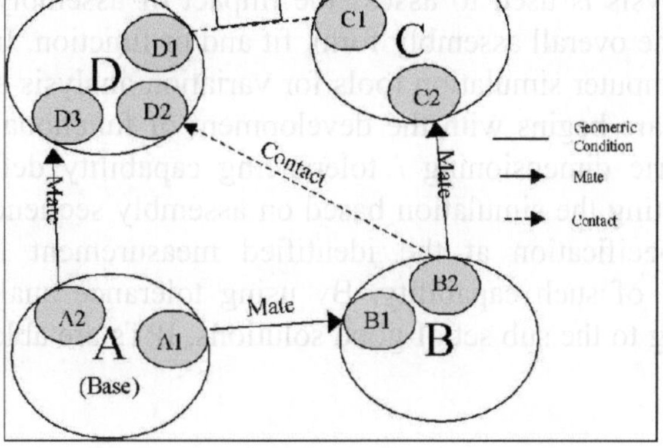

Figure 4. Assembly oriented graph.

As we have seen it before, variations on geometrical conditions depend on variations on mates. All paths on the AOG performed by mate arcs represent then the propagation of geometrical variation through the product. By studying the path going from the base component to a geometric condition we will study the impact of variation on this condition. We call this path of variations, a propagation chain (see figure 5). All links participating to the propagation chain will have an impact on the realization of the geometrical conditions. Then the number of links belonging to the chain increases the magnitude of variations on the geometrical condition. This remark gives us a qualitative criterion to select the best admissible AOG as the shortest propagation chain for a geometrical condition.

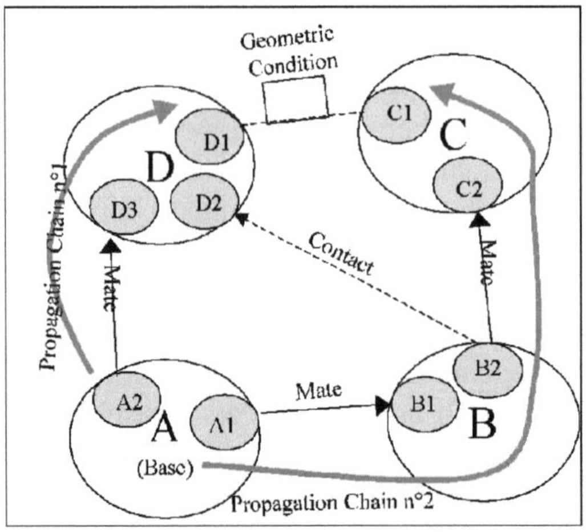

Figure 5. Propagation chain

Based on the propagation chain evaluation for all admissible sequences, the worst assembly sequences can be identified and deleted. The qualitative analysis is then a way for the IPT to select a sub set of good solutions.

4.2 Quantitative Assembly Analysis

In order to select optimal assembly sequence, tolerance analysis has to be realized on each admissible assembly sequence and for each assembly requirement.

Tolerance analysis is used to assess the impact of assembly sequence and detail part tolerances on the overall assembly form, fit and/or function. In the case of complex 3-D assemblies, computer simulation tools for variation analysis are used [Salomons *et al.*, 97]. This software begins with the development of functional feature models with datum and geometric dimensioning / tolerancing capability defined and attached to them. After completing the simulation based on assembly sequence, the tool shows the ability to meet specification at the identified measurement location and provide statistical measures of such capability. By using tolerance analysis on all assembly sequences belonging to the sub set of good solutions, IPTs are able to select the optimal one.

5 IMPLEMENTATION

The integrated design method presented above has been used to improve assemblability of a section on the future Airbus A380. This section is composed by four major components (see figure 6). These major components are in turn decomposed in many sub-components. The first task dedicated to the IPT in charge of this section was to identify the geometrical conditions. About twenty major functional requirements were identifed for this product. Each of these requirements was translated into a geometrical condition. Alignment of the upper and lower sub-sections and minimal gaps between the two door panels and windows panels are two examples of these geometrical conditions.

Based on these geometrical conditions, the IPTs was able to select the KCs related to the design of this section. The selection has been made by a risk analysis and based on the knowledge of similar products. About ten KCs were identified for each geometrical condition. Quality engineers have used these KCs in order to insure that these characteristics will be under control during the manufacturing process.

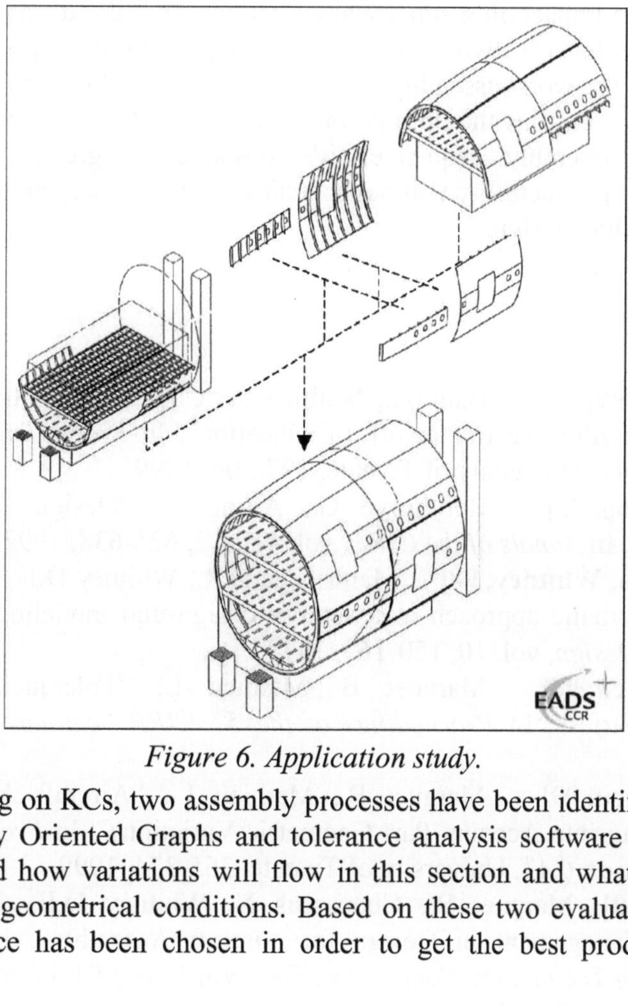

Figure 6. Application study.

After working on KCs, two assembly processes have been identified to realize this section. Assembly Oriented Graphs and tolerance analysis software have allowed the IPT to understand how variations will flow in this section and what variation impact was expected on geometrical conditions. Based on these two evaluations, the optimal assembly sequence has been chosen in order to get the best producibility for this section.

6 CONCLUSION

This paper presented an integrated design method taking into account geometrical variation in order to improve assemblability for complex product. The use of this method intend to select as early as possible in the design cycle, the optimal assembly sequence in order to minimize the propagation of variations on the geometrical condition.

This method is based on two main concepts:

- Focus time and money on product Key Characteristics.
- Importance of assembly sequence in the variation propagation through the product.

This method starts by identifying Key Characteristics for the product. These KCs are the product characteristics, which have the most important impact on major product requirements. Following, our method selects the optimal assembly sequence, which will minimize tolerance impact on geometrical conditions. In order to save time to the IPTs, this selection is realized in two steps. The first step performs a qualitative analysis in order to eliminate the worst assembly sequences. The second step based on quantitative tolerance analysis identifies the optimal one. Based on KC identification and selection of the optimal assembly sequence, it's possible to greatly improve product assemblability and producibility without spending a lot of money on new manufacturing means or new product design.

7 REFERENCES

[Ballu, Mathieu, 99] Ballu A., Mathieu L, "Choice of functional specifications using graphs within the framework of education", In *Proceedings of the 6th CIRP Seminar on CAT*, University of Twente, 197-206, 1999.

[Boothroyd, Alting, 92] Boothroyd G., Alting L., "Design for Assembly and Disassembly", In *Annals of the CIRP*, vol. 41, n°2, 625-638, 1992.

[Mantriparagada, Whitney, 98] Mantripagada R., Whitney D.E., "The Datum Flow Chain: a systematic approach to assembly design and modeling", In *Research in Engineering Design*, vol. 10, 150-165, 1998.

[Marguet, Mathieu, 97] Marguet B; Mathieu L; "Tolerancing Problems for Aircraft Industries", In *Proceedings of the 5th CIRP Seminar on CAT*, Toronto, 1997.

[Marguet, Mathieu, 99] Marguet B., Mathieu L., "Aircraft Assembly Analysis Method Taking into Account Part Geometric Variations", In *Proceedings of the 6th CIRP Seminar on CAT*, University of Twente, 365-374, 1999.

[Marguet et al, 00] Marguet B., Chevassus N, Whitney D.E., Mantripragada R., "Variation Management in Design for Aircraft Assembly", In *SAE Aerospace Manufacturing Technology Conference*, Fort Worth, 200-01-1729,. 2000.

[ReVelle et al., 98] ReVelle J. B., Moran J. W., Cox C. A., "The QFD Handbook", *John Wiley & Sons Editor*, ISBN 0-471-17381-9, 1998.

[Salomons et al, 97] Salomons O. W., Van Houten F., Kals H., "Current status of CAT systems", In *Proceedings of the 5th CIRP Seminar on CAT*, Toronto, 438-452, 1997.

[Yannou, 98] Yannou B, "Analyse fonctionnelle et analyse de la valeur", Conception de produits mécaniques, *Edition Hermes*, ISBN 2-86601-694-7, 1998.

Geometrical tolerancing using linear programming: functional approach and specification of application domain

V. T. Portman*[*], *V. G. Shuster*, and *Y. L. Rubenchik

[*]*Department of Mechanical Engineering, Ben-Gurion University of the Negev, P.O.B. 653, Beer-Sheva, 84105, Israel, Phone: 972-8-6472813, E-mail:portman@menix.bgu.ac.il*

Abstract: A linear programming (LP) approach to the minimal zone (MinZ) based method of assessments of form and position accuracy is investigated. The purpose is to specify domain limitations for this method caused by its own errors and the LP-based assessments uncertainty caused by functional properties of a measured feature. The LP approach has well-established computational advantages. To calculate the linearization-caused error, the second order approximation is used. This type of errors is illustrated by cylindricity assessments when the form and dimension tolerances are specified by the ISO standard system. A distribution of the measurement point positions between the boundaries of the MinZ is analyzed with the aim to estimate the uncertainty of the LP-based assessments. This analysis based on functional approach to the physical sources of the form error. In many instances, a deviation of the result of measurement from the MinZ boundaries may be represented as a sum of limited number of sinusoidal components, which have uniformly distributed phase angles and approximately constant magnitudes.

Keywords: form and position tolerancing, accuracy assessment, linear programming, minimum zone, uncertainty.

1. INTRODUCTION

International and national standards for form and position tolerancing widely use minimal zone (MinZ) for accuracy assessments [ANSI94, ISO83]. When an accuracy of the surface or line is measured, the MinZ is built as follows: a substituted feature (i.e., the feature of the nominal form) is extracted from the measured points; the zone boundaries are built so that all the points of the actual surface lie between or on the boundaries of the zone and the width of the zone is minimal. A procedure of the MinZ construction is reduced to a solution of the minimax optimization problem [Dowling et al., 1995; Choi, et al., 1999]. Various approaches are used for solution of this problem: statistical analysis [Murthy, et al., 1980]; computational geometry techniques such as Voronoi diagrams [Roy, et al., 1995] and convex hull [Dowling, et al., 1995]; non-linear programming [Orady, et al., 1998]; linear programming (LP), etc.

P. Bourdet and L. Mathieu (eds.),
Geometric Product Specification and Verification: Integration of Functionality, 227-236.

The LP procedure has well-established advantages over other minimax optimization methods, such as a rich variety of effective algorithms and software, fast computing, flexibility with respect to on-line data presentation, etc. The principal condition for the application of the LP-base method is the necessity to represent both the optimization criterion and the constraints in linear form. Since size deviations are small in comparison with the nominal dimensions (the ratio of those is in range of 10^{-5} to 10^{-7} for ISO tolerancing grades IT5-IT12), linear presentation of these deviations is a commonly accepted model for regular-sized values.

In this paper, an application domain and an uncertainty of the LP-based method are considered. The application domain is limited because of linearization-caused errors. To estimate these errors, the second order terms in the Taylor series expansion for the matrix-function are used. This type of errors is illustrated by cylindricity assessments when the form and dimension tolerances are specified by the ISO standard system. As a result, the application domain of the LP-based assessments is defined as a function of the nominal size and the IT (International Tolerance) grade number of the measured feature.

The uncertainty associated with measured points depends on the distribution of these points between the boundaries of the MinZ. The distribution is investigated based on functional approach to the physical sources of the form errors. On numerous occasions, an investigated random variable presents a sum of few numbers of sinusoidal components, which have the uniformly distributed phase angles and approximately constant magnitudes. An analytical presentation for the distributions of these sums is obtained through the Bessel functions.

2. PRESENTATION OF THE FITTED FEATURES BY THE MinZ METHOD

A set of data for estimation of the form and position tolerances is defined by two groups of errors:

- errors dq_i, (i =1,...,n) of the ith dimensional parameter q_i of the measured feature, where n is the number of the dimensional parameters; and,
- errors of position and orientation of this feature; $m \leq 6$ is the number of these errors.

The set may be presented in the form of the vector $\boldsymbol{\delta}_b$ of the $p = n+m$ order. In the general case, $m = 6$ and this vector is

$$\boldsymbol{\delta}_b = (dq_1,..., dq_n, \delta_x, \delta_y, \delta_z, \alpha, \beta, \gamma)^T, \tag{1}$$

where δ_x, δ_y, and δ_z are the linear errors (small displacements of the origin of the coordinate system associated with NS) along the X, Y, and Z axes, respectively; α, β, and γ are the small angles of rotations relative to the same axes.

In particular case not all of position error included in the vector $\boldsymbol{\delta}_b$ (for example, in the case of cylindrical surface $\delta_z = \gamma = 0$).

In the linear approximation, the normal deviation $\Delta_{nb}(u, v)$ of the points of actual surface from the points of nominal surface may be presented as follows:

$$\Delta_{nb}(u, v) = \boldsymbol{f}(u, v) \bullet \boldsymbol{\delta}_b \tag{2}$$

where $\boldsymbol{f}(u, v)$ is the $p \times 1$ vector of normal transfer factors for the components of errors vector $\boldsymbol{\delta}_b$; u, v are the curvilinear (Gaussian) coordinates defined on the nominal and actual surfaces.

According to the standard definition, the width Δ_{MinZ} of the MinZ must be minimal, subject to a set of constraints that provide that all the points of the actual (measured) surface $\boldsymbol{r}_a$ are located between two boundaries of the MinZ, i.e.,

$$\Delta_{\text{MinZ}} \rightarrow \min; \tag{3}$$

$$\boldsymbol{f}_k \bullet \boldsymbol{\delta}_b + \Delta_{\text{MinZ}}/2 \geq \Delta_k; \ \boldsymbol{f}_k \bullet \boldsymbol{\delta}_b - \Delta_{\text{MinZ}}/2 \leq \Delta_k; \tag{4}$$

$$k = 1,\ldots,N$$

where $\boldsymbol{f}_k = \boldsymbol{f}(u_k, v_k)$ is the vector $\boldsymbol{f}$ defined for the kth point of the nominal surface; Δ_k is measured deviation in the kth point; N is the number of the measurement points.

Standard method of building of the vector $\boldsymbol{f}$ is known [Portman et al., 1997a]. As an example, the MinZ for the flatness assessment obtained by the *Mathematica* software system [Wolfram, 1996] is shown in Figure 1.

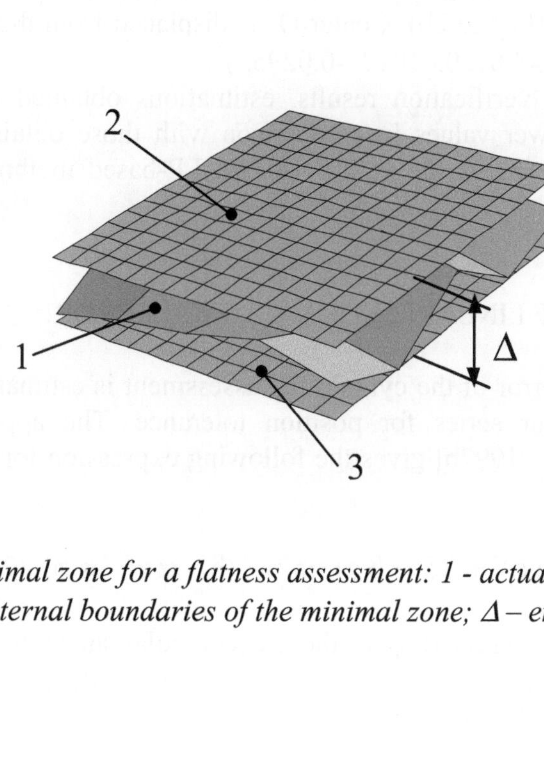

Figure 1; Minimal zone for a flatness assessment: 1 - actual surface; 2, 3 - external and internal boundaries of the minimal zone; Δ – error of the form.

3. METHOD VERIFICATION

To verify the LP-based method, the data of actual measurements on the CMM were used. Roy [Roy, et al., 1995] measured the accuracy of the cylinder in 64 points (a diameter of the cylinder is 10 mm). Using the results of these measurements, the MinZ assessment is calculated by the LP-based method. The obtained cylindricity is 8.8% less than the cylindricity obtained by computational geometry-based algorithm applied by Roy. As another example, data given by Traband [Traband, et al., 1989] is used for straightness and flatness evaluations. The flatness estimations obtained by the LP-based method are 6.7-11.6% less than the convex hull-based estimations and 3.1-6.9% less than the least mean square-based estimations. Assessments, obtained by the LP-based method for the straightness, are approximately equal to the convex hull-based estimations and 3.5-10.4% less than the least mean square-based estimations.

The similar results are obtained when the LP-based approach was applied to sphericity assessments. A set of 100 points given by Huang [Huang, 1999] was used. To generate the points, two spheres with radii 10.0 and 9.0, respectively, were specified; therefore, the MinZ width was prescribed as equal to 1.0. The data points were then randomly generated between the spherical boundaries of the MinZ. The sphericity MinZ obtained by LP method is 0.982379 mm, when 9.9925 mm and 9.01012 mm are the radii of external and inner spheres, respectively. Underestimation of about 1.8% results from a limited number of points. The MinZ is shown in Figure 2a. Each point lying on the meridian φ_i is fixed by the projection onto the horizontal plane by rotation on the opposite latitude angle θ_i (Figure 2b). Center O_1 is displaced from the nominal center O by small vector $\mathbf{OO_1} = (0.00781, 0.01012, -0.02953)^T$.

As is seen from the verification results, estimations obtained by the LP-based method have equal or fewer values in comparison with those obtained by the other methods. This fact characterizes the validity of the LP-based method for the typical applications.

4. ESTIMATION OF LINEARIZATION-CAUSED ERROR

The linearization-caused error of the cylindricity assessment is estimated by the second order terms in the Taylor series for position tolerance. The approach developed previously [Portman, et al., 1997b] gives the following expression for the second order terms:

$$\Delta^{(2)} = (R/2)(\beta \cos \varphi - \alpha \sin \varphi)^2 + [(\delta_x + \beta z) \sin \varphi - (\delta_y - \alpha z) \cos \varphi]^2/(2R), \qquad (5)$$

where R is radius of the cylinder; φ is the current polar angle of the point of the cylinder; and δ_x, δ_y, α, and β are the position and orientation deviations of the cylinder axis (i.e., the deviations of the first order).

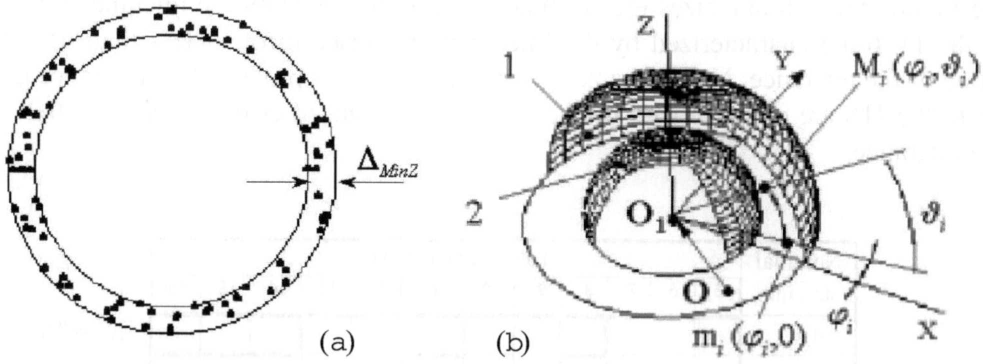

Figure 2; Measurement of the sphere: (a) presentation of the set of the measured points between the zone boundaries; (b) projection m_i of the measured point M_i onto the horizontal plane.

The first term within Eq. (5) presents the error depending on orientation deviation, while the second depends on the eccentricity on the cross-section of the cylinder. Thus, Eq. (5) may be represented in the form:

$$\Delta^{(2)} = (R/2)(\Delta\nu \sin \Phi_1)^2 + [\Delta_{ec}(z) \sin \Phi_2]^2/(2R) \leq (R/2)(\Delta\nu)^2 + [\Delta_{ec}(z)]^2/(2R), \qquad (6)$$

where $\Delta\nu = (\beta^2+\alpha^2)^{1/2}$ is the deviation of the orientation angle; $\Delta_{ec}(z) = [(\delta_x+\beta z)^2 + (\delta_y+\alpha z)^2]^{1/2}$ is the eccentricity at a *z*-height; and Φ_1 and Φ_2 are the current polar angles depending on the values φ, δ_x, δ_y, α, and β.

In accordance with the regular requirements of the international and national standards for geometrical accuracy, the position and orientation errors of the feature must be less than 25%–60% of the size tolerance. These requirements may be interpreted for angular deviations and eccentricity as follows: $\delta\nu \leq 0.6\ T/D$ and $\Delta_{ec}(z) \leq 0.4\ T$, where *D* is the nominal dimension, and *T* is its tolerance. Substitution of these expressions into the right-hand side of inequality (6), taking into account that $R = D/2$, gives:

$$\Delta^{(2)} \leq (D/4)(0.6\ T/D)^2 + [0.4\ T]^2/D = 0.25\ T^2/D\ . \qquad (7)$$

The error caused by LP, thus, depends on the ratio between the tolerance *T* squared and the nominal dimension. To estimate this error, the regular tolerance grade specification was used. In accordance with the ISO (1983) specification, the linearization-caused error of form accuracy estimation of the cylinder is calculated (Figure 3). Three domains, for which the linearization-caused error is less than 0.27%, 1%, and 5%, respectively, of the value of the corresponding tolerance, are built. As is obvious from the Figure, the nominal sizes greater than 1.0 mm produced by the tolerance grades up to the IT16 are characterized by the linearization-caused error of less than 5% of the value of the tolerance. Nominal sizes greater than 1.0 mm produced by tolerance grades up to the IT8 are characterized by the linearization-caused error of less than 0.27% of the tolerance.

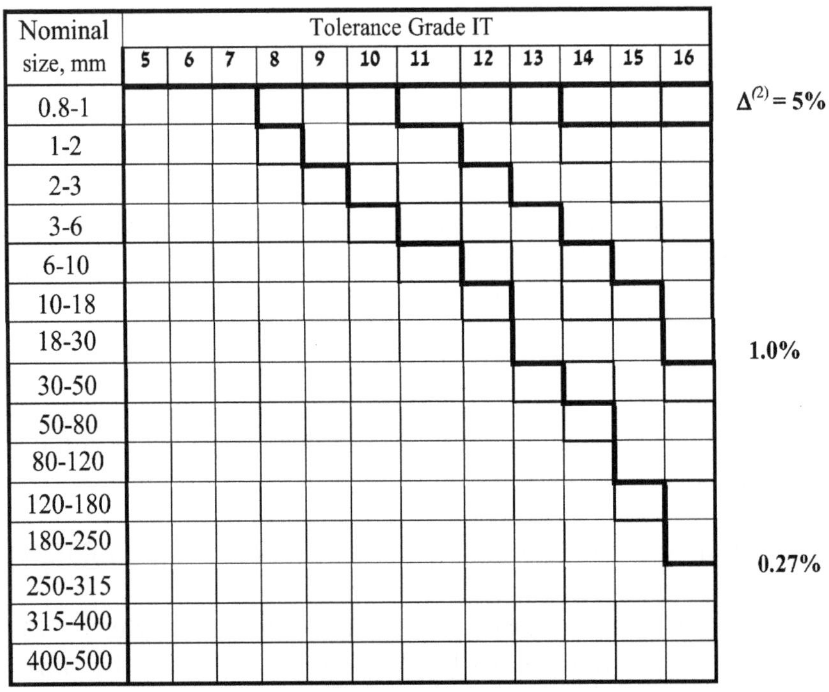

Figure 3; Estimation of the linearization-caused error for cylindricity assessment.

5. TOWARDS UNCERTAINTY OF THE MinZ ASSESSMENTS

For the analysis of the uncertainty problem, a distribution of the deviations of measurements from the boundaries of the MinZ has to be considered. As a rule, two basic models are used: the normal (Gaussian) distribution [Lin et al., 1995; Weckenmann et al., 1998] and the uniform distribution [Choi et al., 1999; Huang, 1999]. Meanwhile, both of these distributions have bad correspondence with majority of the actual measurement data. In the Figure 4, the distributions of the measurements of some real surfaces are shown [Traband et al., 1989, Roy et al., 1995]. All of them significantly different from the normal one: there is no clearly defined maximum in the centre area and the tails of the distribution curve do not tend to zero, as it takes place for the normal distribution (the tails behavior is deciding factor that defines the type of distribution and significantly affects on the quantitative estimations of the uncertainty).

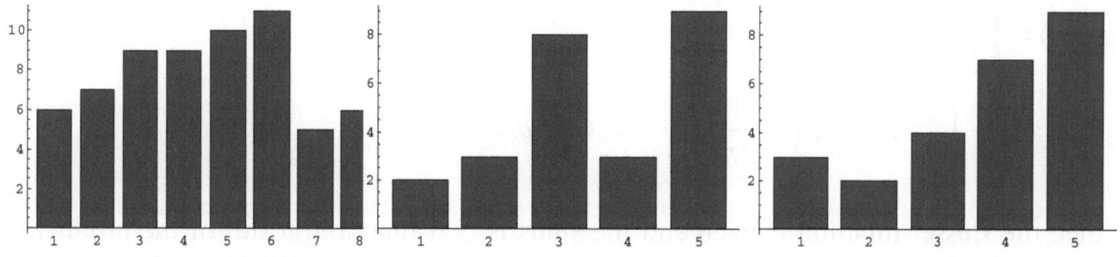

Figure 4; Histograms of experimental data distributions:
(a) measurement of cylinder [Roy et al., 1995], 64 measured points;
(b) measurement of straightness [Traband et al., 1989], 25 measured points;
(c) measurement of flatness [Traband et al., 1989], 25 measured points.

Thus, the assumption about distribution type requires further investigation. This investigation has to be based on a functional approach to the mentioned above distribution. We consider this problem, to form accuracy of cylindrical part. For example, profile of centreless-ground cylindrical parts consists of one or two frequency components affecting on the form accuracy. During turning and cylindrical grinding, there are, as usually, no more than 2–4 dominant harmonics in the profile spectrum. Without rejection the possibility of normal distribution, it is necessary to analyze the case when a spectrum of the profile consists of few number of the harmonics. In this case, form deviation may be defined by sum of 1-3 harmonics, which have the uniformly distributed phase angles and approximately constant magnitudes:

$$X(t) = \sum_{k=1}^{n} a_k \sin(\omega_k t + \alpha_k), \tag{8}$$

where a_k and ω_k are given constants, and the phase angle α_k is a random variable uniformly distributed between 0 and 2π.

Firstly, we will consider a normalized case when there is only one sinusoid with the magnitude $a = 1$, i.e., the unit random value $x = X/a$. In this case, the probability density function $f(x)$ and characteristic function $g(z)$ have the form:

$$f(x) = (1/\pi)\,(1 - x^2)^{-1/2}, \tag{9}$$

$$g(z) = \int_{-\infty}^{\infty} e^{izx} f(x)dx = \int_{-\infty}^{\infty} e^{izx} \frac{dx}{\pi\sqrt{1-x^2}} = J_0(z), \tag{10}$$

where $J_0(z)$ is the Bessel function of order zero. The characteristic function and the probability density function of the sum of n independent random variables with unit magnitudes are given by:

$$g_X(z) = \prod_{k=1}^{n} g_{X_k}(z) = [J_0(x)]^n \tag{11}$$

$$f(z) = \frac{1}{2\pi} \int_{-\infty}^{\infty} e^{-izx} g_X(z) dx = \frac{1}{2\pi} \int_{-\infty}^{\infty} e^{-izx} [J_0(x)]^n dx. \tag{12}$$

Since the Bessel function is even and the sinus is odd function, result of the product of these two functions integrated over x from -∞ to ∞, is zero. So, the final expression for probability density function is:

$$f(z) = \frac{1}{2\pi} \int_{-\infty}^{\infty} \cos zx [J_0(x)]^n dx. \tag{13}$$

The primary problem is related to analysis of the sum of harmonics of type (8) with equal magnitudes. In this case, an analytical solution defined by formula (13) is used. These theoretical distributions for various numbers of harmonics (n = 1, 2, 4, 8) are shown in Figure 5. As indicated in the figure, the distribution curve tended to normal distribution with increase in number of harmonics.

A statistical simulation was applied to fix the dependences between the ratio of the magnitudes and the type of the resulting distribution relating to the sum of two stochastic harmonics. As an example, the histograms of distributions of sum of two sinusoids with different magnitudes are shown in Figure 6, where A_1 and A_2 are the magnitudes of two harmonics. As is seen from the Figure, as the ratio of magnitudes is increases, the distribution tends to distribution that is typical for one sinusoid (Figure 5).

6. CONCLUSIONS

1. The LP procedure for geometrical accuracy assessments using the MinZ approach is considered. The effectiveness of this procedure are demonstrated by examples of actual measurements. The LP-based assessments are equal or less than those obtained by the other methods.

2. The estimation of linearization-caused error associated with the LP-based method application has been formulated and calculated. This error is estimated by the second order terms of the Taylor series approximation of position deviations. It is shown that the nominal sizes greater than 1.0 mm produced in accordance with the tolerance grades up to the IT16 are characterized by the linearization-caused error of less than 5% of the corresponding tolerance.

3. To calculate the uncertainty of the results of measurements, a distribution of the measurement points between the boundaries of the MinZ is analyzed. This analysis is based on the functional approach to the physical sources of the form error and relates to the important case when a spectrum of the profile consists of few numbers of harmonics. The analytical and simulation approaches are applied to obtain the type of distribution.

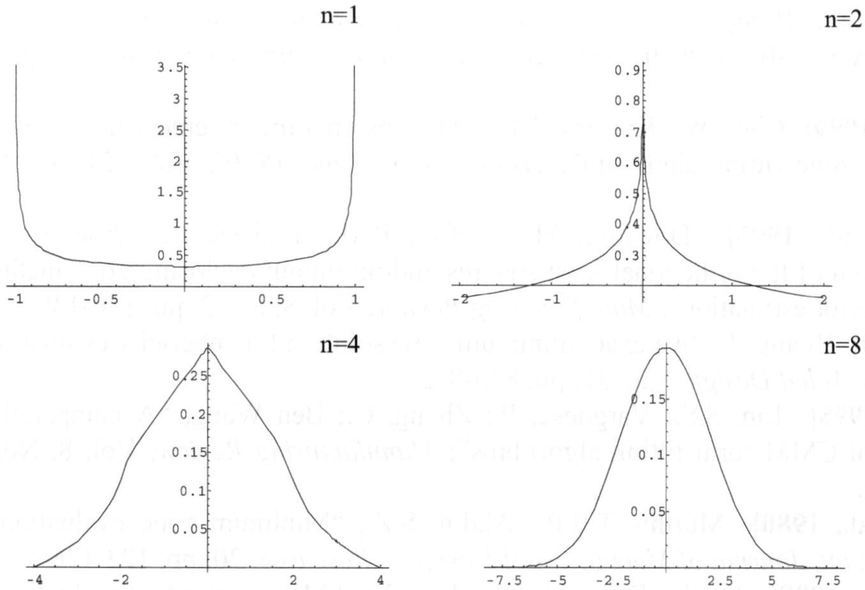

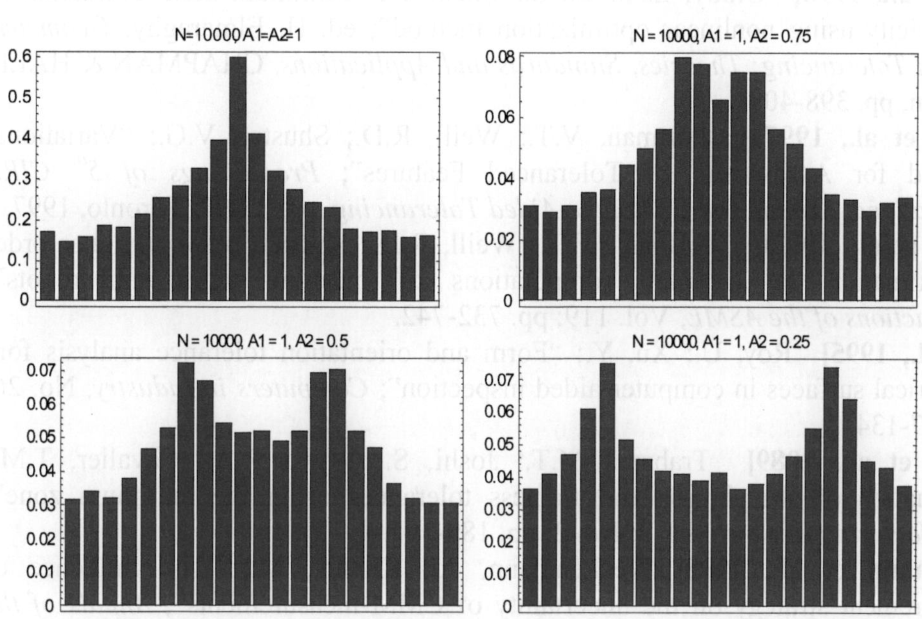

Figure 5; Theoretical distributions of the sum of n sinusoids with uniformly distributed phase angles and equal magnitudes.

Figure 6; Simulated distributions of the sum of two sinusoids with non-equal magnitudes.

REFERENCES

[Anthony et al., 1996] Anthony, G.T.; et. al.; "Reference software for finding Chebyshev best-fit geometric elements"; *Precision Engineering,* Vol. 19; pp. 28-36.

[Choi et al., 1999] Choi, W.; Kurfess, T.R.; "Dimensional measurement data analysis, Part 1: A zone fitting algorithm"; *Transactions of the ASME,* Vol. 121; pp. 238-245.

[Dowling et al., 1995] Dowling, M.; Griffin, P.M.; Tsui, K.-L.; Zhou, C.; "A comparisom of the orthogonal least squares and minimum enclosing zone methods for form error estimation"; *Manufacturing Review,* Vol. 8; No. 2; pp. 120-138.

[Huang, 1999] Huang, J.; "An exact minimum zone solution for sphericity evaluation"; *Computer-Aided Design,* Vol. 31; pp. 845-853.

[Lin et al., 1995] Lin, S.S.; Varghese, P.; Zhang, C.; Ben Wang; "A comparative analysis of CMM form-fitting algorithms"; *Manufacturing Review,* Vol. 8; No. 1; pp. 47-58.

[Murthy et al., 1980] Murthy, T.S.R.; Abdin, S.Z.; "Minimum zone evaluation of surfaces"; *Int. Journal of Machine Tool Design & Research,* 20; pp. 123-136.

[Orady et al., 1998] Orady, E.A.; Li, S.; Chen, Y.; "Minimum zone evaluation of cylindricity using nonlinear optimization method"; ed. H. Elmaraghy; *Geometric Design Tolerancing: Theories, Standards and Applications,* CHAPMAN & HALL, London, pp. 398-409.

[Portman et al., 1997a] Portman, V.T.; Weill, R.D.; Shuster, V.G.; "Variational Method for Assessment of Toleranced Features"; *Proceedings of 5th CIRP International Seminar on Computer-Aided Tolerancing*, pp. 83-97; Toronto, 1997.

[Portman et al., 1997b] Portman, V.T.; Weill, R.D.; Shuster, V.G.; "Higher order approximation in accuracy computations of machine tools and robots"; *Transactions of the ASME,* Vol. 119; pp. 732-742.

[Roy et al., 1995] Roy, U.; Xu, Y.; "Form and orientation tolerance analysis for cylindrical surfaces in computer aided inspection"; *Computers in Industry,* No. 26; pp. 127-134.

[Traband et al., 1989] Traband, M.T.; Joshi, S.; Wysk, R.A.; Cavalier, T.M.; "Evaluation of straightness and flatness tolerances using the minimum zone"; *Manufacturing Review,* Vol. 2; No. 3; pp. 189-195.

[Weckenmann et al., 1998] Weckenmann, A.; Knauer, M.; "The influence of measurement strategy on the uncertainty of CMM-measurements"; *Annals of the CIRP*, Vol. 47; pp. 451-454.

[Wolfram, 1996] Wolfram, S.; "*Mathematica: A System for Doing Mathematics by Computer,*" Addison-Wesley, Redwood City, CA.

Contribution of nonlinear optimization to the determination of measurement uncertainties

Jean Michel Sprauel*, Jean Marc Linares* and Pierre Bourdet**

* I.U.T., Avenue Gaston berger - F 13625 Aix en Provence cedex 1

** E.N.S Cachan, 61 Avenue Président Wilson - F 94230 Cachan

sprauel@iut.univ-aix.fr

Abstract: Much work has already been realized concerning the association of surfaces to points digitized with a Coordinate Measuring Machine (CMM). The developed procedures are usually based on the minimization of the distance between the measured points and the geometric element. They depend on the criterion used for the optimization. The function to minimize is always non-linear. To reduce the computing time, it is therefore transformed usually into linear variational equations, which are solved by an iterative process. Such algorithms require however to define initial parameters close to the final solution. The method presented here is based on a non-linear least squares optimization. It does not require any coordinate transformation and is not sensitive to the selection of initial intrinsic parameters. It applies to all the classical surfaces, i.e. lines, planes, circles, cylinders, cones and spheres. In accordance with quality standards, the uncertainties of the estimated parameters are also defined. These values are derived from the residue between the measured points and the optimized associated surface.

Keywords: CMM, Uncertainties, Nonlinear optimization, Least squares.

1. INTRODUCTION: GEOMETRIC APPROACH OF THE MEASUREMENT.

A lot of work has already been done concerning the association of surfaces to points digitized with a Coordinate Measuring Machine (CMM) [Bourdet and Clément, 1988; Bourdet et al. 1995; Goch et al., 1992]. The developed procedures are generally based on the minimization of the distances δ_k between N measured points M_k and the associated geometric element. Different criteria are used for that purpose, but the function Δ_n to minimize is mostly defined as follows:

$$(1) \qquad \Delta_n = \sum_{k=1}^{N} |\delta_k|^n$$

For n = 1, this leads to the classical algebraic norm criterion. The least squares method corresponds to n = 2. Tchebichev criterion is obtained when $n \to \infty$. The geometric

P. Bourdet and L. Mathieu (eds.),
Geometric Product Specification and Verification: Integration of Functionality, 237-244.

element fitted to the digitized surface is characterized by a set of P parameters a_i: coordinates of the center of the measured surface, cosines of the normal to a plane or of the direction vector of an axis, radius, angle of a cone, etc. Δ_n is however a nonlinear function of these variables a_i to optimize. To reduce the computing time, the coordinate system is therefore usually translated and rotated in order to obtain a small displacement formulation of the problem. This allows a linear approximation of the distance δ_k. The parameters a_i are then computed through an iterative process. Such algorithms require however to define initial parameters close to the final solution [Kruth et al. 1992].

This geometric approach only allows estimating mean values of the parameters a_i. However, according to quality standards, the uncertainties of the estimated parameters should also be defined.

The set of digitized coordinates forms a statistical sampling of the true analyzed surface. It contains therefore some mixed information about the distribution of the matter around the ideal geometric element and about the quality of the measurement. This information is not yet treated. A new procedure has therefore been developed in our laboratory to analyze the statistical properties of the digitized coordinates. The error bars on the estimated parameters can thus be estimated.

2. STATISTICAL APPROACH OF THE MEASUREMENT.

We will demonstrate first that each classical optimization criterion assumes a specific statistical distribution of the distance δ_k between the digitized points M_k and the associated perfect geometric element. For a given exponent n, the corresponding distribution function $f_n(\delta)$ will be:

$$(2) \qquad f_n(\delta) = \frac{1}{a_n\,(b_n\sigma)} \operatorname{Exp}\left[-\left|\frac{\delta}{b_n\sigma}\right|^n\right]$$

where σ is the standard deviation of the random variable δ,
a_n and b_n are two constants.

In fact, according to the maximum likelihood criterion, the best statistical estimator of the parameters a_i will maximize the conditional probability of all the realized independent measurements:

$$(3) \qquad \phi_n = \prod_{k=1}^{N} f_n(\delta_k) = \left[\frac{1}{a_n\,(b_n\sigma)}\right]^N \operatorname{Exp}\left[-\frac{\Delta_n}{(b_n\sigma)^n}\right]$$

The likelihood function ϕ_n becomes maximum when Δ_n is minimum. This shows clearly that the criteria used in the actual coordinate measuring machine software's implicitly assume a given distribution of the digitized points around the perfect geometric element.

The coefficients a_n and b_n can be derived from the normalization condition of the probability density $f_n(\delta)$ and from the mathematical definition of the standard deviation:

$$(4) \qquad \int_{-\infty}^{+\infty} f_n(\delta)d\delta = 1 \qquad \text{and} \qquad \sigma^2 = \int_{-\infty}^{+\infty} \delta^2 f_n(\delta)\, d\delta$$

This leads to following values :

$$(5) \qquad a_1 = 2\,,\ b_1 = 1/\sqrt{2}\ ;\ a_2 = \sqrt{\pi}\,,\ b_2 = \sqrt{2}\ ;\ a_\infty = 2 \ \text{and}\ b_\infty = \sqrt{3}$$

From these results, we deduce that the algebraic norm criterion corresponds to a symmetrical exponential probability density. The least squares method assumes a normal law. The Tchebichev criterion is obtained for a uniform distribution (Figure 1).

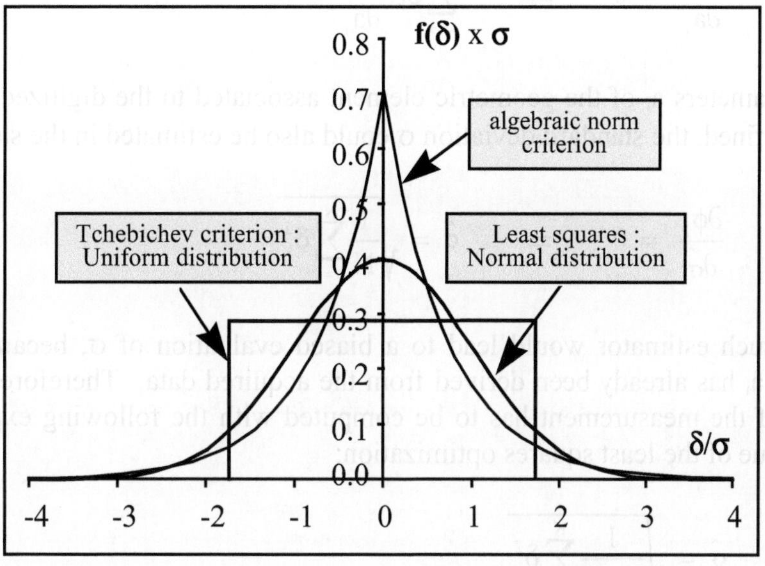

Figure 1; Probability density associated with the classical criteria used in coordinate measuring machine software's.

The deviations included in the experimental data result from the convolution between statistical perturbations of the coordinate measuring machine and the distribution of the matter around the ideal geometric element. Since the instrument depends on a great number of uncontrolled parameters, the first deviation is associated with a normal law.

The second component is more difficult to define. However, its probability density will correspond neither to an exponential function, nor to a uniform distribution. As a first approximation, the global experimental deviations will therefore be described by a normal law. With this assumption, the parameters a_i of the mathematical surface fitted to the experimental data will be defined through a least squares optimization.

3. UNCERTAINTY ESTIMATED FROM LEAST SQUARES.

The likelihood function ϕ_2 of the least squares optimization criterion is:

$$(6) \qquad \phi_2 = \left[\frac{1}{\sigma\sqrt{2\pi}}\right]^N \mathrm{Exp}\left[-\frac{1}{2}\sum_{k=1}^{N}\left(\frac{\delta_k}{\sigma}\right)^2\right]$$

The best-unbiased estimators $\hat{a}_i$ of parameters a_i are thus derived from the following optimization conditions:

$$(7) \qquad \frac{\partial\phi_2}{\partial\hat{a}_i} = 0 \quad \Rightarrow \quad \sum_{k=1}^{N}\delta_k\frac{\partial\delta_k}{\partial\hat{a}_i} = 0$$

If the P parameters a_i of the geometric element associated to the digitized surface were perfectly defined, the standard deviation σ could also be estimated in the same way:

$$(8) \qquad \frac{\partial\phi_2}{\partial\hat{\sigma}} = 0 \quad \Rightarrow \quad \hat{\sigma} = \sqrt{\frac{1}{N}\sum_{k=1}^{N}\delta_k^2}$$

However, such estimator would lead to a biased evaluation of σ, because a set of P parameters $\hat{a}_i$ has already been derived from the acquired data. Therefore, the standard deviation of the measurement has to be computed with the following expression, also called residue of the least squares optimization:

$$(9) \qquad \hat{\sigma} = \sqrt{\frac{1}{N-P}\sum_{k=1}^{N}\delta_k^2}$$

This deviation $\hat{\sigma}$ can be propagated to deduce the covariance matrix of the estimated parameters $\hat{a}_i$, using equation (7) and classical differential expressions of the uncertainties [BIPM 1993, Arri et al. 1995, Tuninski et al. 1999]. From the diagonal components of the covariance matrix, the error bars of $\hat{a}_i$ are then easily calculated, since the statistical distribution of these random variables corresponds to a Fisher-

Student law. Moreover, the covariance matrix is also useful to propagate the uncertainties of the fitted surfaces to any derived geometric element.

It has to be pointed out that the propagation process of the standard deviation $\hat{\sigma}$ requires a precise definition of the derivatives $\partial\delta_k/\partial\hat{a}_i$. For that reason, a nonlinear algorithm, based on a Gauss-Newton type method [Dennis et al., 1983], has been implemented to solve the optimization conditions (7). It does not require any coordinate transformation and is not sensitive to the selection of initial intrinsic parameters. It allows proceeding all the classical surfaces, i.e. lines, planes, circles, cylinders, cones and spheres.

Usually, the measurement uncertainties are assumed the same for all the digitized coordinates. Nevertheless, this hypothesis in not required to apply the maximum likelihood criterion. In fact, if the standard deviation σ_k of each distance δ_k is well known, a weighted least squares criterion can be minimized to estimate the parameters of the fitted surface.

4. EXPERIMENTAL RESULTS.

In order to test the influence of some acquisition parameters onto the resulting uncertainty, the method has been applied to the measurement of a machined hole. Such geometrical element is characterized by seven parameters: the three coordinates (x, y, z) of the center of the cylinder, the three direction cosines (v_x, v_y, v_z) of its axis and its diameter D. The analyzed surface has a diameter D of about 20 mm, a height of 25 mm and his axis parallel to the direction $\vec{z}$ of the coordinate-measuring machine.

4.1. Effect of the number of digitized points.

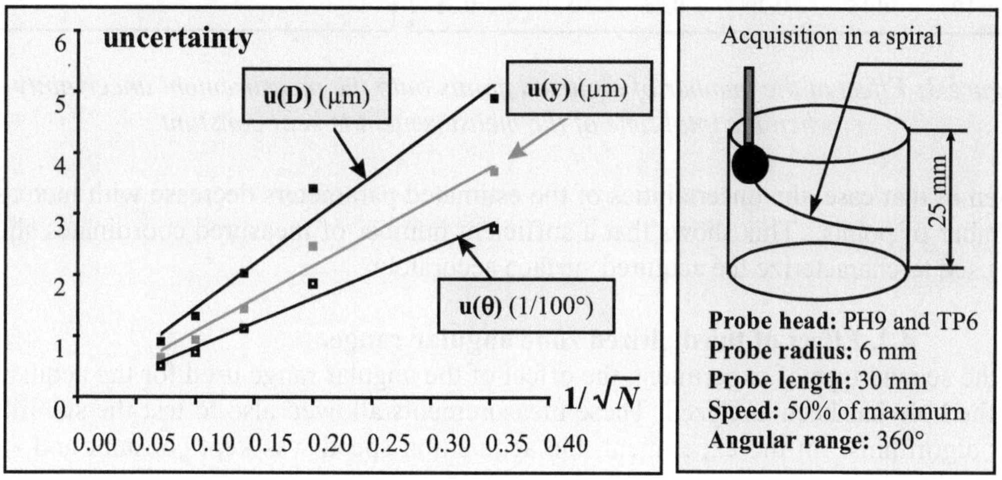

Figure 2; Effect of the number of digitized points onto the measurement uncertainty.

In a spiral acquisition mode, the influence of the number N of digitized points has been tested first. In this experiment, the acquisition speed was kept constant and set to half the maximum sampling rate of the coordinate-measuring machine. The results are summarized in figure 2. Only the error bars on diameter D, on coordinate y and on the angle θ between the axis of the hole and the direction $\vec{z}$ have been plotted, since these parameters are the most significant. As it could be expected from the statistical properties of sampling, these uncertainties decrease with increasing point number and are proportional to $1/\sqrt{N}$.

As already pointed out, the previous experiment was conducted with a constant acquisition speed. Increasing the number of digitized points leads thus to great acquisition times and consequently to high measurement costs. The experiment was therefore carried out again, adjusting the acquisition speed in order to keep the total acquisition time constant. The results are plotted in figure 3.

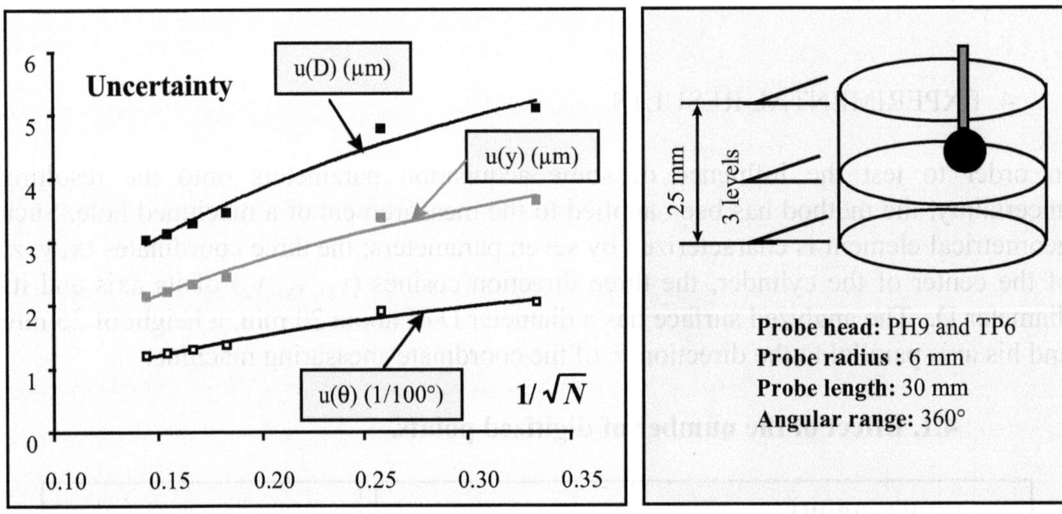

Figure 3; Effect of the number of digitized points onto the measurement uncertainty when the total time of the measurement is kept constant.

Even in that case, the uncertainties of the estimated parameters decrease with increasing number of points. This shows that a sufficient number of measured coordinates should be used to characterize the acquired surface accurately.

4.2. Effect of the digitized zone angular range.

In the second type of experiment, the effect of the angular range used for the acquisition of the hole has been analyzed. These measurements allowed also to test the stability of our algorithms. In the experiment, the acquisition speed was kept constant and set to half the maximum sampling rate of the coordinate-measuring machine. The number of digitized points was fixed to 30. The results are shown in figure 4.

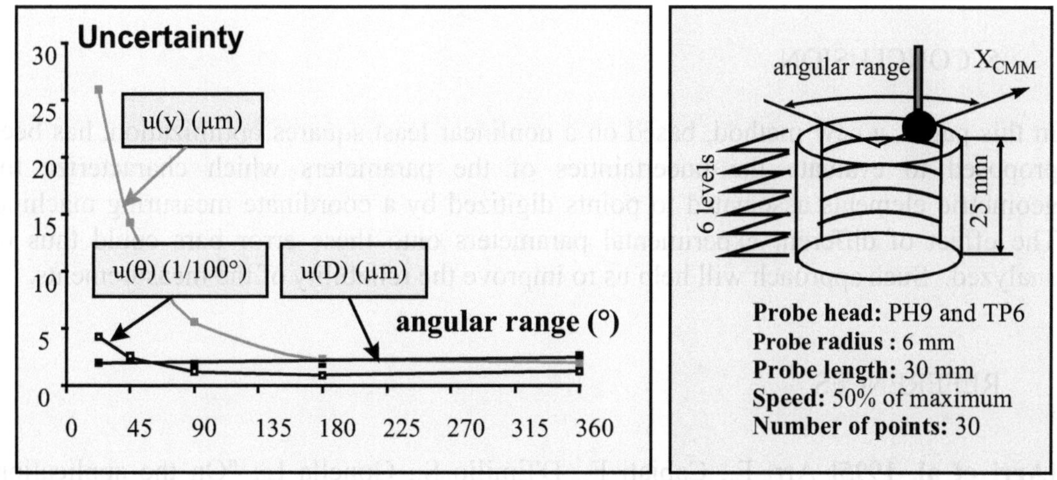

Figure4; Effect of the angular range onto the measurement uncertainty.

A decrease of the angular range greatly affects the precision at which the center of the hole is localized. On the contrary, the diameter of the cylindrical surface is accurately defined, even with a very small digitized zone.

4.3. Effect of the digitized zone height.

In the last type of experiment, the effect of the height of the digitized zone has been defined. As previously, the acquisition speed was set to half maximum and the number of points was fixed to 30. The results are presented in figure 5.

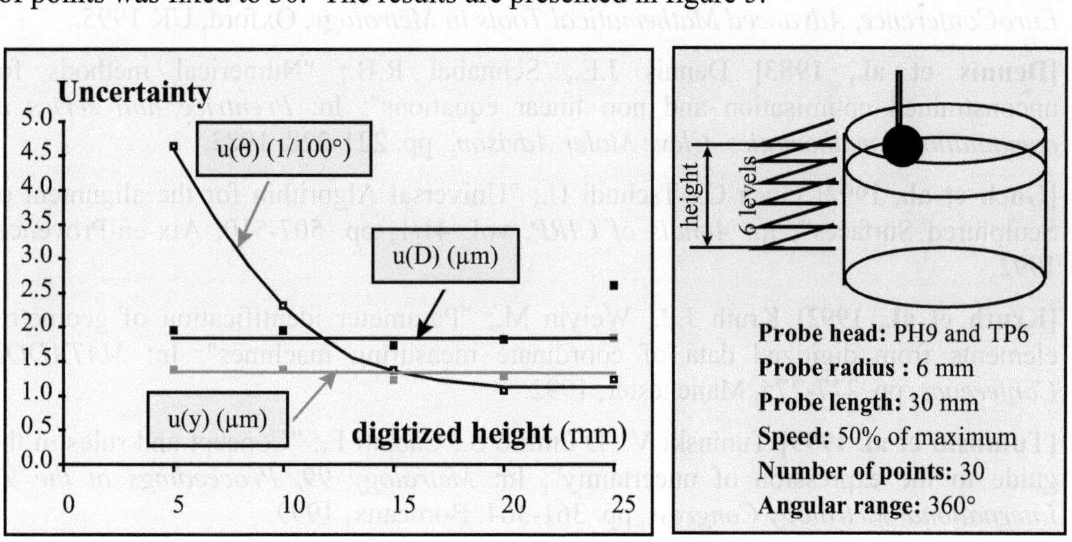

Figure5; Effect of the height of the digitized zone onto the measurement uncertainty.

A reduction of the height of the digitized zone greatly affect the precision at which the direction of the hole axis is defined.

5. CONCLUSION.

In this paper, a new method, based on a nonlinear least squares optimization, has been proposed to evaluate the uncertainties of the parameters which characterize the geometric elements associated to points digitized by a coordinate measuring machine. The effect of different experimental parameters onto these error bars could thus be analyzed. Such approach will help us to improve the reliability of the measurements.

REFERENCES

[Arri et al. 1995] Arri E., Cabiati F., D'Emilio S., Gonella L.; "On the application, Guide to the expression of the uncertainty in measurements to measuring instruments", In: *Measurement*, vol. 16, pp. 51-57, 1995.

[BIPM, 1993] BIPM, IEC, ISO, IUPAC, IUPAP, OIML; "Guide to the expression of the uncertainty in measurement, First Edition". 1993, ISBN 92-6710188-9.

[Bourdet and Clément, 1988] Bourdet P., Clément A.; "A study of optimal-criteria identification based on the small displacement screw model"; In: *Annals of CIRP*, vol. 37, pp. 503-506, 1988.

[Bourdet et al., 1995] Bourdet P., Mathieu L., Lartigue C., Ballu A.; "The concept of the small displacement torsor in metrology"; In: *Proceedings of the International EuroConference, Advanced Mathematical Tools in Metrology*, Oxford, UK 1995.

[Dennis et al., 1983] Dennis J.E., Schnabel R.B.; "Numerical methods for unconstrained optimisation and non linear equations"; In: *Prentrice hall series in computational mathematics. Cleve Moler Advison*, pp. 221-228, 1983.

[Goch et al., 1992] Goch G., Tschudi U.; "Universal Algorithm for the alignment of Sculptured Surfaces"; In: *Annals of CIRP*, vol. 41/1, pp. 507-510, Aix-en-Provence, 1992.

[Kruth et al., 1992] Kruth J.P., Weiyin M.; "Parameter identification of geometric elements from digitized data of coordinate measuring machines"; In: *MATADOR Conference*, pp. 277-275, Manchester, 1992.

[Tuninski et al. 1999] Tuninski V., D'Emilio S., Cabiati F.; "Concept and rules in the guide to the expression of uncertainty"; In: *Metrology 99, Proceedings of the 9th International Metrology Congress*, pp. 361-364, Bordeaux, 1999.

Probabilistic evaluation of invariant surfaces through the Parzen's method

Paolo Chiabert, Mario Costa
Department of Economics and Production, Politechnic of Turin
C.so Duca degli Abruzzi, 24 –10129 Torino (Italy)
chiabert@lep.polito.it

Abstract: Clouds of points are often obtained by measuring manufactured parts both in presence of sculptured and simple surfaces. A cloud of points can be imagined as a set of random variables centred on the (unknown) corresponding nominal points.
This paper proposes a statistical approach to the extraction of a nominal model from the cloud of measured points. In particular, the Parzen's method is employed to obtain a model-independent, non-parametric estimate of the unknown Probability Density Function (PDF) of the measurements from a finite number of samples. The nominal models are defined according to the classification based on surface invariance with regard to the rigid motions recently submitted to ISO/TC213 Technical Committee.
Some experimental results show the performance of the method when applied to sparse sampling of simple surfaces.

Keywords: Surface recognition, Non-parametric statistical analysis.

1. INTRODUCTION

The Geometrical Product Specification (GPS) language, proposed by ISO/TC213 in the field of definition, control, and verification of dimensional and geometrical characteristics of parts, is focused on the logical process that associates a nominal design feature to the measurement of the corresponding surface on the actual part.

The extraction of data from actual objects is naturally affected by errors: each step along the path moving from the measurement of actual part to the identification of nominal shape and geometrical errors, introduces uncertainty.

This paper proposes a non-parametric approach where the measurement of actual shape is interpreted as a probabilistic representation of a nominal surface not necessarily described by a finite set of parameters. The focus of the paper is centred on the analysis of single surfaces although in a near future the partition problem could be a privileged application field.

The probabilistic interpretation of the clouds of points does not require an explicit parametric model of the nominal surface. On the other side it does not condense the information if some additional data are not inferred from the PDF. Moreover, the three-dimensionality of the PDF represents a hard computational problem that can be

P. Bourdet and L. Mathieu (eds.),
Geometric Product Specification and Verification: Integration of Functionality, 245-254.

simplified by employing some tricks in case the PDF should present some peculiar characteristics. As a general case, the PDF requires a complex description often based on estimators.

2. MATHEMATICAL FOUNDATION

The proposed method uses two powerful mathematical tools. The Parzen method provides a model-independent, non-parametric estimate of an unknown PDF, while the classification of subsets of Euclidean space according to invariance under rigid motions constitutes a criterion to investigate on the nominal surface hidden in the cloud of measured points.

2.1 Parzen method in the non-parametric estimation of PDF

When the parametric form of a density function is unknown, it is necessary to employ *non-parametric estimation.* One of the classical non-parametric approaches is Parzen Density Estimation, which works best in situations where data are the only source of information about the density.

Suppose $p(x)$ is a density for the random variable x. The range of x is given by the region $D \subseteq \Re^n$. The likelihood for x, lying in some regions $\Re \subset D$, is $P(x \in \Re) = \int_{\Re} p(\zeta) d\zeta$. We know $P(x \in \Re)$ is just a smoothed version of p and it is to say that P gives an estimation of p.

Consider a set $\{x_1, .., x_n\}$ of independent and identically distributed observations of x, according to the distribution $p(x)$.

The probability that k sampling fall into the region $\Re$ is a binomial distribution, which peaks sharply about the mean $E[k] = nP(x \in \Re)^K$ and drive a random sample S. Let consider $p(x)$ smooth, and $\Re$ small enough for $p(x)$ doesn't have big variation. If the volume of $\Re$ is V, is reasonable to estimate $\int_{\Re} p(\zeta) d\zeta \approx p(x \in \Re) V$. Considering $\int_{\Re} p(\zeta) d\zeta = P(x \in \Re) \approx k_{\Re} / n$ the estimation $p(x) \approx (k_{\Re} / n) / V$ is derived and it represents a space average estimation of $p(x)$ on the region $\Re$: $p_{\Re}(x) = \int_{\Re} p(\zeta) d\zeta \, / \int_{\Re} d\zeta$.

The estimation of $p(x)$ is better approaching the volume of $\Re$ to zero. In this case, if the number of samples remains constant and $\Re$ becomes so small to enclose no samples, we have $p(x) \approx 0$, so this is useless. But if an infinite number of samples are available estimating $p(x)$, it is possible to operate defining a sequence of regions $\Re_1$, $\Re_2$, .., $\Re_n$ (each of them must contain x), extracting n samples of the distribution, using $\Re_n$ to estimate the n'th $p(x)$ for each value of n, computing k_n (number of samples from the sample population of size n) which are contained in $\Re_n$, computing V_n (volume of $\Re_n$):

the n'th estimation to $p(x)$ is given by $p(x) = \frac{k_n / n}{V_n}$. Assuring the convergence to $p(x)$, the following conditions are required:

$$\lim_{n\to\infty} V_n = 0, \quad \lim_{n\to\infty} k_n = \infty, \quad \lim_{n\to\infty} k_n / n = 0.$$

Now our attention is concentrated on the generation of the regions $\Re_n$: the strategy consists on shrinking $\Re_n$ ensuring that the volume V_n has been expressed by a function of n, for example $V_n = 1/\sqrt{n}$, and the random variables k_n and k_n/n behave respecting the conditions above. This target is gained by application of Parzen Window method.

2.2 Classification of invariant subsets in $\Re^3$

In the Euclidean space a rigid motion is an operator m: $\Re^3 \to \Re^3$ that transforms a geometric subset $S \subseteq \Re^3$ in a congruent copy $S' = m(S)$. From a mathematical point of view, the set RM of rigid motions in $\Re^3$ forms a group because:

- successive applications of rigid motion m_1, which transforms S in S', and m_2, which transforms S' in S'', is equivalent to the application of a rigid motion m_3, which transforms S in S'', i.e. $m_1(m_2(S)) = m_3(S)$ for $m_1, m_2, m_3 \in RM$;
- a null rigid motion i exists in RM which leaves the geometrical set unchanged, i.e. $i(S)=S$ for $i \in RM$;
- an inverse rigid motion m^{-1} exists for each motion m which moves the geometrical object back to its original place, i.e. $m^{-1}(m(S))=S$ for $m^{-1}, m \in RM$.

Number	Rigid motion	Surface $S \subseteq \Re^3$	$Aut(S)$	Reference element $\supset \Omega_i$	dim $(Aut(S))$
1	Spherical rotations	Sphere	R(3)	Point	3
2	Axial rotations and unidirectional translations along the same axis;	Cylinder	T(1)xR(1)	Straight line	2
3	Planar translations	Plane	T(2)xR(1)	Plane	3
4	Helical motions of pitch ε	Helical	T(1)xR(1) pitch ε	Helix	2
5	Axial rotations;	Revolution	R(1)	Point, Straight line	1
6	Unidirectional translations	Prismatic	T(1)	Straight line, Plane	1
7	Identity	Complex	i	Point, Straight line, Plane	0

Table I. Classes of invariance or symmetry.

The rigid motions in $\Re^3$ can be grouped in twelve different subgroups and seven of them contain operators that leave a proper set $S \subset \Re^3$ invariant. These operators are called automorphisms and are mathematically described as: $Aut(S)= \{m \in RM: m(S)=S\}$. It is possible to show that $Aut(S)$ is a group and, in more detail, that it is a subgroup of RM:

- if $m_1, m_2 \in Aut(S)$ then $m_1(m_2(S))=m_1(S)=S$ and $m_1 m_2 \in Aut(S)$;
- $i(S)=S$ and $i \in Aut(S)$;
- since $m \in RM$ there is $m^{-1} \in RM$ such that $m^{-1}(S)= m^{-1}(m(S))= i(S)$ and $m^{-1} \in Aut(S)$.

Automorphisms capture the symmetry of S and the invariance of S can be restricted to some peculiar geometric elements used to define the location of the set. In Table I are summarised the results on invariance classification.

3. METHODOLOGICAL APPROACH

Given a set $S \subset \Re^3$, let i_S: $\Re^3 \rightarrow \{0,1\}$ be the indicator function of S (i.e. $\forall P \in \Re^3$, $i_S(P)=1$ iff $P \in S$). In case S has finite volume (as it should be the case for all practical purposes), we are allowed to normalise i_S so as to define the PDF $p_S(P) = i_S(P) / \int i_S(Q) d^3Q$ which, by construction, is uniform on its support S. Of course, given any rigid motion $g \in Aut(S)$ we have $p_S(P)=p_S(gP)$ $\forall P \in \Re^3$. A measured point $P \in \Re^3$ can be viewed as a "noisy" version of some unknown point $Q \in S$. In case S is sampled uniformly, the related PDF is $\hat{p}_S(P) = \int p_n(P|Q) p_S(Q) d^3Q$, being $p_n(P \mid Q)$ the conditional PDF that accounts for the noise.

The proposed method does not assume any explicit parametric model for the set S but founds its efficiency on the following two aspects:

1. it looks for invariance in the measured data in order to extract practical results from $\hat{p}_s$;
2. it employs techniques to evaluate object having different complexity degrees

The former aspect can be afforded by resorting to the invariance in the set S and, consequently, in the measured data. The preservation of invariance between nominal surface and measured data implies precise conditions on the noise structure.

Given any rigid motion $g \in Aut(S)$ is $p_S(P)=p_S(gP)$ $\forall P \in \Re^3$, so we'll demonstrate that $\hat{p}_S(P)= \hat{p}_S(gP)$. This can be accomplished noting $\hat{p}_S(gP)$ = $\int p_n(gP \mid gQ) p_S(gQ)\, d^3(gQ)$, where $d^3(gQ)=|\mathbf{J}| d^3Q$ and $|\mathbf{J}|=1$ because $g \in RM$. In other words $\hat{p}_S(P)$ = $\hat{p}_S(gP)$ iff $p_n(gP|gQ) = p_n(P|Q)$, i.e. if the noise is invariant under the rigid motion g. These conditions are naturally satisfied when the noise only depends on the distance $\|P-Q\|$ between the nominal and the measured points. Under these conditions it can be asserted that $\hat{p}_s$ reflects the incidental invariance of S and vice versa.

In spite of the conditions imposed on the noise structure, it is impossible to describe the three-dimensional PDF in a closed form because, among those assumptions, there isn't an explicit definition of noise. However, the invariant hypothesis allows the use of a semi-parametric model M_i whose parameters Ω_i take into account the symmetry of the invariant surface and are part of the set of restricted invariant elements that locate the measured set.

Moreover, the invariant hypothesis defines a subset $E_i \subseteq \Re^3$ containing the elements belonging to equivalence relationship imposed by the symmetry. In mathematical terms, let Gi the symmetry group associated to model Mi, the equivalence relationship $E_i=\{(P,Q)\in \Re^3 \times \Re^3 : Q=gP$ for some $g\in G_i\}$ simplifies the original PDF $\hat{p}_s$ by projecting it onto the quotient set $\Re^3/E_i$ through a suitable parametric function $Z_i(\cdot;\Omega_i)$. The projected PDF is then approximated by a consistent estimate $\tilde{p}$ provided by Parzen methods.

Replacing a direct comparison of the models' likelihood with a model ranking method based on the Leave One Out technique explains the ability of this method in evaluating different level of complexity. Each hypothesis on the structure of the measured surface will be compared to all the available possibilities creating a classification that establishes the most confident solution. The comparison is performed evaluating the probability that the "left out point", not used in the evaluation of a particular hypothesis, belongs to the nominal surface described in the hypothesis.

From a mathematical point of view, if D is a set of n measurements $D=\{(x_1,y_1,z_1),(x_2,y_2,z_2),\ldots,(x_n,y_n,z_n)\}$, $D_j=D\backslash\{(x_j,y_j,z_j)\}$ $\forall j\in\{1,2,\ldots,n\}$, M is a set of m alternative models $M=\{M_1,M_2,\ldots,M_m\}$ where M_i may involve a set of parameters Ω_i, and $Z_i(\cdot;\Omega_i)$ is the parametric function that modifies the original three-dimensional PDF according to the invariance hypothesis, the direct ranking based on the evaluation of models' likelihood

$$\max_{\Omega_i} \tilde{p}(D \mid M_i, \Omega_i, Z_i(D;\Omega_i))$$

is replaced by the Leave One Out strategy which does not penalise the simplest model:

$$L(M_i) = \prod_{j=1}^{n} \tilde{p}(x_j, y_j, z_j \mid M_i, \Omega_{ij}, Z_i(D_j;\Omega_{ij}))$$

being Ω_{ij} the maximum likelihood estimate of Ω_i based on data set D_j. That is:

$$\Omega_{ij} = \arg\max_{\Omega_i} \tilde{p}(D_j \mid M_i, \Omega_i, Z_i(D_j;\Omega_i))$$

3.1 Spherical invariance G_1

In case of spherical invariance the Parzen method is used to estimate the unknown PDF $\hat{p}(\rho)$. In detail, defined $\Omega_1=\{a,b,c\}$, being $(a,b,c)\in\Re^3$ the centre of the sphere, in order to project raw measurements onto the quotient set $\Re^3/E_1=\Re_+$ it is worth resorting to polar co-ordinates:

$$(x,y,z)\rightarrow(\rho,\phi,\theta)=$$
$$=\left(\sqrt{(x-a)^2+(y-b)^2+(z-c)^2},\arctan\left(\frac{y-b}{x-a}\right),\arcsin\left(\frac{z-c}{\sqrt{(x-a)^2+(y-b)^2+(z-c)^2}}\right)\right)$$

so that $Z_1(D_j;\Omega_1)=\{\rho_1,\rho_2,\ldots,\rho_{j-1},\rho_{j+1},\ldots,\rho_n\}$.

The invariance of $\tilde{p}(x,y,z\,|\,M_1,\Omega_1,Z_1(D_j;\Omega_1))$ with regard to rotations about the point (a,b,c) is then enforced as follows:

$$\tilde{p}(\rho,\phi,\theta\,|\,M_1,\Omega_1,Z_1(D_j;\Omega_1))=\frac{\cos(\theta)}{4\pi}\tilde{p}(\rho\,|\,Z_1(D_j;\Omega_1))$$

where the PDF $\tilde{p}(\rho\,|\,Z_1(D_j;\Omega_1))$ is supplied by the Parzen method. Going back to cartesian co-ordinates we finally obtain :

$$\tilde{p}(x,y,z\,|\,M_1,\Omega_1,Z_1(D_j;\Omega_1))=\frac{\tilde{p}\left[\sqrt{(x-a)^2+(y-b)^2+(z-c)^2}\,|\,Z_1(D_j;\Omega_1)\right]}{4\pi\left[(x-a)^2+(y-b)^2+(z-c)^2\right]}$$

3.2 Cilindrical invariance G_2

Model M_2 states that there exists an axis ***u*** such that $\hat{p}(x,y,z)$ is invariant with regard to rotations about ***u***. It is therefore convenient to define a suitable transformation of co-ordinates $(x,y,z)\rightarrow(x',y',z')$ so that z' lies along ***u***. Such a transformation can be described through a rotation around the ***x*** axis followed by a rotation around the **current *y*** axis and then by a translation in the **current *x-y*** plane. That is:

$$\begin{bmatrix}x'\\y'\\z'\end{bmatrix}=\begin{bmatrix}\cos(\beta)&0&-\sin(\beta)\\0&1&0\\\sin(\beta)&0&\cos(\beta)\end{bmatrix}\begin{bmatrix}1&0&0\\0&\cos(\alpha)&\sin(\alpha)\\0&-\sin(\alpha)&\cos(\alpha)\end{bmatrix}\begin{bmatrix}x\\y\\z\end{bmatrix}-\begin{bmatrix}a\\b\\0\end{bmatrix}=$$
$$=\begin{bmatrix}x\cos(\beta)+y\sin(\beta)\sin(\alpha)-z\sin(\beta)\cos(\alpha)-a\\y\cos(\alpha)+z\sin(\alpha)-b\\x\sin(\beta)-y\cos(\beta)\sin(\alpha)+z\cos(\beta)\cos(\alpha)\end{bmatrix}$$

Thus $\Omega_2=\{a,b,\alpha,\beta\}$. If we now adopt cylindrical co-ordinates:

$$(x',y',z')\rightarrow(\rho,\phi,z')=\left(\sqrt{x'^2+y'^2},\arctan\left(\frac{y'}{x'}\right),z'\right)$$

the above invariance can be stated as follows:

$$\hat{p}(\rho,\phi,z')=\frac{1}{2\pi}\hat{p}(\rho)\hat{p}(z')$$

The unknown PDFs $\hat{p}(\rho)$ and $\hat{p}(z')$ are estimated through the Parzen method upon projection of each triplet (x_k,y_k,z_k) in D_j onto the corresponding couple (ρ_k,z'_k) of transformed co-ordinates. Let $\tilde{p}[\rho,|Z_2(D_j;\Omega_2)]$ and $\tilde{p}[z'|Z_2(D_j;\Omega_2)]$ be the Parzen estimate of $\hat{p}(\rho)$ and $\hat{p}(z')$, where:

$$Z_2(D_j;\Omega_2)=\{(\rho_1,z_1),(\rho_2,z'_2),\ldots,(\rho_{j-1},z'_{j-1}),(\rho_{j+1},z'_{j+1}),\ldots,(\rho_n,z'_n)\}.$$

We have:

$$\tilde{p}(x,y,z\,|\,M_2,\Omega_2,Z_2(D_j;\Omega_2))=\frac{\tilde{p}[f_\rho(x,y,z;\Omega_2),f_{z'}(x,y,z;\Omega_2)\,|\,Z_2(D_j;\Omega_2)]}{2\pi f_\rho(x,y,z;\Omega_2)}$$

$$f_\rho(x,y,z;\Omega_2)=\{[x\cos(\beta)+y\sin(\beta)\sin(\alpha)-z\sin(\beta)\cos(\alpha)-a]^2+[y\cos(\alpha)+z\sin(\alpha)-b]^2\}^{1/2}$$

$$f_{z'}(x,y,z;\Omega_2)=x\sin(\beta)-y\cos(\beta)\sin(\alpha)+z\cos(\beta)\cos(\alpha)$$

3.7 Identity invariance G_7

Stated $\Omega_7=\varnothing$ and $\Re^3/E_7=\Re^3$, there is little to project here. The Parzen method is directly applied on the raw measurements and provides the PDF:

$$\tilde{p}(x,y,z\,|\,M_7,\Omega_7,Z_7(D_j;\Omega_7))=\tilde{p}(x,y,z\,|\,D_j)$$

4. EXPERIMENTAL TESTS

The algorithm developed on the basis of the illustrated methodology, can be schematised as follows:

1. Acquisition of input data $\mathbf{D}=\{P_1, P_2, \ldots, P_N\}$
2. Selection of invariance hypothesis G_k k=1, ..., 7
3. Loop A: For j=1, ..., N
   4. Leave Out P_j
   5. Loop B: While Ω_k not satisfying
      6. Optimise Ω_k with regard to $\mathbf{D_j} = \mathbf{D} / P_j$
   7. Evaluate $\tilde{p}$ (P_J | M_k, Ω_{kj}, $Z_k(D_j; \Omega_{kj})$)
8. End

A preliminary test of this algorithm has been performed in the recognition of clouds of points sampled from the sphere, cylinder and generic surface represented in Figure 1.

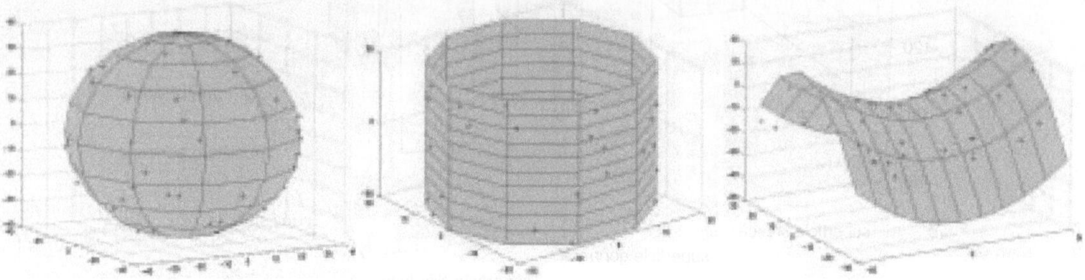

Figure 1 Graphic representation of sampled surfaces

Sampling Points =10	*Invariance hypothesis*		
	Spherical invariance	**Identity invariance**	**Cylindrical invariance**
Sphere	- 87.6031	-141.7135	-138.778
Generic surface	-128.5544	-126.7122	-128.7103
Cylinder	-140.7322	-143.1779	- 87.8604

Table II. Results of matching between sampled data and invariant models

Table II reports the numerical values of log *L(Mi)* to rank the sampling of sphere, cylinder and generic surface against the corresponding models. Figure 1 shows a graphic representation of data collected in Table II, and demonstrates that algorithm correctly assigns the clouds of points to the right classes.

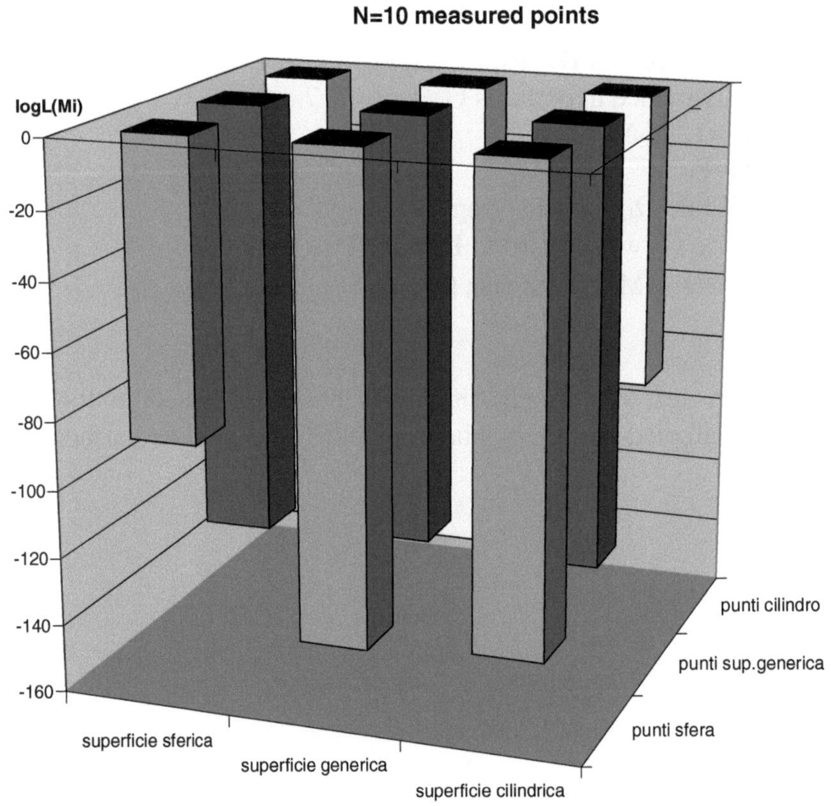

Figure 2. Graphical representation of data collected in Table II

Results in Table II are obtained using a very sparse sampling over the unknown surface: ten points over a sphere of diameter 80 mm.

In order to evaluate the impact of the sampling dimension on the discriminatory capability, the test has been repeated increasing the number of points. Figure 2 shows the results of this test. It is possible to point out that the recognition of the right solution is shaper with higher number of points.

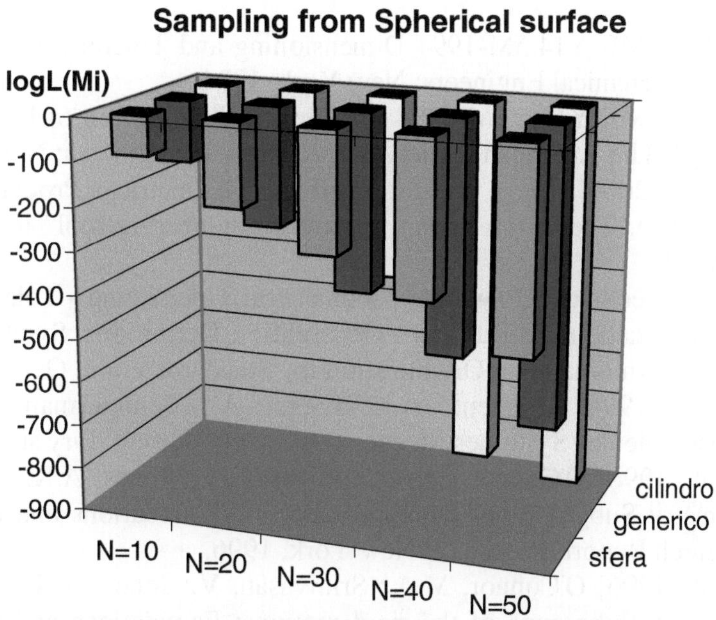

Figure 3. Graphical representation of results from different number of sampled points

It is noteworthy that identity invariance, corresponding to generic surfaces, has a smoothed behaviour in comparison to spherical and cylindrical invariance. Obviously a generic surface comprehends the particular cases of spherical and cylindrical surfaces, and although with a slower convergence, it is able to match every kind of surface. This means that the identity invariance model is likely to rank first only if no other invariance class is able to gain a good matching against the sampled points.

The computational effort is quite relevant, especially when a robust optimisation of invariant parameters is requested. The computing time excludes at the moment industrial applications of this algorithm.

5. CONCLUSION AND REMARKS

In this paper an innovative and robust approach to the recognition of sampled surface is presented. At the time of paper publication, the methodology is not ready for an industrial application, but this fact does not diminishes its merits that, according to the authors' opinion, consist of the robust mathematical foundation and the wide generality and applicability. Further development of this work could interest the partition of complex shapes as well as the evaluation of geometrical defects of measured surfaces.

REFERENCES

[ASME, 1994] "ASME Y14.5M-1994 Dimensioning and Tolerancing"; The American Society of Mechanical Engineers; New York, 1994.

[ASME, 1994] "ASME Y14.5.1M-1994 Mathematical Definition of Dimensioning and Tolerancing"; The American Society of Mechanical Engineers; New York, 1994.

[ISO/TC213N355, 2000] "Next generation of the Geometrical Product Specification (GPS) language. The vision for an improved engineering tool" in ISO/TC 213 N 355 Annexe 1.

[ISO1101, 1983] "Technical Drawings - Geometrical Tolerancing - Tolerances of Form, Orientation, Location and Run-out - Generalities, Definitions, Symbols, Indications on Drawings"; International Organization for Standardization; Geneve, 1983.

[Clement et al., 1994] Clement, A.; Riviere, A.; Temmerman, M.; "Cotation Tridimensionelle des Systemes Mecaniques"; PYC Edition, Ivry sur Seine 1994.

[O'Connor et al., 1996] O'Connor, M.A.; Srinivasan, V.; Jones, A.K.; "Connected Lie and Symmetry Subgroups of the rigid motions: Foundations and classifications"; IBM Research Report, RC20512; New York, 1996.

[O'Connor et al., 1996] O'Connor, M.A.; Srinivasan, V.; Jones, A.K.; "Connected Lie and Symmetry Subgroups of the rigid motions: Foundations and classifications"; IBM Research Report, RC20512; New York, 1996.

[Fukunaga, 1990] Fukunaga, K.; "Introduction to Statistical Pattern Recognition"; Academic Press; Boston, 1990.

[Karger et al., 1985] Karger, A.; Novak, J.; "Space Kinematics and Lie Groups"; Gordon and Breach Science Publishers; New York, 1985.

[Srinivasan, 1999] Srinivasan, V.; "A Geometrical Product Specification Language Based on a Classification of Symmetry Groups"; IBM Research Report, RC21384; New York, 1999.

[Bourdet et al., 1988] Bourdet, P.; Clement, A.; "A Study of Optimal- Criteria Identification Based on Small Displacement Screw Model"; In: CIRP Annals 1988 Manufactuting Technology, vol 37/1/1988, pp.503-506, 1988.

Automated Seam Variation and Stability Analysis for Automotive Body Design

Rikard Söderberg and Lars Lindkvist
Machine and Vehicle Design
Chalmers University of Technology
SE-412 96 Göteborg, Sweden
Email:riso@mvd.chalmers.se

Abstract: The spatial relations between parts in assembled products are often critical. In the automotive industry, the relations between the doors, hoods and panels are important quality characteristics. Functionally it must be possible to open the door without interference with other parts, but the esthetical aspect is also important. Today, the quality appearance of a vehicle is judged by the quality of the gaps between the body panels, i.e. doors, hoods, fenders and panels. This paper presents three basic types of analysis that can be used to evaluate design concepts and predict the final variation: *stability analysis*, *seam variation analysis* and *quality appearance evaluation*. The stability analysis focuses on the locating schemes in the assembly and aims at making the concept as insensitive to variation as possible. Seam variation analysis focuses on the relation between parts in an assembly. A *seam*, as introduced here, is a relation between two parts over a distance. Typically, seam variation is measured and evaluated in two directions, the gap and flush directions. The paper introduces two goodness values for automotive body variation evaluation. The *instability index* reflects the total locating scheme stability for the whole assembly while the *quality appearance index* rates the total variation in all seams of the body. The latter is calculated as the mean variation in all defined seams of a body and allows for evaluating the final appearance of the body, with one measure, already in the early concept phase. The analyses are implemented and performed in the CAT software, RD&T, presented numerically and visualized by color coding. An auto body example is used to illustrate the analyses. The analyses are however general and valid for most assembled products.

Keywords: tolerance analysis, CAD, robust design

1. INTRODUCTION

The spatial relations between parts in assembled products are often critical. In the automotive industry, the relations between the doors, hoods and panels are important quality characteristics. It must be possible to open the doors and hoods without

P. Bourdet and L. Mathieu (eds.),
Geometric Product Specification and Verification: Integration of Functionality, 255-264.

interference with other parts. There is also an esthetical aspect. Today, the quality appearance of a vehicle is often judged by the gaps between doors, hoods, fenders and panels. Tolerance analysis aims at predicting the variation in one or several critical dimensions of a design with respect to expected manufacturing variation. Typically, a critical dimension of an assembled product is the relation between features of two parts. Tolerance analysis is an important link between design and manufacture and has also been an area of academic interest over the years. Tolerance chain detection was treated for early configuration stages in [Söderberg and Johannesson, 1999] and robustness evaluations and coupling analysis were reported in [Söderberg and Lindkvist, 1999]. [Lindkvist and Söderberg, 1999] described concurrent engineering with focus on locating scheme definition and tolerance analysis in a CAT environment. [Carlson *et al.*, 1999] described multi-fixture assembly system diagnosis. Flexible assemblies are treated in [Liu *et al.*, 1996] and [Hu, 1997]. [Ceglarek and Shi, 1997] used a beam-based model for tolerance analysis in sheet metal assemblies. [Sellem and Rivière, 1998] use FEA for the same purpose. [Hu, 1997] present the "stream-of-variation" theory for automotive body assembly. [Maxfield *et al.*, 2000] and [Wickman and Söderberg, 2001] discuss quality appearance and the use of VR (virtual reality) techniques for evaluation.
This paper will describe CAT functionality for supporting two important stages of the geometry design process typical for the automotive industry, namely locating scheme definition and evaluation of exterior panel relations, i.e. quality appearance.

1.1. Scope of the paper

[Söderberg and Lindkvist, 1999] presented a general goodness measure for the degree of geometrical coupling in assemblies. In this paper, the two quality characteristics, robustness and variation as described in the previous section, are analyzed and evaluated using two new goodness values, the *instability index* and the *quality appearance index*. The instability index, described in chapter 2, can be used during the concept phase to judge and evaluate the goodness of the locating schemes controlling the final position of all parts in an assembly. The seam variation analysis described in chapter 3 calculates the flush and gap variation along a relation between two parts, a seam, based on assumed tolerances on parts and fixtures. The quality appearance analysis described in chapter 4 ranks flush and gap for a group of seams and can be used to be used to judge expected external body quality.

2. STABILITY ANALYSIS

2.1 Stability Analysis

A locating scheme such as the 3-2-1 locating scheme, six theoretical points are used to lock six degrees of freedom for a part. In reality, the six theoretical locating points are represented by physical geometrical features such as holes, planes and slots.

By varying each locating point in a locating scheme by a small increment, *Δinput*, one at a time, *Δoutput/Δinput* may be determined in the X, Y and Z directions for a number of output points, *n*, representing the geometry. The *RMS* values corresponding to variation in each of the six locating points, *i*, are then determined for all points. The total *RSS* influence of all six locating points, *i*, is then calculated in each direction, as is the total *RSS* magnitude. The *RSS* sensitivity value shows how well a certain locating scheme controls the position stability of a certain part. In an assembly consisting of a number of parts with individual locating schemes, the individual *RSS* values are presented in the *stability matrix*. Stability analysis, described originally in [Söderberg and Lindkvist, 1999], is done to find the most important positioning schemes, i.e. the most important parts, in an assembly and the most important individual locators in each locating scheme.

Figure 1 shows the stability matrix for the side of a vehicle with front fender, front door and rear door, located on the body with one locating scheme each. The list in figure 1 shows the individual stability index (amplification factor) for each locating point. As is seen in the matrix in figure 1, the amplification factor for the front door locating scheme (the RSS value of the individual amplification factors in the locating scheme) is the highest. This means that a unit variation applied to the door via its locators is amplified by the positioning scheme more than for the other parts. The color coding also shows what areas have the highest amplification. The analysis result and the color coding can be shown in each direction, X, Y, Z, separately or, as in figure 1, absolute (MAG).

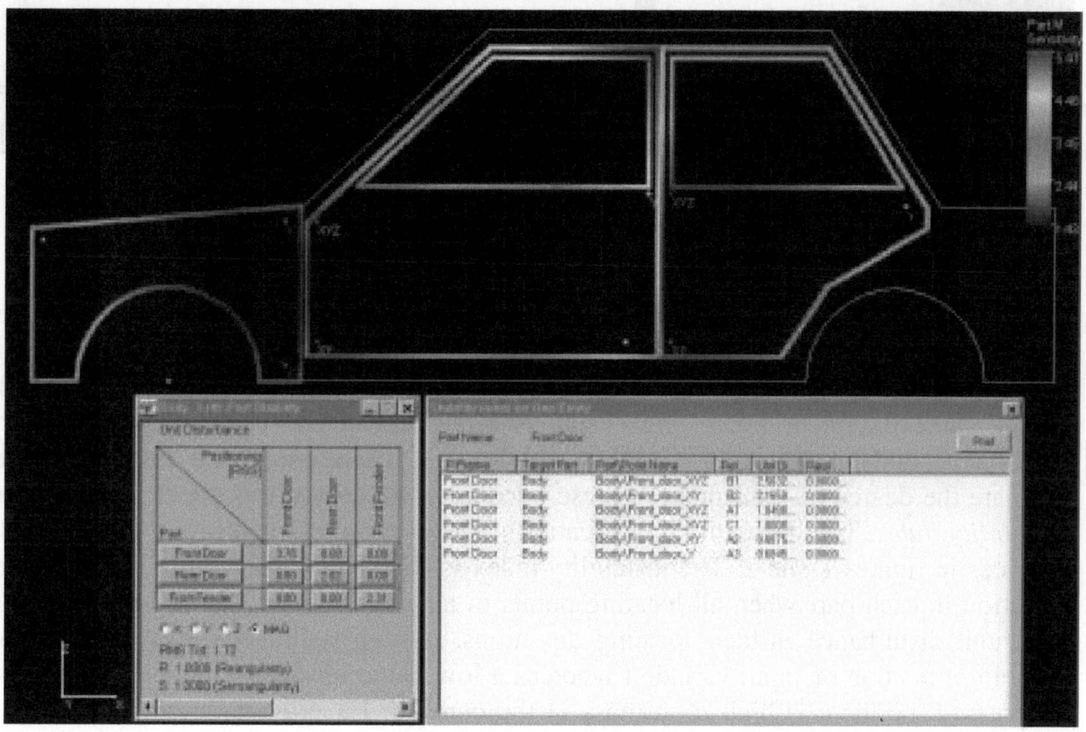

Figure 1; Stability analysis for body side

As is seen, the XYZ and the XY locators are the ones that amplifiy variation most. In figure 2 these points are moved from the left side to the bottom of the door to improve the stability. As can be seen from the matrix, the amplification factor for the front door is now lower as a result of lower individual amplification factors for the XYZ and XY locating elements (the two B points).

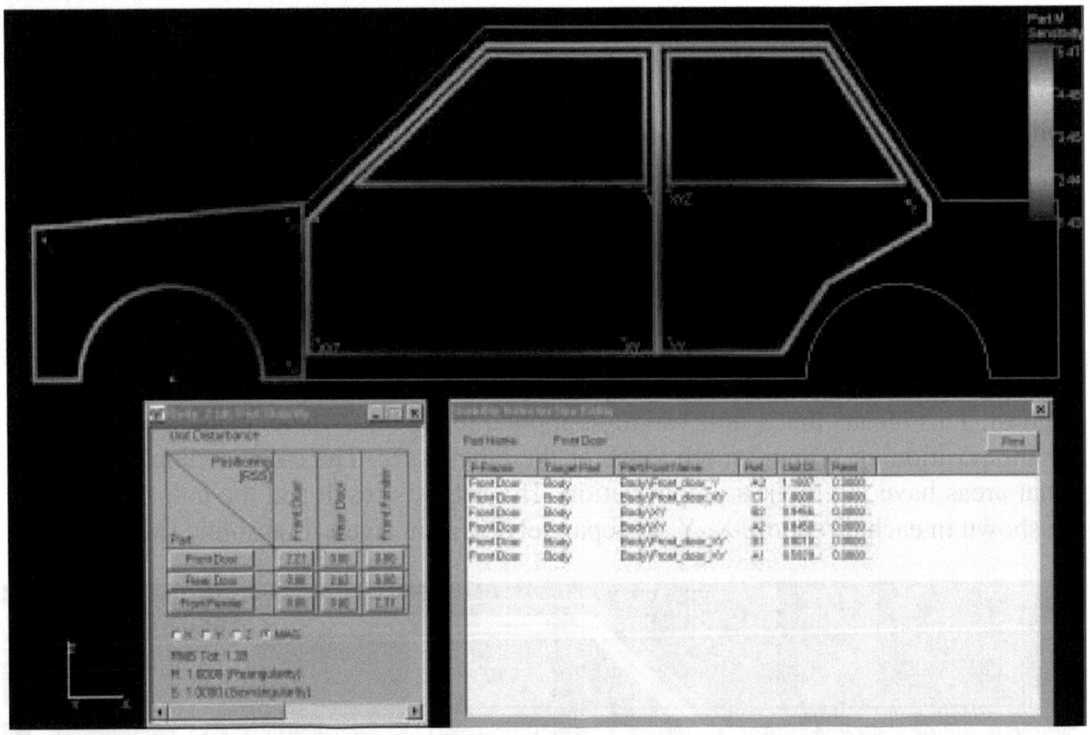

Figure 2; Stability analysis – modified locating scheme for front door

2.2 Instability index

[Söderberg and Lindkvist, 1999] used the two goodness values R and S to evaluate the degree of geometrical assembly coupling, which assumes an effort toward uncoupled concept solutions, see also [Suh, 1990]. For exterior automotive panels or parts located on the same fixture or frame, the design is often uncoupled. This is shown by the diagonal stability matrices and the fact that R=S=1. Therefore, to be able to further evaluate the degree of robustness in these uncoupled situations, this paper presents the *instability index*. The instability (amlification) indexes are presented below the stability matrices in figures 1 and 2. The instability index is calculated as the RMS value of the variation in each part when all locating points in all positioning schemes are disturbed by a unit disturbance in their locating directions. A high stability index value means high amplification of input variation whereas a low value means that input variation is suppressed, i.e there is high robustness in the concept. Equation 1 describes how the instability index is calculated:

$$RMS_{Tot} = \sqrt{\frac{1}{n \cdot m} \sum_{i=1}^{n} \sum_{j=1}^{m} a_{ij}^2} \qquad (1)$$

The instability index represents the total sensitivity (amplification) in an assembly when all locators are disturbed equally in their respective locating directions. As can be seen in figures 1 and 2, improving the stability of the front door also improves the total stability of the assembly, i.e. decreases the instability index.

3. SEAM VARIATION ANALYSIS

This section introduces the *seam* as the relation between two parts over a specified distance and describes the most frequently used quality characteristics for evaluations of geometrical variation in automotive body design.

3.1 Seam variation

Figure 3 describes the two most commonly used relations between exterior automobile panels, *gap* and *flush*. While the gap refers to the distance between two parts in a common plane, the flush refers to the distance between two parts, perpendicular to a plane or a surface. Often these two critical dimensions are measured or calculated for specified point pairs, with one point at each part in a specified 2D cut.
There is often a need to evaluate the gap or the flush variation over the whole distance of the critical dimension, not only in two points. Figure 4 describes the door of a vehicle positioned in relation to the rest of the body.

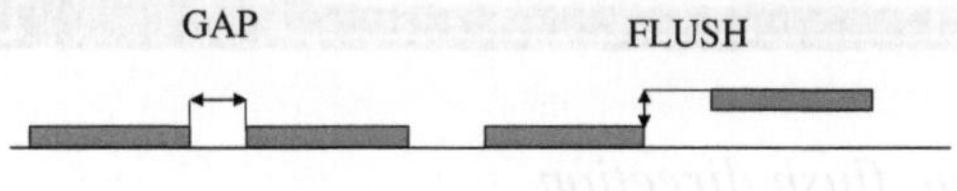

Figure 3; Gap and flush

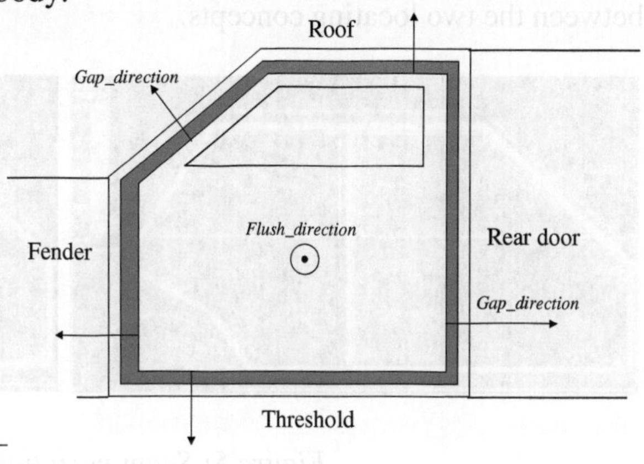

Figure 4; Door seams

It is of interest in terms of the door in figure 4 to be able to predict both gap and flush variation around the whole door edge to evaluate its relation to all surrounding parts. In this paper we refer to a *seam* as the relation between two parts over a specified distance. The gray area in figure 4 shows the seams between the door and its neighbor parts. Seam variation is evaluated in both the gap and flush directions.

3.2 Automatic Seam Generation

To be able to efficiently evaluate flush and gap along the seams of an automotive body, they must be generated more or less automatically in the CAT environment. For that purpose an algorithm for automatic seam generation is developed is used for seam variation evaluation in the following subsections. The algorithm uses points and lines to create seams. Several chaining options and criteria are then available. In this paper, fairly simple geometry is used. However, real tests have been done with good results on real automotive parts. The models have then been generated in CATIA, translated to IGES and imported to RD&T for seam generation. The IGES format is therefore not a prerequisite for the algorithm but was used for this implementation. The algorithm is described in pseudo-code in [Söderberg and Lindkvist, 2001].

3.3 Simulation and Visualization of Seam Variation

Figures 5 and 6 visualise the door relations for the example used in section 2. Three door seams are used. Seam_1 describes the relation between the front door and the body, Seam_2 describes the relation between the front door and the rear door and Seam_3 describes the relation between the rear door and the body. This Monte Carlo based analysis assumed 1mm variation in the locators (mating part or fixture) and 1mm surface profile tolerance for the doors. Figure 5 shows the variation in the flush direction for the two different front door locating concepts described in section 2. As seen from in figure 5, there is not a large difference in variation in the flush direction between the two locating concepts.

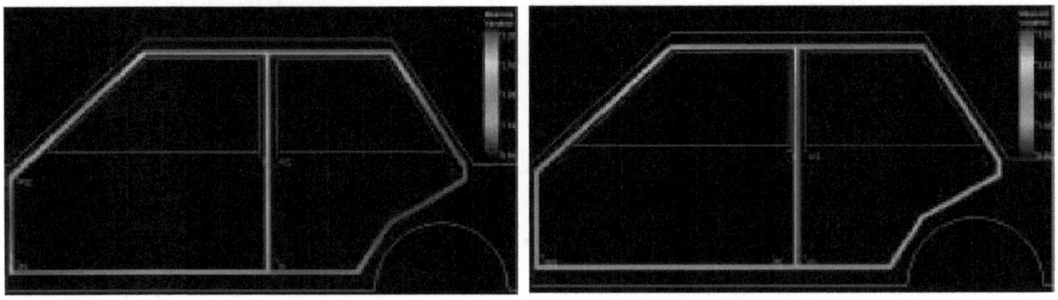

Figure 5; Seam variation, flush direction

Figure 6 shows the variation in the gap direction for the two locating concepts. Here, there is quite a remarkable difference between the concepts. The color coding scale shows that shifting from locating concept one to concept two decreases the variation by approximately a factor 2.

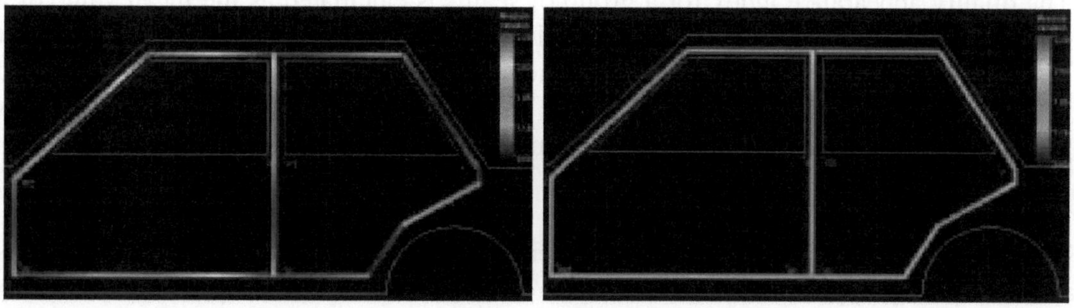

Figure 6; Seam variation, gap direction

Figures 5 and 6 visualise seam variation by using color coding. In many situations, underlying calculated data and complementary information are needed for evaluation.

4. QUALITY APPEARANCE EVALUATION

In this section we present the *quality appearance index* to evaluate how variation and mean flush and gap values affect the visual quality appearance of a vehicle.

4.1 Setting the requirements

[Wickman and Söderberg, 2001] describes the geometry design process in an automotive company and relate it to a general design process description. In the *conceptual design phase,* styling requirements are set, often by the styling department. These requirements are based mainly on competition analyses of other brands in the same segment and in agreement with the quality philosophy of the company. Styling requirements describe the relation and interaction between different features and functions that have a great impact on the overall design quality impression of the vehicle. [Wickman and Söderberg, 2001] describes how these requirements can be defined and evaluated using integration between variation simulation and virtual reality techniques.

Two aspects are important for the quality appearance of a vehicle: i) the variation in flush and gap *within a seam* (comp. parallelism) and ii) the variation *between seams* in the same visual focus. This means that, viewing the vehicle from the side for instance, all seams between doors, fenders and roof must, beyond having small flush and gap variation, also be of equal size. To evaluate these aspects of visual quality appearance we introduce the *quality appearance (QA) index* in this section.

On the basis of a predefined scale, the flush and gap requirements for a group of seams may be translated into an index value representing the required visual quality appearance. The translation table translates the mean seam variation (other quality characteristics than mean value may be used) in flush and in gap for one seam into a quality appearance index.

The quality appearance index for a group of seams is calculated as the mean value of a number of individual seam values, see [Söderberg and Lindkvist, 2001]. Since different areas are visually sensitive to different extents, a weighting system may also be used when calculating the total QA Index. In the requirement definition phase, the QA index is calculated on the basis of styling requirements.

4.2 Evaluating the requirements

Today, the first physical prototype often serves as an evaluation model of the esthetics and the quality appearance of the vehicle. However, one physical prototype does not verify the concept with respect to geometrical variation. When locator positions and expected variation in parts and fixtures are available, the QA index may be calculated on the bases of expected process variation and compared to the one calculated according to styling requirements. Simulated seam variation for a group of seams is translated to a quality appearance index using a company-specific translation table. In our test implementation it is possible to customise a scale with up to ten levels, see [Söderberg and Lindkvist, 2001]. Figure 7 shows the results of the QA analysis and presents the index based on styling requirements and the simulated resulting index, based on selected tolerances and locating schemes. Figure 7 also presents the simulation results in

Figure 7; Total QA Index for the group Doors

The underlying simulation results used to calculate the total QA index may also be analyzed seam by seam. Figure 8 shows the results of the simulation of a group of seams, consisting of three seams, as described earlier in this paper, and is evaluated in both the gap and flush directions. The distribution of maximum, minimum and mean variation along the seam, and the range variation (parallelism) are presented for each seam together with the individual QA index. The total QA index for the group is calculated as the mean of all individual indices.

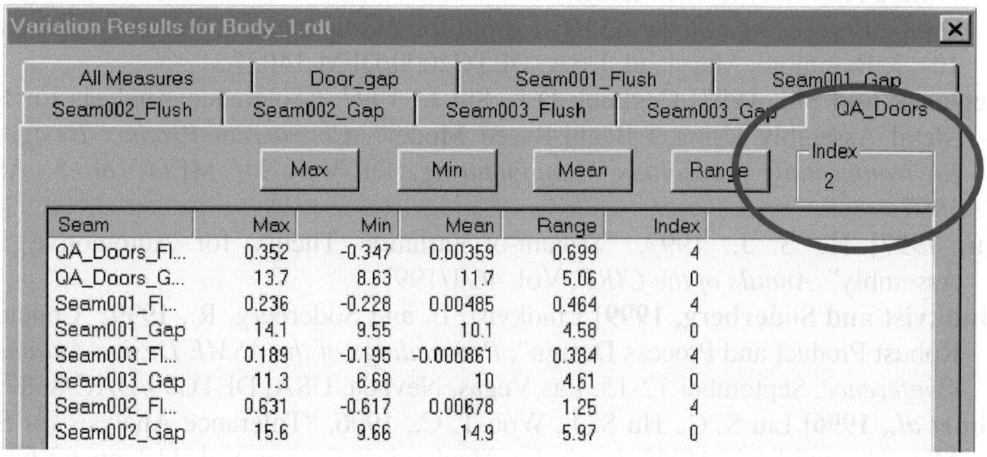

Figure 8; QA Index for group (QA_Doors) and individual seams

5. CONCLUSIONS

This paper has presented a CAT tool that supports the geometry design process in very early stages when locators are determined and robustness is built into the concept as well as the setting and evaluation of visual requirements with respect to expected part and process variations. The functionality presented is specially adapted in this paper to the automotive area and is described using a side door example to illustrate three basic functionalities:

1. *Stability analysis*, used in early design phases to elaborate and optimize the selection of locators for exterior automobile panels. The effect of geometrical variation amplification is shown with color coding and numerical sensitivity matrices. The *instability index* allows two similar concepts to be compared with respect to geometrical robustness.
2. *Seam variation analysis*, used to evaluate expected variation in relations between exterior panels, based on assumed variation in individual parts and fixtures. The seam between two parts is automatically defined using curve tracing and evaluated in the flush and gap directions. The expected variation over a seam is shown with statistical data as well as with color coding.
3. *Quality appearance evaluation*, used to evaluate the total quality appearance of a group of seams, typically the front end, the rear end or the side with the doors, with respect to expected variation. The quality appearance index summarizes the variation in a group of seams and allows benchmarking against other brands on the market.

6. REFERENCES

[Carlson *et al.*, 2000] Carlson J. S., Lindkvist, L. and Söderberg, R., 2000, "Multi-Fixture Assembly System Diagnosis Based on Part and Subassembly Measurement Data", *Proceedings of the ASME Design for Manufacture Conference*, September 10-13, Baltimore, Maryland, USA, DETC2000/DFM-14037.

[Ceglarek and Shi, 1997] Ceglarek D.J., Shi J., 1997, "Tolerance Analysis for Sheet Metal Assembly Using a Beam-Based Model", *Concurrent Product Design and Environmentally Conscious Manufacturing*, DE-Vol. 94/ MED-Vol. 5. ASME 1997.

[Hu, 1997] Hu S. J., 1997, "Stream-of-Variation Theory for Automotive Body Assembly". *Annals of the CIRP*, Vol. 46/1/1997.

[Lindkvist and Söderberg, 1999] Lindkvist, L. and Söderberg, R., 1999, "Concurrent Robust Product and Process Design", *Proceedings of the ASME Design Automation Conference*, September 12-15, Las Vegas, Nevada, USA, DETC99/DAC-8687.

[Liu *et al.*, 1996] Liu S. C., Hu S. J., Woo T. C., 1996, "Tolerance Analysis for Sheet Metal Assemblies", *ASME Journal of Mechanical Design*, Vol. 118, pp. 62-67.

[Maxfield et al., 2000] Maxfield, J., Dew, P.M., Zhao, J., Juster, N. P., Taylor, S., Fitchie, M., Ion, W. J., Thompson, M., 2000a "Predicting Product Cosmetic Quality using Virtual Environment", *Proceedings of the ASME Computers and Information in Engineering Conference*, September 10-13 2000, Baltimore, Maryland

[Sellem and Rivière, 1998] Sellem E., Rivière A., 1998, "Tolerance Analysis of Deformable Assemblies", *proceedings of ASME Design Engineering Technical Conference*, September 13-16, 1998, Atlanta, USA.

[Söderberg and Johannesson, 1999] Söderberg, R., and Johannesson, H. L., 1999, "Tolerance Chain Detection by Geometrical Constraint Based Coupling Analysis", *Journal of Engineering Design*, Vol. 10, No. 1, pp. 5-24.

[Söderberg and Lindkvist, 1999] Söderberg, R. and Lindkvist L., 1999, "Computer Aided Assembly Robustness Evaluation", *Journal of Engineering Design*, Vol. 10, No. 2, pp. 165-181.

[Söderberg and Lindkvist, 2001] Söderberg, R. and Lindkvist L., 2001, "CAT Supported Automotive Body Design ", submitted to *Journal of Engineering Design*

[Wickman and Söderberg, 2001] Wickman, C. and Söderberg, R., 2001, "Towards Non-Nominal Virtual Geometric Verification by Combining VR and CAT Technologies", accepted for the 7th *CIRP Seminar on Computer-Aided Tolerancing*, April 24-25 2001, Cachan, France

Functional assistance for computer-aided tolerancing software

H. Toulorge, D. Buysse*, A. Bellaccico**, A. Rivière***
**Renault, direction de la mécanique,*
78 rue des bons raisins, 92500 rueil malmaison
e.mail : harald.toulorge@renault.com
e.mail : didier.buysse@renault.com
*** Laboratoire de Mécatronique, ISMCM – CESTI*
3, rue Fernand Hainaut, 93407 Saint Ouen Cedex, France
e.mail : ariviere@ismcm-cesti.fr
e.mail : alain.bellacicco@worldonline.fr

Abstract: A great deal of scientific research and industrial development working the field of C.A.T is aimed at providing user assistance with semantics and syntax. However, little or no assistance is offered to ensure the functionality of this type of specification.

The objective of this paper is to indicate the functional assistance identified by the world automobile group, Renault, and then to specify the first elements of response envisaged for the computerisation of this method.

Keywords: Functional Conditions - Tolerancing - link graph - TTRS

1. FUNCTIONAL SPECIFICATION DETERMINATION PROCESS

The first stage of the functional characterisation process identified by RENAULT consists of the identification and description of entry data: Functional Conditions or Use Aptitude Conditions (UACs) and the contacts between the various components of the system under consideration.

The second part of the assistance process consists of describing the "minimum link" positional chain and the isolation of the surfaces (elementary or functional group) to be associated by the tolerancing process.

The last stage consists of deducing the schema of a functional tolerancing plan from previously identified data and then proposing and validating values for nominals and tolerances.

1.1. The identification of entry data

The function-assisted design process starts with the identification and description of UACs and then the various contacts between components.

P. Bourdet and L. Mathieu (eds.),
Geometric Product Specification and Verification: Integration of Functionality, 265-274.

Identification and description of UACs
UACs contain geometric and qualitative information :
<u>geometric information</u>
The geometric aspect affects the choice of the type of standardised geometric tolerancing ensuring efficient operation of the mechanism. Initial work to determine the various types of UACs normally encountered in mechanical engineering has shown that the UACs of a mechanism are often of two kinds. The geometric aspect of UACs can either be expressed in the form of:

- gap and functional adjustments,
- geometric relations between geometric entities belonging to the same part or to different parts: for example, the UAC for the correct meshing of two cogs can be represented by the parallelism of the 2 axes of the cogs.

<u>qualitative information</u>
Once the geometric aspect of UACs has been ascertained, information regarding their qualitative aspect now has to be determined. We will use this information to propose and validate the values of the nominals and tolerances by means of a dimensional chain calculation.

Identification and description of contacts between components
In this stage, we identify the surfaces and the functional groups of surfaces which form the contacts between the various components of the mechanism. We also identify the types of links connecting these components.

This stage provides us with information on the geometric nature of the surfaces involved in making the contact and the nature of the link between them.

1.2. Position control

Once the UACs and contacts have been described, the third stage consists of determining the minimum link positional chain. It allows the surfaces or interfaces to be associated by tolerancing to be identified.

Once the positional loop of each UAC has been defined, we estimate the impact of each link by calculating the contribution to the UAC and establishing a hierarchy system.

1.3. Tolerancing plan

Tolerancing schema
Once all entry data (UAC and contact) have been compiled, the positional loop determined for each UAC and the pertinent links isolated, all the UACs must now be grouped into every possible state in order to deduce the minimum standard tolerancing schema.

This stage involves intrinsic definition of the surfaces of which each recorded interface is composed. This plan is deduced by means of:

- translation of links into dimensional and geometric tolerancing with standardised position and orientation specifications,
- the intrinsic definition of the components: surfaces or groups of surfaces of each interface (diameter of bores, etc.),
- the establishment of the hierarchy of the elements of each interface,
- the definition of the toleranced element-reference element direction,
- assembly capability requirements,
- the management of hyperstaticity within the interfaces

Quantification of tolerances

The last stages of the function-assisted design process consists of proposing values for the nominals and tolerances, then calculating the values of UACs resulting from these proposals and, finally, validating these choices.

The values of the nominals and tolerances can eventually be optimised, by maximising either the sum of the TIs (Tolerance Intervals) or one TI in particular, by steering it by the impact of these TIs on the UACs.

validation

Once the values of the nominals and tolerances have been validated, the last stage consists of eliminating inclusions and redundant elements.

2. ELEMENTS OF RESPONSE CONCERNING THE COMPUTERISATION OF THIS METHOD

We have endeavoured to determine responses by studying a test mechanism. This mechanism, a balancing cartridge, is used for the dynamic balancing of a crankshaft. It is composed of 6 main parts:

- a base plate and a cover forming the cartridge
- two shafts with their respective pinions constituting the input and output shafts.

The input pinion is activated by a fretted drive wheel on the crankshaft and drives the output pinion.

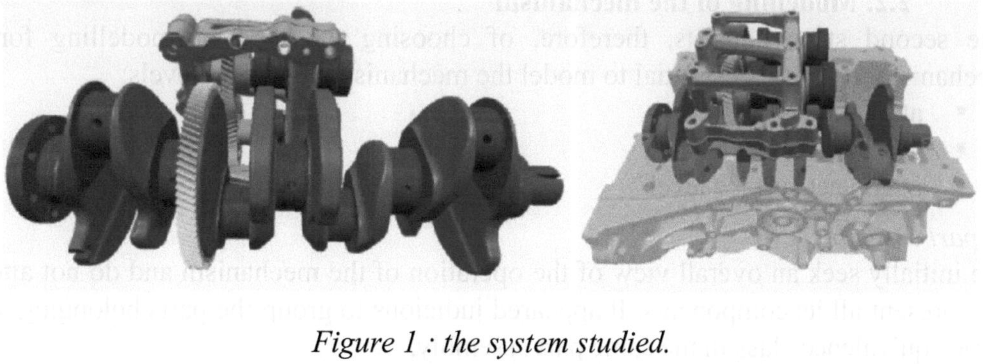

Figure 1 : the system studied.

2.1. Formalisation of a UAC

To ensure correct identification and representation, we felt it would be judicious to use ISO language to express the UACs. ISO language offers the following advantages:

- It is a structured language
- The use of ISO language to express the UACs ensures they are consistent with one another and with the dimensional and geometric specifications applied to the parts

Use of ISO language may seem a roundabout means and does pose limits on the structuring of the language: the formalisation of a UAC is actually achieved:

- without hierarchy system criteria,
- without observing the independence principle, and
- can involve two fictive elements.

We can give the example whereby certain French standards express a functional requirement for geometric tolerancing between two different parts [CLE 99].

The NF ISO 1328-1 standard defines the condition for the correct meshing of two cogs in terms of the parallelism between the two axes, while NF E 48-371 defines the requirement for a tight seal between a shaft and a structure in terms of the coaxiality between the contact surface of the seal and the transmission shaft.

In our test case we have chosen to formalise the correct meshing of the UAC input and output pinions by the parallelism between the axes of these pinions instead of applying a condition relating to the contact surfaces. This UAC is thus formalised by:

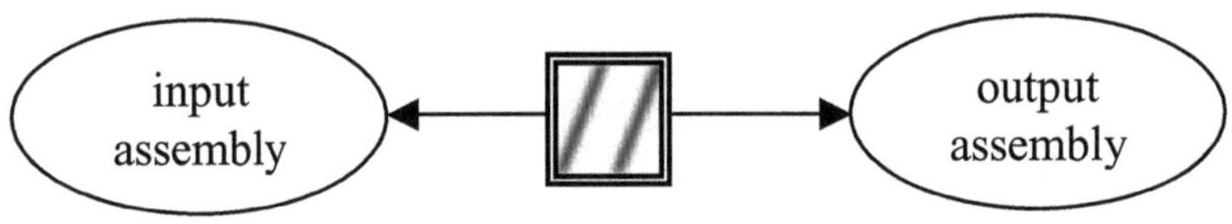

Figure 2 : the system studied.

The experiment shows that expressing a UAC by a simple ISO specification is not sufficient. This formalism leads to the loss of a substantial amount of information. This is why it appeared necessary to add a linkage or contact sub-graph to the formalism of the UACs.

2.2. Modelling of the mechanism

The second stage consists, therefore, of choosing the type of modelling for the mechanism. It appears essential to model the mechanism at several levels:

- at part assembly level
- at part level
- at contact surface level

At part assembly level

We initially seek an overall view of the operation of the mechanism and do not attempt to represent all its components. It appeared judicious to group the parts belonging to the same equivalence class in the same part assembly.

In our example, the mechanism can be represented by five part assemblies corresponding to the five classes of equivalence determined. These different assemblies are linked by lines representing the mechanical links between the assemblies. Moreover, we have chosen to represent the UAC under consideration on the graph. We obtain the following for the cartridge:

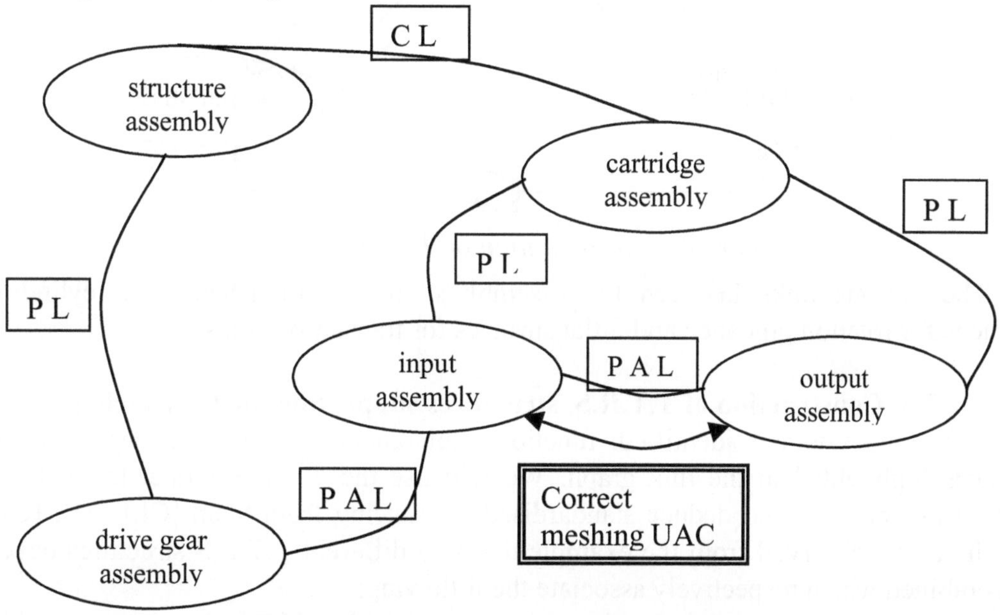

Figure 3: representation of the mechanism

It then appeared necessary to work by sub-system. Working on the mechanism as a whole would, in fact, cause too much dispersion. Please refer to the work of Ballu and Mathieu [BAL 99] on the notion of "Key parts" and "Key surfaces". The various possibilities envisaged to determine the pertinent assemblies are as follows:

- user assistance
- use of a functional selection
- use of geometry; small displacement equations could be written, the coefficient of the influence of each parameter on the UAC calculated and the pertinent sub-system identified.

Once the sub-system corresponding to the meshing UAC is isolated, it will be modelled at contact surface level:

At surface level

We will be using the "link graph", defined and used by Desrochers [DES 91], [SEL 99] and apply it to the surface entities defined. The parts assemblies constitute the nodes of the graph whereas the contacts (pertinent or not) correspond to the arcs or link of the same graph.

The meshing UAC can be represented as follows in figure 4:

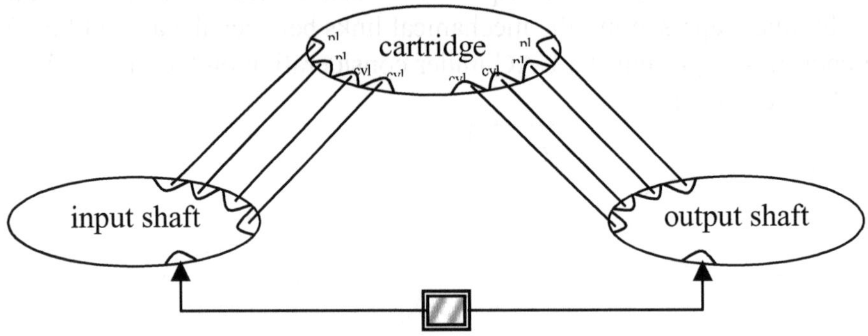

Figure 4 : representation of the meshing UAC

The various links between the assemblies are accomplished by 2 cylindrical surfaces for rotation guidance and 2 flat surfaces for translation stops.

2.3. Construction of T.T.R.S. structures supporting the tolerancing

Once the sub-system is identified, functional tolerancing has to be generated on the surfaces highlighted in the link graph. We will use the T.T.R.S model to model the surface associations and deduce standardised tolerancing from them [CLE 94], [GAU 94]. It can be observed from the example that two different T.T.R.S structures have to be combined which respectively associate the following:

- all the services used to position one parts assembly with another parts assembly.
- the surfaces within a parts assembly which are in contact with those making another contact.

Generating the T.T.R.S structure associated with link constraints

Several stages are involved in the generation of tolerancing specification:

- the generation of all possible loops.
- the choice of associations, consistent with the functional requirement.
- the choice of the type of consistent associations.

While it is easy to computerise the first stage, the other two stages are more difficult because of the choice of consistent associations and type.

The possibilities currently envisaged for the construction of the T.T.R.S structure corresponding to contact constraints are:

- to allow the user to guide us in the construction of this structure
- to have professional rules which allow us to associate this set of surfaces with a standard tolerancing specification. The user would then seek out the case relevant to his model in a professional library.
- to have simulation calculation results for this structure [LEP.94]

In the example, we will concentrate on the contact between the cartridge assembly and the input assembly. The key link identified brings 8 elementary surfaces into play. This link is a classic rotation guidance link which can be associated with a standard tolerancing specification and a corresponding T.T.R.S. structure. The following tolerancing specifications and T.T.R.S.s are then obtained for the 4 surfaces belonging to the cartridge assembly:

For the two cylinders:

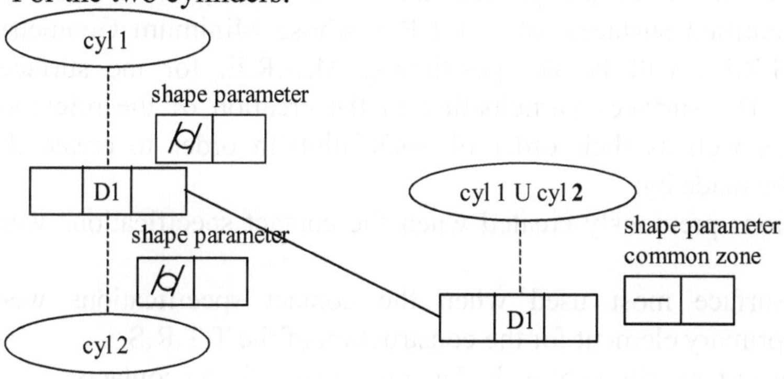

For the planes:

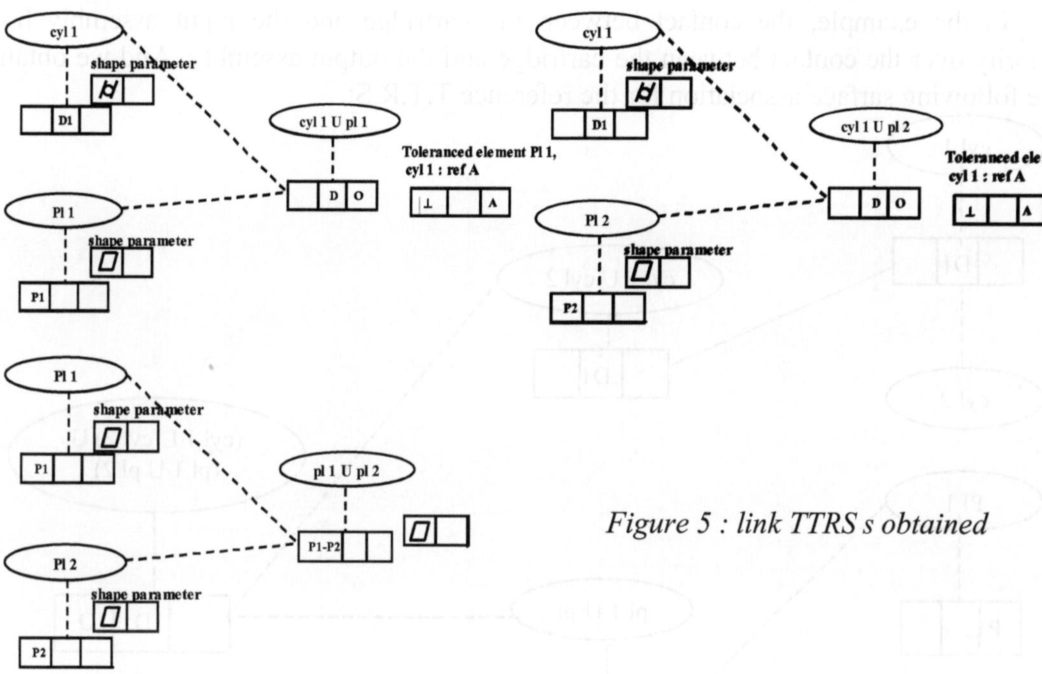

Figure 5 : link TTRS s obtained

We note that the structure obtained cannot be arborescent: the same surface participates in the creation of several T.T.R.S.s.

Generation of the T.T.R.S associating the contacts

The second stage consists of associating the 2 units of 4 surfaces belonging to the cartridge and making the key links with the input and the output assemblies. This association is achieved in two stages. The reference surfaces must be chosen, therefore a priority contact, and the Reference and Positioning System (R.P.S) constructed. The non-priority contact surfaces are then associated in relation to the R.P.S. constructed .

The first stage is the choice of the priority link. The second consists of the construction, with the identified surfaces, of a T.T.R.S whose Minimum Geometric Reference Element (M.G.R.E.) will be the positioning M.G.R.E. for the surfaces defining the second link. The surfaces participating in the creation of the reference system must be chosen as well as their order of association in order to create the T.T.R.S. This choice can be made by:

- using the associations previously created when the contact specifications were determined.
- considering the surface most used when the contact specifications were determined as the primary element for the construction of the T.T.R.S.
- using the hierarchy system criteria already dawn up to specify the contacts.
- calling on user assistance.

In the example, the contact between the cartridge and the input assembly has priority over the contact between the cartridge and the output assembly. And we obtain the following surface association for the reference T.T.R.S:

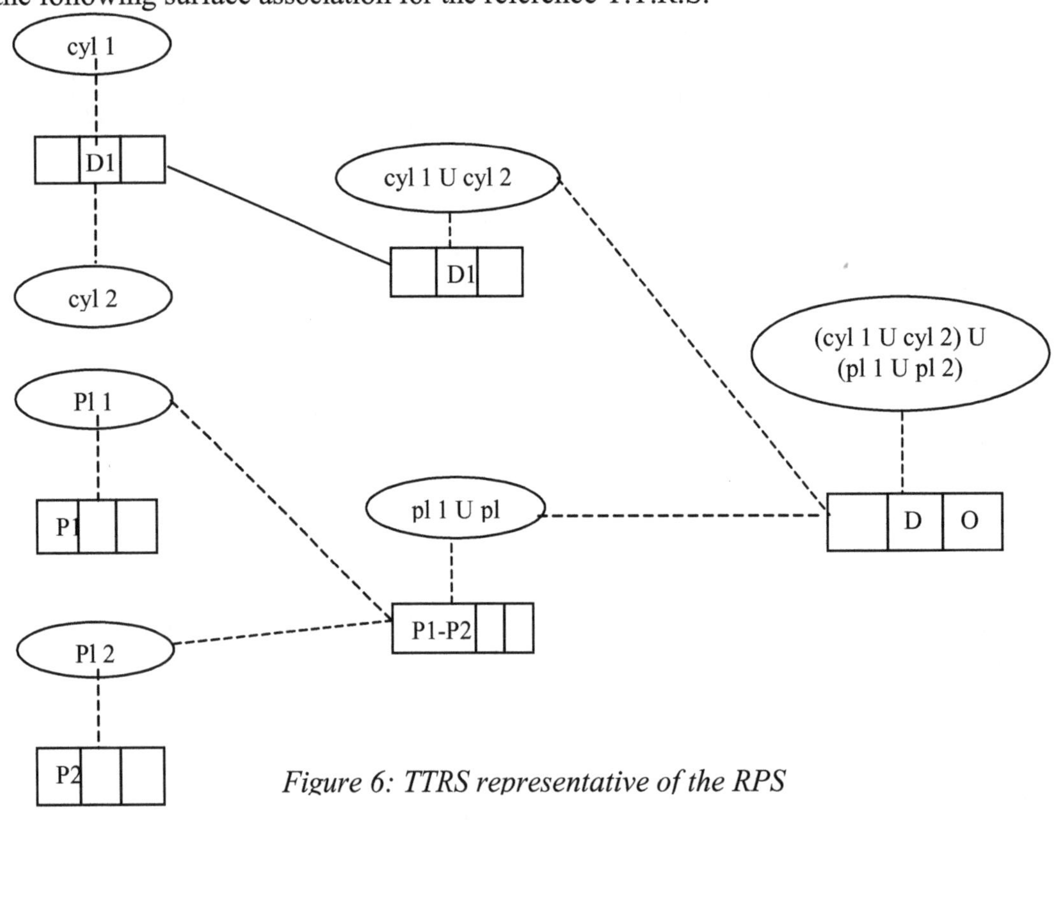

Figure 6: TTRS representative of the RPS

Once the R.P.S. is defined, we then specify the surfaces or surface associations for the non-priority contact, defined in relation to the T.T.R.S constructed. From the surfaces and the surface associations defining the link we have to choose the one to associate within a T.T.R.S. The choice criteria of these surfaces are:

- non-observance of ISO standard tolerancing specifications
- the absence of specification duplication on a surface by favouring its use in an association rather than by itself.

For the cartridge assembly, the specifications linking the surfaces defining the link with the output assembly with the surfaces defining the link with the input assembly are as follows:

For the cylinders: for the planes:

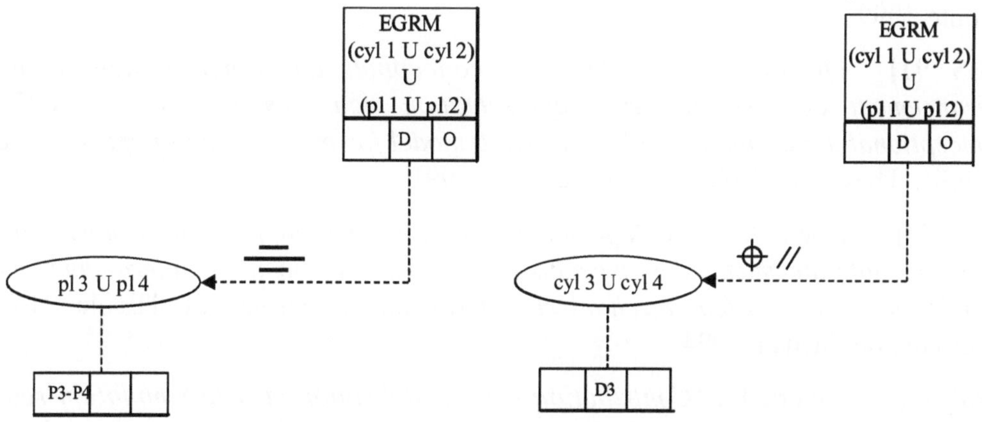

Figure 7 : possible specifications

2.4. Conclusion

The initial methods enabling digital functional assistance put forward in this paper are based on good dialogue with the user. In addition to obvious user participation in the definition and writing of professional rules, which will certainly be used by the software, the assistance of the user will enable:

- to specify UACs correctly.
- to isolate the sub-systems corresponding to the UACs to be studied.
- to guide the choice of T.T.R.S. structures by his/her professional knowledge.
- It should also be noted that a few difficult points require further research, including:
- The transcription of UACs onto fictive geometric entities and the creation of pseudo-T.T.R.S.s.
- Compensating for the shortcomings of the ISO language to describe UACs appropriately.

REFERENCES

[BAL 99] Ballu, A.; Mathieu, L.; "Choice of functional specifications using graphs within the framework of education"; In: *Proceeding of the 5th CIRP seminar on computer aided*; p197-206; Twente 1997

[CLE 94] Clément, A.; Rivière, A.; Temmerman, M.; *"Cotation tridimensionelle des systèmes mécaniques. Théorie et pratique. (Three-dimensional functional tolerancing of mechanical systems. Theory and practice)";* PYC édition,*1994*

[CLE 97] Clément, A.; Rivière, A.; P. Serre; "A declarative information model for functional requirement"; *Proceeding of the 6th CIRP seminar on computer aided*; Toronto 1999

[DES 91] Desrochers, A.; *"Modèle conceptuel du dimensionnement et du tolérancement des mécanismes. Représentation dans les systèmes de CFAO - (Conceptional dimensioning and tolerancing model for mechanisms. Representation in CFAO)";* Thesis of l'Ecole Centrale de Paris; 1991

[GAU 94] Gaunet, D.; *"Modèle formel de tolérancement de position. Contribution à l'aide au tolérancement des mécanismes - (Formal position tolerancing model. Contribution to assisted mechanism tolerancing)";* Thesis of l'École Normale Supérieure de Cachan, 1994.

[LEP 94] Le Pivert, P.; *"Contribution à la modélisation et à la simulation réaliste des processus d'usinage - (Contribution to the modelling and realistic simulation of machining processes)";* Thesis of l'Ecole Centrale de Paris; 1998

[SEL 99] Sellakh, R.; *"contribution à l'intégration de la définition multiniveaux de la géométrie des systèmes mécanique pour le tolérancement - (Contribution to the integration of multilevel definition of mechanical system geometry for tolerancing)";* Thesis of l'INSA de Lyon, 1999

NF ISO 1328-1 *Engrenages cylindriques - Système ISO de précision - Partie 1 : définitions et valeurs admissibles des écarts pour les flancs homologues de la denture (Cylindrical meshing – ISO precision system – Part I: definitions and admissible deviation values for equivalent denture faces)*; AFNOR 1997

NF E 48-371 *Transmissions hydrauliques - Etanchéité par bagues à lèvres pour arbres tournants - Dimensions et caractéristiques du logement et de l'arbre (Hydraulic transmission – Sealing with lip seals for turning shafts – Dimensions and characteristics of receptacle and shaft);* AFNOR 1994

Towards a Method for Early Evaluations of Sheet Metal Assemblies

Stefan Dahlström, M.Sc., Ph.D. Student
Volvo Car Corporation
Complete Vehicle
Dept. 91340, PVD3:2
SE-40531 Göteborg, Sweden
E-mail: dahste@mvd.chalmers.se

Rikard Söderberg, Ph.D., Associate Professor.
Chalmers University of Technology
Dept. of Machine and Vehicle Design
SE-412 96 Göteborg, Sweden
E-mail: riso@mvd.chalmers.se

Abstract: During sheet metal assembly, part and process variations influence the final geometry of the assembly. Normally, the information needed to make realistic analysis of the final geometric variation is extensive and in the early stages of the development process still unknown. Tolerance analysis, as conducted in most industries today, is normally based on the assumption of rigid parts and is therefore not always valid for sheet metal assemblies. Lately, research on using FEA to predict the influence of flexibility has been reported on by some authors. However, FEA is still a timeconsuming activity for experts, requiring a lot of information.

This paper presents a structured method for early evaluation of sheet metal design and assembly concepts with respect to different evaluation criteria. The method is based on a structured, hierarchical product description and constraint decomposition. It identifies, captures and utilizes critical information needed for sheet metal concept evaluations. An inventory of the key parameters and groups that control the final assembly geometry and variations, an identification of the evaluation criteria used to evaluate different concepts, and a mapping matrix to identify what key parameters need to be defined to analyze the concepts against different evaluation criteria are presented.
Keywords: assembly, tolerancing, variation, evaluations, compliance

1. INTRODUCTION

Shortening of the product development time and concurrently improving the design of new products makes it necessary to evaluate concepts in early development phases. The problem with early evaluation of concepts is the lack of needed information. Most of the needed information will not be available until late in the design process. Many assumptions therefore have to be made to analyze the concept. One important aspect to evaluate in product concepts is how the final product will fulfil the geometrical requirements.

Since all manufacturing processes are afflicted by variation the nominal value of a key characteristic or critical dimension may not be expected at all times. Variation in a

P. Bourdet and L. Mathieu (eds.),
Geometric Product Specification and Verification: Integration of Functionality, 275-286.

geometrical key characteristic (KC) [Söderberg 1998] of an assembly product typically results from a number of different sources, as shown in Figure 1. There is variation in the individual component geometry, which results from machine precision itself, but there is also process variation over time. Similarly, the assembly process will contribute variation, related to the way in which parts are assembled, which may also vary over a period of time. The geometrical robustness of an assembly concept can be evaluated by its ability to minimize the effect of geometric variation in the end assembly product. Low robustness means that our assembly concept (the components and processes used) adds variation to the final assembly. High robustness on the other hand means that the system is able to suppress the variation in individual components and processes so that they do not affect the quality of the end product to any great extent.

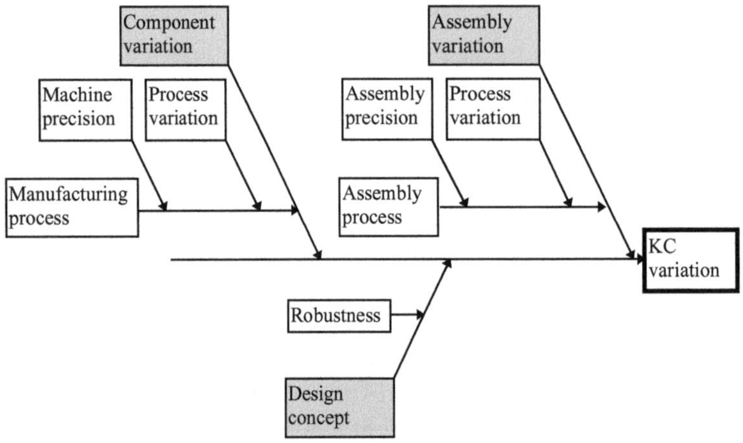

Figure 1: Geometrical KC variation contributors [Söderberg 1998]

The industrial need for geometrically robust concepts increases rapidly as the development and production ramp-up time decreases. Since a concept change during pre-production is often very costly, geometry-related production problems must be avoided already in early design stages.

1.1. Sheet Metal Assemblies

Sheet metal assemblies are used to manufacture many different products such as automobiles and airplanes. These products often use the sheet metal assembly as their supporting and functional structure, so it is a key issue to control or predict the final geometry and variation of the assembly. The assembly process for sheet metal assemblies is often conducted in three steps;

- positioning and clamping the parts in a fixture,
- welding or joining the parts and
- finally releasing the parts from the fixture.

During assembly, parts and process variations will influence the final geometry of the assembly proportional to the robustness of the assembly concept. Normally, the information needed to make realistic analysis of the final geometric variation is extensive and in early stages of the development process still unknown. Tolerance analysis, as conducted in most industries today, is normally based on the assumption of rigid parts and is therefore not always valid for sheet metal assemblies. In early phases of the product development process, a prediction or evaluation of the final geometry of the assembly is necessary to ensure that the final geometrical requirements for the assembly are fulfilled.

1.2. Variation source and assembly criteria for sheet metal assemblies

The variations introduced during the assembly process can be divided into two main sources;

- *part variation*, mainly shape, size deviations and material property variations
- *process variation*, introduced to the assembly by clamping, welding and joining forces or by variation in the fixtures used.

These variations give different effects to the sheet metal assembly. The difficulty in analyzing a sheet metal assembly is when the variation listed above, forces the sheet metal parts to bend in order to fix them in their nominal position. In order for the above listed variation sources to influence a critical dimension of the assembly, two aspects of the assembly concept (design and production) are critical:

- whether the concept allows variation to propagate through the assembly structure, which is an aspect that is determined in the early design stages and
- whether the assembly sequence allows for spring-back, which is an aspect that is mainly determined during production preparation

1.3. Methods for analyzing sheet metal assemblies

The difficulty in predicting sheet metal assemblies is to simulate the quantity of variations that can occur during the assembly process. FEA is used in many methods for the evaluations, since the flexibility of the parts must be taken into consideration. One way of predicting the final geometry is to conduct a *Monte Carlo simulation* [Liu and Hu 1997]. This method is very time-consuming if many types of variations are to be simulated. Furthermore, the random numbers generated must be correlated to make realistic distributions, and in many cases this is difficult to achieve. A faster way of doing the simulation is to establish a linear relationship between the parts deviation and the assembly spring-back deviations by using the *method of influence coefficient* [Liu and Hu 1997]. Defining the sensitivity matrix, which describes how the assembly spring-back is influenced by initial geometrical variations in the parts, does this. Lately, a robustness evaluation method for compliant assembly systems has been presented [Lee *et al.,* 2000].

A method developed by Sellem and Rivière [Sellem and Rivière 1998], [Sellem *et al.,* 1999] takes into account three different kinds of variation in the simulation: positioning, conformity and shape variabilities. The output of the simulation produces the influence of these variabilities on: distributions at control points, the force required for the clamping process and the residual stresses near the fastening points. The theoretical basis for this method is based on the influence matrices.

The use of *beam-based modeling* to describe a sheet metal assembly has been presented by [Ceglarek and Shi 1998] and [Shiu *et al.,* 1997]. The method identifies different joint types and beam elements to achieve a simplified model for the assembly. The use of transformation vectors to describe variations and displacement of features has been presented by Chang and Gossard [Chang and Gossard 1997]. The method represents the interaction between parts and tooling by contact chains which are later used in vector equations.

1.4. Scope of the paper

This paper presents a method for early evaluation of potential flexibility-related geometry problems. The method is based on a structured, hierarchical product description and constraint decomposition. It identifies, captures and utilizes critical information needed for sheet metal concept evaluations. When necessary, FEA could be used to define the part and joint types for analysis of propagation of the direction of the variations. Because of the time limit in the product development and production startup process, often the concept with minimum risk will be chosen. Thus, the best concept concerning other design criteria is not always chosen, due to lack of information or difficulties in conducting a realistic simulation. A mapping matrix between key parameters that control the final assembly variation and the evaluation criteria is presented. The method is an attempt to evaluate concepts with little information by simple assumptions based on experience and simulations.

2. EARLY EVALUATION OF SHEET METAL ASSEMBLIES

The objective in the early design phases is to achieve a desired sheet metal design; (Figure 2). Due to the limitations of the sheet metal forming process often the desired design has to be assembled by feasible parts.

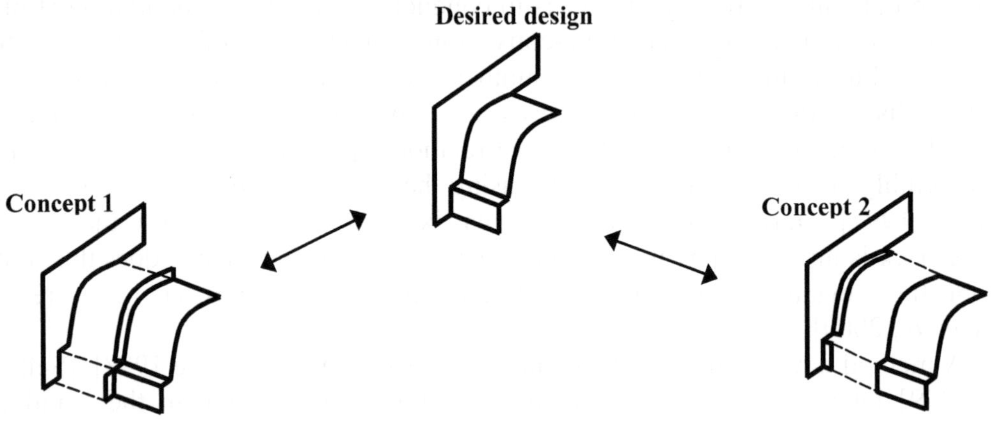

Figure 2; Example of two concepts for a desired design

Both the dividing of the desired design and the assembling of the parts can be conducted in numerous different ways and will affect the final geometry differently. The key issue is then to evaluate which concept fulfills the geometrical requirements best. A concept is analyzed by evaluating different characteristics depending on the development phase in the project. To conduct the evaluation, an identification of the key parameters that control the final geometry needs to be done.

2.1. Key parameters

This section presents a set of controlling key parameters. The key parameters have been collected from present research results but also completed by industrial experience. The key parameters are divided into three groups; *variations, design concept* and *process layout.*

The *variation* group consists of two kinds of variation: part and process variations. Part variations include all variations that are associated with variations that occur on the single sheet metal part. The major part of the variations comes from the sheet metal forming process or in some cases from the transportation and the stacking of the parts. The process of sheet metal forming is a well-known process and the variation type tends to occur more frequently in some areas of the part. The process variations are defined as including all variations introduced during the process. Variations from the welding process, such as the alignment, force and location of the spot welds, fixture errors and clamping forces, are the most important. Experience can be used to sort out the most common variation types and their frequency.

The *design concept* group includes the parameters that control how sensitive the concept is to variations. The parameters in this group are:

- *geometry of the parts:* depending on size, shape and how complex the geometry is, it controls how sensitive the parts are and what kind of variations they are expected to have.
- *joint type:* there are two basic joint types [Ceglarek and Shi 1998], the lap joint and the butt joint; (Figure 3). The joint type allows the variations to propagate in defined directions.
- *stiffness condition:* depending on how the variations are introduced to the part, it will have different characteristics depending on the stiffness condition in the variation direction of the part. Furthermore, the stiffness condition between parts has a major impact on how the variations are transferred between the parts. For example, assembly variation is dominated by the variation of the more rigid parts, [Liu *et al.*, 1995].
- *number of parts*. This will effect to the complexity of the assembly. Increased number of parts will most likely increase the number of variations sources.

The *process layout* controls how the variations are introduced to the assembly. The parameters in this group are:

- *welding configuration:* by experience, this parameter introduces the most variations and geometrical defects to the assembly. The most important configurations are: *welding type, location, alignment, sequence* and *welding forces*.
- *assembly sequence:* by different sequences, the variations can moderate or increase each other depending on the sequence (serial or parallel).

- *fixture design:* the layout of the master location points affects how the variations are amplified when afflicted to part and process variations.
- *clamping layout:* sheet metal parts often need to be clamped by additional clamps to hold the part against forces introduced during assembly. For large parts, several clamps are needed to reduce the influence of gravity.

Figure 3; Butt and lap joint

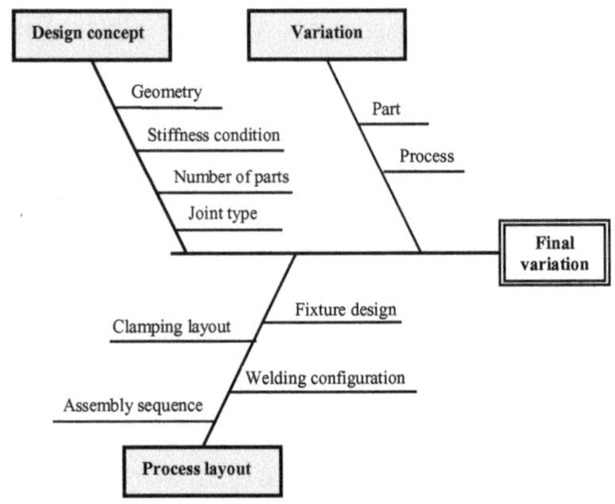

Figure 4; Key parameters

All these three groups (Figure 4) interacts with each other in a complex way, i.e. the final geometry and the sensitivity to variations is depended on the internal relationship between the parameters. When improving a concept regarding the final geometry, the groups to control and adjust the parameters are the *design concept* and *process layout.* The *variation* group is compared to the others very difficult or expensive to make improvements of. Information of these parameters in early design phases is limited and assumptions and experience is necessary to obtain the relevant information.

2.2. Evaluation criteria

Sheet metal assembly concepts are evaluated by different criteria. Depending on the stage of the development process and the amount of available information, different criteria are used to evaluate different concepts. The different evaluation criteria used during the development process are:

• *analysis requirement level* for a concept. Since lead time in product development process is a very important factor, this has the consequence that the concept with minimum risk regarding the required evaluation analysis need is often chosen. This concept may not always be the best in terms of geometrical requirements but allows for fairly precise tolerance analysis. There are three scenarios in industry for failure to replace an old, well-known concept with a new one: 1) the new concept is not proven to give any new benefits, 2) the analysis to identify the benefits is too extensive and 3) there is no time to implement the new concept, even though it offers improved benefits. It is therefore desirable to estimate how extensive the evaluation is going to be further down the line. One way to measure this is by analyzing the differences between number of parts, variation type, location and frequency of the variations for the concepts.

• the degree of *robustness/coupling*. Normally, an uncoupled relationship between critical measures and parts and process variations indicates a concept that is easy to tune in production. Often this also results in a concept that is less sensitive to variations and where the flexibility of the parts does not add more variation to the final geometry of the concept, i.e. the concept is more robust. Unexpected variations have less influence on the final geometry if the concept is considered to be robust and has uncoupled relationships [Söderberg and Lindkvist, 1999].

• the degree of *complexity*. Number of parts, assembly tolerance chains [Söderberg and Johannesson 1999], contact areas between parts, assembly sequence and the geometry of the parts affects the degree of complexity of the concept. A DFA analysis could be used to measure the degree of complexity. This is a critical criteria for choosing concepts for further improvements.

• *propagation direction of the variations*. When the information for a complete tolerance analysis of the concepts is not available, a prediction of potential propagation direction of the variations is desirable. These directions could be estimated by the use of experience of the assembly characteristics regarding welding configuration and part and joint type. [Mantripragada and Whitney, 1998], [Söderberg and Johannesson, 1999]

• *numerical distributed variations*. When most of the information needed is available, an analysis of predicted final geometry could be conducted by the use of FEA or a CAT tool that simulates non-rigid parts. [Lee *et al.*, 2000], [Chang and Gossard, 1997], [Sellem and Rivière, 1998]

	Evaluation criteria				
Key parameters	Analysis requirement level	Robustness /coupling	Complexity	Propagation direction	Numerical distributed variations
Part variation	x	x	x	x	x
Process variation	x	x	x	x	x
Geometry of the parts	x		x	x	x
Joint type		x	x	x	x
Stiffness condition				x	x
Number of parts	x		x	x	x
Welding type	x		x	x	x
Welding location					x
Welding alignment					x
Welding sequence			x	x	x
Welding force					x
Assembly sequence	x	x	x	x	x
Fixture design	x	x	x	x	x
Clamping layout	x	x	x	x	x

Figure 5; Mapping matrix

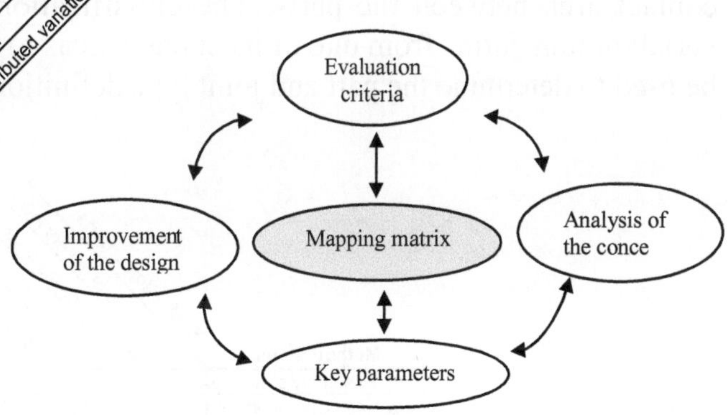

Figure 6; Mapping between key parameters and evaluation criteria

The different evaluation criteria require a number of the key parameters to be defined. The mapping matrix shown in Figure 5 identifies the coupling between the key parameters and the evaluation criteria. This matrix is used during the evaluation and improvement of concepts, (Figure 6), by identifying what key parameters control an evaluation criteria. The matrix can be extended to include more key parameters or evaluation criteria.

2.3 Evaluation of sheet metal design and assembly concepts

Sheet metal design and assembly is a complex task and requires a structured method to evaluate and improve different concepts. In this section we suggest a method that simplifies the evaluation by the use of a structured definition of necessary steps. These definition steps are still under development but they are in general based on present research and industrial experience. The method consists of five steps; 1. *Definition of an assembly structure,* 2. *Part and joint type definition,* 3. *Equation system definition*, 4. *Evaluation matrix formulation* and 5. *Numerical simulation*. The first two steps are always conducted independently of the evaluation criteria being used. The definition of the first two steps are:

1. Definition of an assembly structure. The assembly is decomposed into a hierarchical assembly structure. The assembly structure is used to define the assembly sequence and subassemblies, to identify serial or parallel assemblies and to find potential tolerance chains between the parts in the assembly.

2. Part and joint type definition. Depending on the joint type between two sheet metal parts, the variations will propagate through the assembly in different ways. Regardless of the kind of variation in the sheet metal parts, these will only be able to propagate the variation to other parts through their contact areas. The definition used here for a joint type is based on the directions in which the joint can transfer a variation through the contact area between the parts. The classification of the part types defines how a variation transforms from one of its contact areas to others. When necessary FEA could be used to determine the part and joint type definition.

Method steps / Evaluation criteria	Analysis requirement level	Robustness /coupling	Complexity	Propagation direction	Numerical distributed variations
Assembly structure	x	x	x	x	x
Part/joint type definition	x	x	x	x	x
Equation system		x	x	x	x
Compairision matrix		x	x		
Numerical simulation					x

Figure 7; Coupling between the method steps and the evaluation criteria

Depending on the current evaluation criteria different steps are conducted. Figure 7 shows which steps to conduct depending on the evaluation criteria, and the mapping matrix in figure 5 is then used to identify what key parameters need to be defined.

3. *Equation system definition*. After the definition of the part and joint type, a logical equation system can be formulated and used to analyze the propagation directions of the sheet metal assembly. The equation system is used to find potential propagation directions of the variations compared to critical measures and also find what effect a certain variation has on the final assembly variation. The equation system is based on vector equations to describe and represent the interaction between propagation directions of the variations and part or joint type.

4. *Evaluation Matrix formulation*. This evaluation matrix is used to evaluate different concepts regarding the degree of *analysis requirement level*, *complexity* and *robustness/ coupling*. The results from the previous steps can be compared with different concepts in an evaluation matrix much like the Kesselring [Pahl and Beitz 1999] or the Pughs matrixes [Pugh 1991].

5. *Numerical simulation*. When sufficient or estimated information is available, numerical calculations of the final assembly variation and geometry can be conducted. FEA or a CAT tool which can simulate non-rigid parts is used to calculate displacement and the contribution from the flexibility in the parts, [Lee *et al.*, 2000], [Chang and Gossard, 1997], [Sellem and Rivière, 1998]

These steps are used to compare different concepts regarding the evaluation criteria and to facilitate and point out with help of the mapping matrix what adjustments to implement to the key parameters. A detailed implementation of these five steps and validation of the definitions will be presented in later research results.

2.4. Including experience in the method

In most industries where the development time for new products is several years, it is difficult to carry over the experience gained in the previous project. Furthermore, it is very valuable to derive information about the influence a key parameter has on an evaluation criteria and to map how frequently variations occur under certain conditions. One way of including the experience in the method is to save the analysis results from previous evaluations steps with respect to:

- the different behavior of a joint between parts
- the influence of welding sequence and the location of the spot welds
- characteristics of a similar geometry.

These analysis results are then used improve the definitions in the method in order to adapt these to empirical definitions and to:

- support the formulation of *design guidelines* in order to support early concept improvements

- obtain the *percentage* contribution of a key parameter or a group in the final assembly geometry
- support *root cause analysis* in order to detect production problems.

3. CONCLUSIONS

Sheet metal design and assembly is a complex task and requires a structured method to evaluate and improve different concepts. Lack of information and the difficulty of conducting realistic simulations are the most important reasons why sheet metal design and assembly is a complex task. Due to shortening lead time it is of major interest to evaluate, as early as possible in a project, the complexity and the need for deeper analysis, in a concept. Often, the concept that minimizes the risk of a failure has to be chosen.

In this paper we have presented an inventory of the key parameters and groups that control the final assembly geometry and variations, and a have identified the evaluation criteria used to evaluate different concepts. A mapping matrix has been presented to identify the key parameters that need to be defined to analyze the concept against different evaluation criteria. Furthermore, a proposal for a structured method to evaluate sheet metal design and assemblies with respect to different evaluation criteria is presented.

4. FUTURE WORK

This evaluation method is under development and still needs to be refined and verified. In specific we will i) develop the joint and part definition, and further ii) adapt these to experience of the real behavior of sheet metal assemblies and to realistic variations, and iii) study the influence of the welding sequence on final assembly variations. Furthermore, we will conduct a study to obtain the percentage by which the key parameter or group influences the final geometry of the assembly.

REFERENCES

[Author et al., 2001] Author1, F.N.; Author2, S.A.; Author 3, T.A.; " i.Title of the publicationl"; In: *Proceedings of the7th International Seminar on Interesting Topics*, pp. 507-518; Cachan 2001; ISBN 1-12345-123-1

[Ceglarek and Shi 1998] Ceglarek, D.; Shi, J. “Design Evaluation of Sheet Metal Joints for Dimensional Integrity” , Journal of Manufacturing Science and Engineering, Transactions of the ASME v 120 n 2 May 1998 p 452-460

[Chang and Gossard 1997] Chang M., Gossard D.C., 1997, "Modeling the Assembly of Compliant, Non-ideal Parts", Computer-Aided Design, Vol. 29, No. 10, pp. 701-708.

[Lee *et al.,* 2000] Lee J., Long Y., Hu S.J., "Robustness Evaluation for Compliant Assembly Systems", Proceedings of The 2000 ASME International DETC and CIE: Design for Manufacturing Conference, September 10-13, 2000, Baltimore, Maryland

[Liu and Hu, 1997] Liu S. C., Hu S. J., 1997, "Variation Simulation for Deformable Sheet Metal Assemblies Using Finite Element Methods". ASME Journal of Manufacturing Science and Engineering. Vol. 119. August. pp. 369-374.

[Liu *et al.,* 1995] Liu, S., H.-W. Lee, and Hu.S., 1995. " Variation Simulation for Deformable Sheet Metal Assemblies Using Mechanistic Models". Transactions of the North American Manufacturing Research Institute of SME. May 1995.

[Mantripragada and Whitney 1998] Mantripragada R., Whitney D.E., 1998, "The Datum Flow Chain: A Systematic Approach to Assembly Design and Modeling" Proceedings of the ASME Design Theory and Methodology Conference, September 13- 16, Atlanta, GA, USA

[Merkley *et al.,* 1996] Merkley K. G., Chase K. W., Perry E., 1996. "An introduction to Tolerance of Flexible Assemblies". MSC 1996 World Users' Conference Proceeding Newport Beach.

[Pahl and Beitz 1999] Pahl G., Beitz W., 1999 "Engineering Design", Springer-Verlag London, ISBN 3-540-19917-9

[Pugh 1991] Pugh S., 1991 "Total Design", Addison-Wesley Publishing Company, ISBN 0-201-41639-5.

[Sellem and Rivière 1998] Sellem E., Rivière A., 1998, "Tolerance Analysis of Deformable Assemblies". Proceedings of ASME Design Engineering Technical Conference., September 13-16, 1998, Atlanta, USA.

[Sellem *et al.,* 1999] Sellem E., Rivière A., Clement A., 1999, "Validation of The Tolerance Analysis of Compliant Assemblies", Proceedings of DETC99: 1999 ASME Design Engineering Technical Conference, September 12-15, 1999, Las Vegas, Nevada.

[Shiu *et al.,* 1997] Shiu, B. W., Ceglarek, D., and Shi, J., 1997. "Flexible Beam_based Modeling of Sheet Metal Assembly for Dimensional Control," Trans. of NARMI, Vol. XXV, pp. 49-54.

[Söderberg, 1998] Söderberg, R., 1998, "Robust Design by Support of CAT Tools", Proceedings of the ASME Design Automation Conference, September 13-16, Atlanta, GA, USA, DETC98/DAC-5633.

[Söderberg and Johannesson 1999] Söderberg R., Johannesson H., 1999, ”Tolerance Chain Detection by Geometrical Constraint Based Coupling Analysis” Journal of Engineering Design, Vol. 10, No. 1, 1999.

[Söderberg and Lindkvist 1999] Söderberg R., Lindkvist L., 1999, ”Computer Aided Assembly Robustness Evaluation” Journal of Engineering Design, Vol. 10, No. 2, 1999.

An assisted method for specifying ISO tolerances applied to structural assemblies

R.Sellakh, A.Rivière*, N.Chevassus**, B.Marguet***
** Laboratoire de Mécatronique, ISMCM – CESTI*
3, rue Fernand Hainaut, 93407 Saint Ouen Cedex, France
sellakh@ismcm-cesti.fr, ariviere@ismcm-cesti.fr
*** EADS, Centre Commun de Recherche*
4, bis rue du Val d'Or, 92152 SURESNES
nicolas.chevassus@eads-nv.com, benoit.marguet@eads-nv.com

Abstract:. The method for generating assembly plan oriented tolerancing presented in this paper enables the systematic breakdown of a functional requirement, given in dimensional, shape and position specifications. It uses graphs to choose the geometric specifications and the Technologically and Topologically Related Surface (TTRS) concept to define the model for the parts. This method is composed of two principal stages. Functional requirements are identified by assembly level from the path of a DFC (Datum Flow Chain) oriented graph, representing part positioning principles. They are then associated with TTRS modelling in order to achieve the expression of vectoral tolerancing and translated to comply with ISO standards. A specific case from the aircraft industry is presented to validate our approach.
Keywords: Assemblies – Assembly plan – TTRS - Tolerancing – Geometric specifications.

1. INTRODUCTION

The tolerancing of mechanical parts is a delicate operation requiring expert knowledge of ISO standards together with a good professional grounding. Various methods of generating functional tolerancing specifications exist which are not all equivalent.

However, they may turn out to be complementary in compound processing. We will first present the principles of three methods, give details of the various stages and tools of the method developed and then illustrate it with an industrial example which has enabled it to be validated.

2. REVIEW OF THE TOLERANCE ASSISTANCE METHODS

P. Bourdet and L. Mathieu (eds.),
Geometric Product Specification and Verification: Integration of Functionality, 287-299.

2.1. TTRS method

The purpose of the TTRS method (Technologically and Topologically Related Surfaces) [Clément et al 94] is to obtain a design assistance tool which allows the full tolerancing specifications of a mechanism to be obtained in compliance with the designer's requirements. The modelling offered by this method integrates technological data into geometric information, within the same model.

A basic element of the TTRS theory is the notion of successive relations of surfaces or series of surfaces. These relations are preserved as independent entities which have their own intrinsic characteristics. These characteristics include dimensions, tolerances, a partial referential called MRGE (Minimum Reference Geometric Elements) and a pointer towards a corresponding TTRS class. The tolerancing synthesis of a mechanical system using the TTRS model is made in stages, as follows: "Stage 1 consists of "The identification of the functional contact surfaces", stage 2 of the "Creation of TTRSs" and, finally, stage 3 of the "Functional Parameterisation of TTRSs (dimensioning and tolerancing)". The type of contact surfaces are identified from a model of the mechanism, according to 7 surface classes. This information is transcribed in the form of a link graph in order to determine related surface couples or those brought into relationship.

The relationship graph allows the possible relation scenarios to be summarised on arborescent graphs. The creation of TTRS is a recursive process, enabling several surfaces to be related in succession. These related surfaces must belong to the same kinematic loop or cycle in a mechanism. 28 classes of TTRS exist which are the 28 relative positioning possibilities. The surface resulting from the union of two surfaces belongs to one of the 7 classes. This results in 44 possibilities of reclassification , i.e. the determination of the resulting class corresponding to already identified tolerancing cases.

2.2. Graph analysis method

The graph analysis method was developed by A.BALLU and L.MATHIEU [Ballu 99] with the initial objective of formalising the specification of requirements for teaching tolerancing. Use of the method is in three stages:

The first consists of the study of the mechanism structure, the second of the search for "influential" elements and, finally, the third enables the tolerancing plan to be set up. The authors started from the assumption that functional conditions are independent. Thus all the conditions of a mechanism must be studied independently of one another. The structure of a mechanical assembly is modelled by a graph of the contacts, composed of the parts, identified by a vertex and functional surfaces identified by a pole of the part vertex to which they belong, and the elementary links, represented by the edges between the vertices. Once the link graph is established, the authors propose to install the functional conditions referring to the geometry of the two parts. Influential variation (corresponding to the variation of the surfaces on which the functional condition is dependent) is determined by using graph simplification rules. They consist of eliminating parts which do not belong to a cycle going through the condition or an influential cycle, as well as eliminating surfaces linked to non-

influential parts. Graphs are then reduced either by reducing the serial links which enable a part to be eliminated that only has contact with independent parts or by reducing parallel links which consists of replacing links by an equivalent link.
The authors did not use any specific method to choose the specifications; however, they propose a few rules to specify the components of a mechanism. It should thus be kept in mind that the specifications are only installed on influential surfaces and the references of a specification by tolerance zone should be applied to the fixed contact surfaces.

2.3. Indeterminates method

Developed by E.Ballot [Ballot 99], the purpose of this method is to model a mechanism with a set of equations using small displacement torsors modelling geometric faults and clearance in the links. The use of substitution surfaces the same as the nominal surfaces enables actual surfaces to be modelled. Geometric faults and clearance in the links are modelled thanks to a small displacement torsor. The author uses several types of torsors. The variation torsor expresses the variation between the nominal surface and the substitution surface, the fault torsor which is the composition of the variation torsors of two surfaces belonging to a part and the clearance torsor which expresses the clearance between two surfaces nominally in contact with two parts. Thus for parts connected by links, the displacement of one in relation to the other corresponds to the union of the displacements induced by each of the links. By a system of equation and, after resolution, the position of the parts and constraints between the variation and indeterminates are obtained. The consistency of a specification can be verified by means of the equation system obtained and by the use of the indeterminates. However, this work is only possible once the specification has been translated into an equation.

3. METHOD DEVELOPED TO ASSIST ISO TOLERANCING

We will now use the principal results of this research work for our method, as follows:
- The rigorous structure of the TTRS method,
- The graph analysis techniques developed by A.Ballu and L.Mathieu,
- The idea of using a simulation method to guide our tolerancing synthesis.

All the functional conditions to be observed by the mechanism were determined from the technical functional analysis. Our purpose, therefore, is to find ISO specifications to be indicated on the documents for the various parts for each functional condition. This approach is based on the classical tolerancing approach. It consists of seeking positioning loops for each condition in order to determine all the parts concerned and their surfaces and then to find the tolerancing specifications for the parts.

The method developed permits a functional requirement to be systematically broken down into specifications of various kinds (dimension, shape and position). It is principally based on the use of graphs to choose the geometric specifications whilst integrating the TTRS concept [Clément et al 94].

3.1. Assumptions formulated

A functional requirement generally covers the geometry of two or several parts. Thus an independent study will be made for each of these conditions. The latter will be processed by assembly level.

3.2. The tools implemented

To assist with our approach, we will use several tools. The first is the link graph or assembly graph showing the positioning links (contact between 2 parts which serves to position them) and the fastening links (contact between 2 parts which serves to maintain them in position). They are oriented with the following differentiation: arrow side, Referenced surface, start side (opposite the arrow); Reference surface. .
In addition, the assembly plan will assist with the choice of reference surfaces. Finally, the functional requirements are represented by labels linking the surfaces they refer to.

The second is the contact graph. It stems from the link graph and, on this graph, each part is represented by a vertex, the surfaces are represented by a vertex pole and the links by an edge between two poles.

3.3. Approach

Our approach is broken down into several stages which can be summarised in two principal phases. The first involves the representation of the mechanism by the assembly sequence and the second the search for ISO specifications influencing a functional requirement. The detail of the stages can be summarised as follows:

PHASE I: Representation of the mechanism.

- Establishment of the link-oriented graph for each assembly sequence,
- Placing of labels corresponding to the functional requirements on the final link graph,
- Reduction of the link graph by functional requirement,
- Establishment of the simplified contact-oriented graph,
- Simplification of the contact graph to the influential parts and surfaces for the functional requirement processed,

PHASE II: Specification of functional requirements.

- Identification and recording of chains involving a requirement,
- Regrouping and classification of associations backed by the TTRS concept,
- Analysis and search for relations meeting technological constraints and ISO tolerancing,
- Inscribing of the specifications on each part for each functional requirement,
- Regrouping of tolerancing and elimination of specification redundancy.

Moreover, during the creation of the contact graph, we will not take into account surfaces involved in fastening links, given that these links do not participate in the positioning of the parts. The elementary surfaces or set of surfaces formed, likely to be references, are those which belong to parts located on the opposite side to the arrows (outgoing links) on the graph path. The same applies to those which are to be tolerances; they belong to the parts which are situated on the arrow side (incoming links).

Relations are composed from the likely sets of surfaces obtained, preferably in the direction of assembly sequence. The set of surfaces are identified with the assistance of the TTRS concept applied during the first stage. From this, the paths obtained with respect to assembly sequences and ISO tolerancing cases are analysed. If the tolerancing of the components of the assembly satisfies these constraints (technological and standards constraints), a tolerancing specification is proposed using the various cases of relation and reclassification noted when the TTRS theory was applied. If this is not the case, the hierarchical tree of relations is reconsidered in order to construct a new tolerancing structure.

4. APPLICATION TO A SPECIFIC CASE IN AERONAUTICAL ASSEMBLY

The example (see Figure [1]) to which the method presented is applied stems from a concrete case in aeronautical assembly, the bottom panel of the A319. This panel is composed of several parts, as follows:

- a panel (2), placed on tooling equipment,
- L-shaped edge cleats (3), located on either side of the edges of the panel,
- internal cleats (4), the same shape as the edge cleats, located between the latter,
- a ruler (5), to align the cleats along the same straight line,
- L-shaped string (7), fixed simultaneously on the panel and cleats.

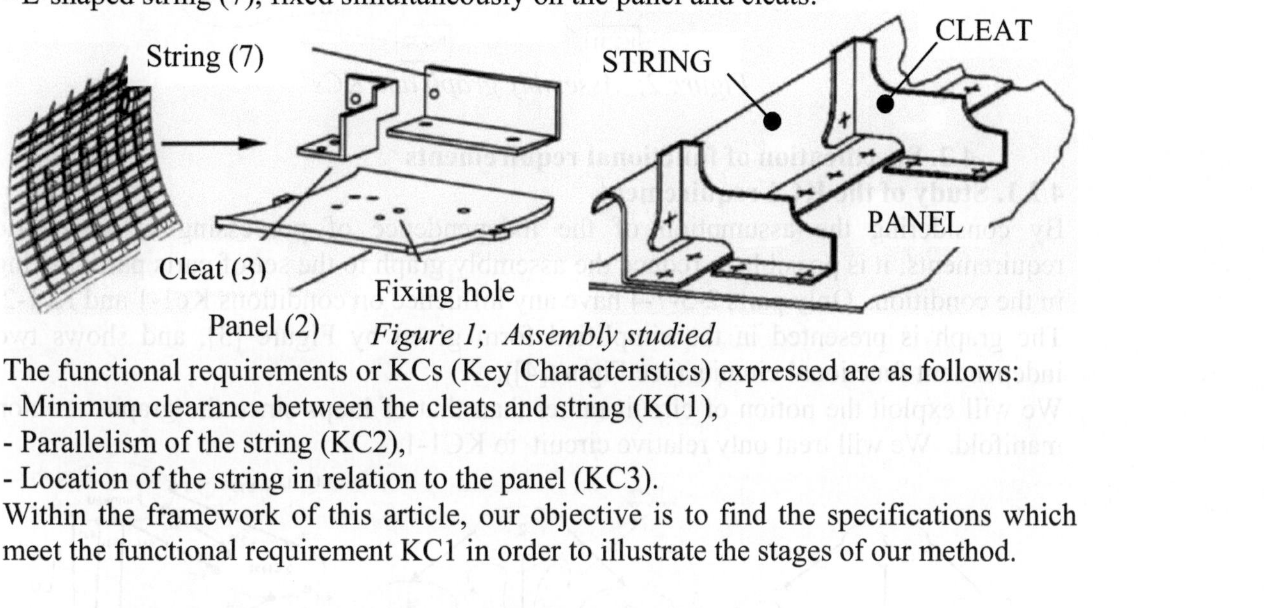

Figure 1; Assembly studied

The functional requirements or KCs (Key Characteristics) expressed are as follows:

- Minimum clearance between the cleats and string (KC1),
- Parallelism of the string (KC2),
- Location of the string in relation to the panel (KC3).

Within the framework of this article, our objective is to find the specifications which meet the functional requirement KC1 in order to illustrate the stages of our method.

4.1. Representation of the mechanism: assembly graph

The assembly sequences are described as follows:

SEQUENCE I Assembly of Tooling (1) /Panel (2)

SEQUENCE II Assembly of Panel (2) /Cleat row1(3-4) & Cleat row 2(3'-4'):
There are two cleats per row corresponding to the external cleats,

SEQUENCE III Alignment Rules (5 et 5')/(Cleat_rg1 & Cleat_rg2 & panel (2)),

SEQUENCE IV Placing of the Cleat_int(6)/Rule (5) & Panel (2),
SEQUENCE V Assembly of beam (7)/(Cleat1 (3) & Cleat2 (4) & Panel(2)).
The functional requirements or KCs are reflected on the link graph by links represented by dotted lines (see Figure [2]). The following references are used:
KC1-x: The minimum clearance between cleats and string, indicated by an arc linking the following part couples: (7/3), (7/4),(6/7'), (6'/7'),(3'/7''), (4'/7''),
KC2-x: The parallelism between the string, indicated by an arc linking the couples (7/7'), (7'/7''),(7/7''),
KC3-x: The location of the string in relation to the panel supports of the same row, indicated by arcs linking the following couples: (7/2), (7'/2), (7''/2).

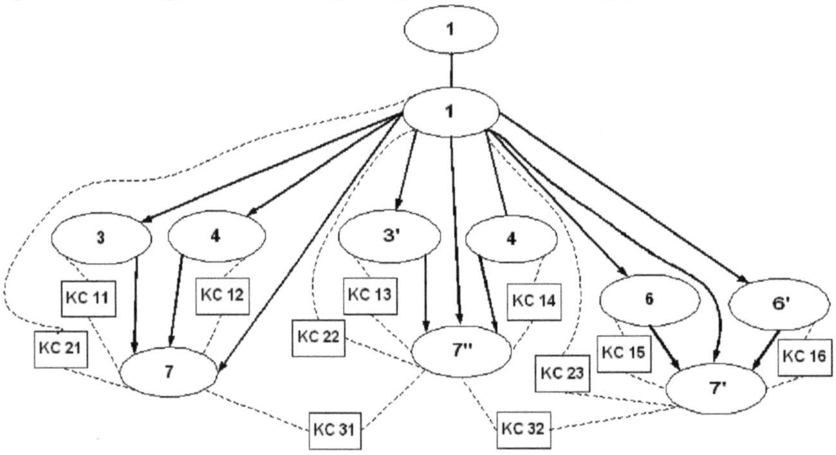

Figure 2; Assembly graph and KCs

4.2. Specification of functional requirements

4.2.1. Study of the KC1 requirement

By considering the assumption of the independence of processing of geometric requirements, it is possible to reduce the assembly graph to the set of parts participating in the condition. Only parts 2-3-7-4 have any influence on conditions Kc1-1 and Kc1-2. The graph is presented in the simplified form given by Figure [3], and shows two independent functional circuits (see Figure[4]).

We will exploit the notion of circuit rather than that of loops since the graphs are non manifold. We will treat only relative circuit to KC1-1.

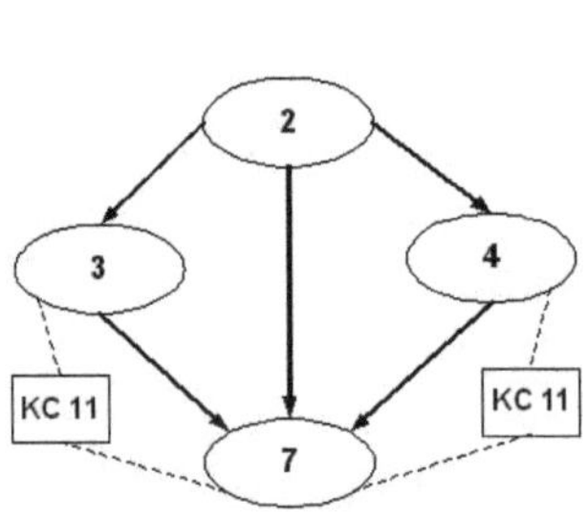

Figure 3 ; Condition graph

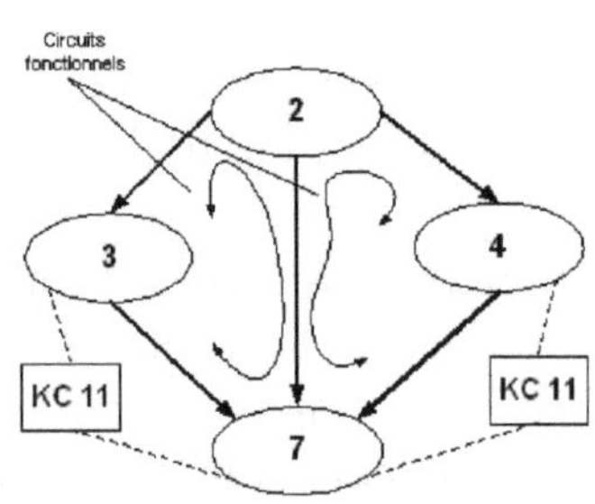

Figure 4; Functional circuits KC1

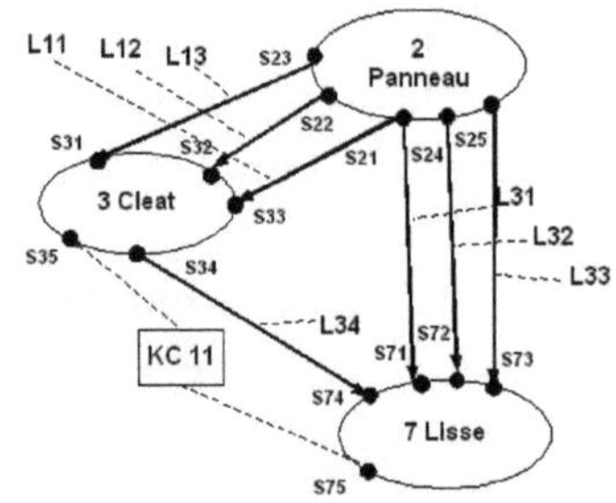

Figure 5; Contacts graph KC1

The link graph has highlighted the structure of the assembly influencing the condition. The latter is transformed into the contacts graph. It shows the elementary contacts at the level of the intervening functional surfaces. After taking stock of the set of surfaces of the various parts and their TTRS classification type, we establish the contacts graph for KC1-1, by corresponding assembly level. The elementary link arcs are given the following references:
L ij, where i represents the assembly level index and j the order of establishment within this level. The contacts graph is presented in the following form (see Figure[3]). The surfaces (S21, S22, S23, S24, S25) for the panel, (S31, S32, S33, S34, S35) for the cleat, and (S71, S72, S73, S74, S75) for the beam, participate in the creation of elementary chains which have an influence on condition Kc1-1. The calculation of the cyclomatic number $\delta = m - n + 1$, where n represents the number of vertices (parts) and m the number of links (contacts) between these vertices, gives $\delta = 8 - 3 + 1 = 6$ independent cycles.

4.2.1.1. Strategy of Choice of relation creation path

We will move towards the need for ISO standard tolerancing of characteristic groups of surfaces for a given task (positioning, fastening or simply professional culture). Thus the only usable objects are isolated surfaces, a pair of planes and a pair or group of cylinders. During the creation of relations we will keep the idea of the creation of technological features. According to the TTRS approach, once the independent loops have been determined, the next stage consists of carrying out the corresponding relations.

We will retain this idea by seeking this time the most technological chains in the sense of the assembly sequence, which pass through a link and therefore influential edges, in order to relate surfaces to them and form the TTRSs on each solid.

The following principal criteria will be adopted:

- The type of contact. This is an important element which can influence the choice of specifications. The type of contacts to be considered are fixed contacts (without any clearance) and floating contacts (with clearance)
- The hierarchy for positioning neighbouring parts: In concrete terms, this is the hierarchy of the reference system which, via the preponderant surfaces, gives the surfaces that will be declared within the reference system as primary, secondary and/or possibly tertiary.

In order to identify these criteria more rapidly at assembly and part level, we will use two types of tables.

- The assembly table: This gives the list of parts participating in the assembly sequences with indications on their levels, the number of incoming and outgoing links and their respective contact surfaces.

For our example, it is presented in the form given in table [1].

- The part table: It gives indications concerning the types of functional surfaces influencing a link with a neighbouring part, the level of the link where it occurs within the assembly, the order of positioning, the neighbouring part index and the type of contact.

To study KC1-1 the tables of parts 2,3 and 7 are given by tables [2], [3], and [4] below. With the assistance of these tools (cyclomatic number, relations graph and tables), it is possible to sketch out possible path chains and bring in technological justification to establish the tree of relations for each part.

PIECE	NIVEAU	Liaison Entrante et contacts	Liaison Sortante et contacts
2	1	0	2: S21, S22, S23
			S24, S25, S26
3	2	1: S31, S32, S33	1: S34, S35
7	3	3: S71, S72, S73,	0
		S74, S75, S76, S77, S78	

Table 1; Assembly Table

Pièce2	S21	S22	S23	S25	S26
Classe surface	PL	CY	CY	CY	CY
Niveau	1/3	1	1	3	3
Liaison					
Ordre de					
Mise en Position	1	2	3	2	3
Pièce Voisine	3/7	3	3	7	7
Type	Fixe	Fixe	Flottant	Fixe	Flottant

Table 2; Table Part 2

Pièce3	S31	S32	S33	S34	S35
Classe surface	PL	CY	CY	CY	PL
Niveau	1	1	1	3	3
Liaison					
Ordre de					
Mise en Position	1	2	3	1	2
Pièce Voisine	2	2	2	7	7
Type	Fixe	Fixe	Flottant	Fixe	Flottant

Table 3; Table Part 3

Pièce7	S71	S72	S73	S74	S75
Classe surface	PL	CY	CY	CY	PL
Niveau	4	4	4	4	4
Liaison					
Ordre de					
Mise en Position	1	2	3	1	1
Pièce Voisine	2	2	2	3	3
Type	Fixe	Fixe	Flottant	Fixe	Flottant

Table 4; Table Part 7

The choice criteria that we will choose to establish our relations are:

1. The level of the link in the assembly plan,
2. The order of positioning of the surfaces and the neighbouring part,
3. The type of contact.

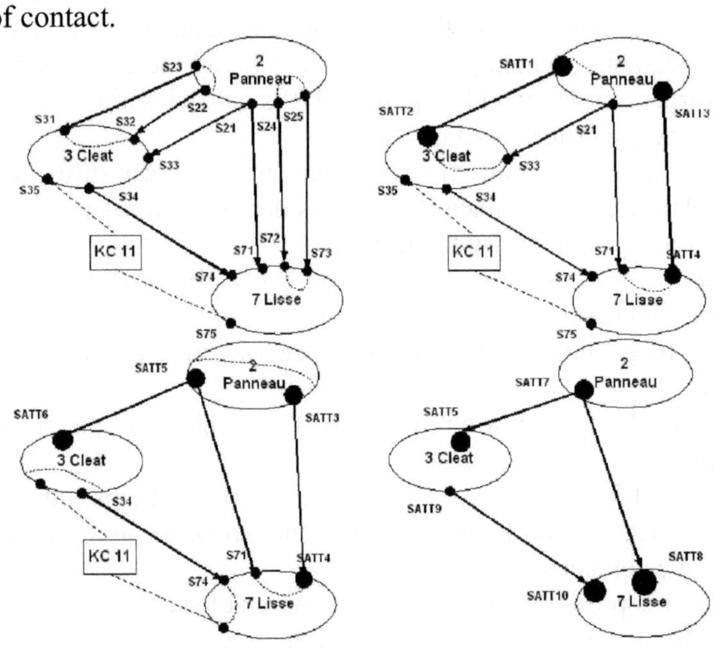

Figure 6; Chains path

Initially, we apply the TTRS approach in order to record possible relations. The calculation of the cyclomatic number for the circuit paths (see Figure [6]) relative to condition KC1-1 indicates 6 chains:

Chain 1 gives for: panel 2, TTRS1= S23 u S24,
cleat 3, TTRS2= S32 u S33,
Chain 2 gives for: panel 2, TTRS3 = S24 u S25,
beam 7, TTRS4= S72 u S73.
Chain 3 creates for: panel 2: TTRS5= TTRS1 u S21,
cleat3: TTRS6= S31 u TTRS2.
Chain 4 creates for: panel 2: TTRS7= TTRS5 u TTRS3, giving the final_TTRS for the part, beam 7: TTRS8= S71 u TTRS4,
Chain 5 creates for cleat 3: TTRS9= S34 u S35,
beam 7: TTRS10= S74 u S75.

Finally, chain 6 generates the final TTRSs for the parts (see Figure [7]):
cleat 3, TTRS_final= TTRS6 u TTRS9,
beam 7, TTRS_final= TTRS8 u TTRS10.

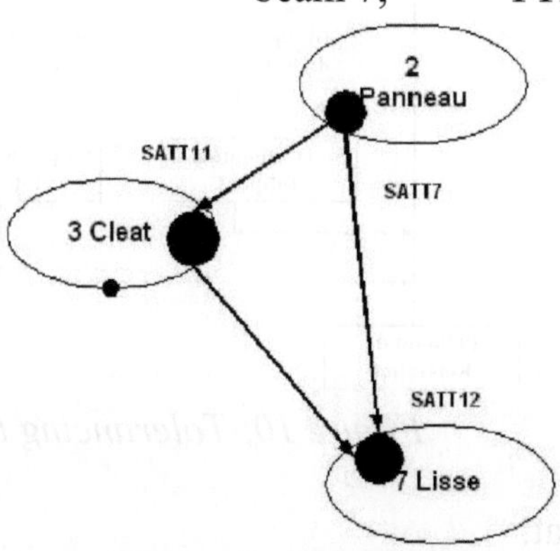

Figure 7; Final TTRSs

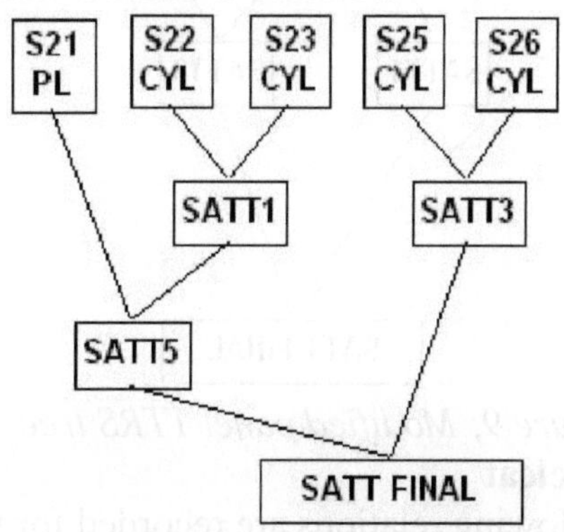

Figure 8; Final TTRSs

a. The Panel

At this stage, we have the hierarchical tree of the TTRSs (see Figure [8]) for the panel. This enables the set of surfaces to be toleranced to be recorded in relation to one another. We note the appearance of a typical TTRS (relation of a plane, a cylinder without clearance and a cylinder with clearance), corresponding to a classic positioning (surface bearing, centring with clearance and centring without clearance) commonly used in aeronautics with a classic tolerancing structure. We will call this TTRS_mecano.

However, this arborescent structure leads to non-compliant tolerancing as regards ISO standards. In fact, for the panel-cleat link, we can see that the first TTRSs position surface S23 (hole with clearance) in relation to the surface S22 (hole without clearance). This group is itself positioned in relation to reference surface S21. We find a similar architecture for the panel/beam link, referencing on the previous structure. This construction does not correspond to any case of tolerancing defined by the standard.

We will redefine a construction tree, therefore, by respecting our principal technological constraint (use of TTRS_mecano) and the standard. This hybrid usage takes us from the classic hierarchy model towards a closely related model. Figure [9] illustrates the shape of the tree of relations of the newly created panel. Indirectly, two TTRS_mecanos are identified, each composed of a preponderant plane surface (common to both) and a group of holes, each composed of holes with and without clearance.

The sequence order (see Figure[10]) is established in accordance with the standard which permits the positioning of a group of holes with different diameters relative to one another in relation to an external reference. In this case, S21 becomes the only reference common to both groups. The preponderance of links (L1j in relation to L3j) does not occur here and, moreover, it does not lead to a hierarchy between the two groups of holes.

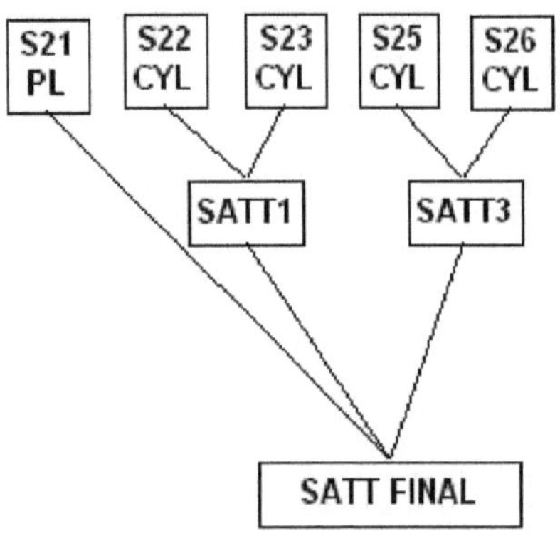

Figure 9; Modified panel TTRS tree

Classe surface	PL	CY	CY	CY	CY
Niveau Liaison	1/3	1	1	3	3
Ordre MP	1	2	3	2	3
Pièce voisine	3/7	3	3	7	7
Type	Fixe	Fixe	Flottant	Fixe	Flottant

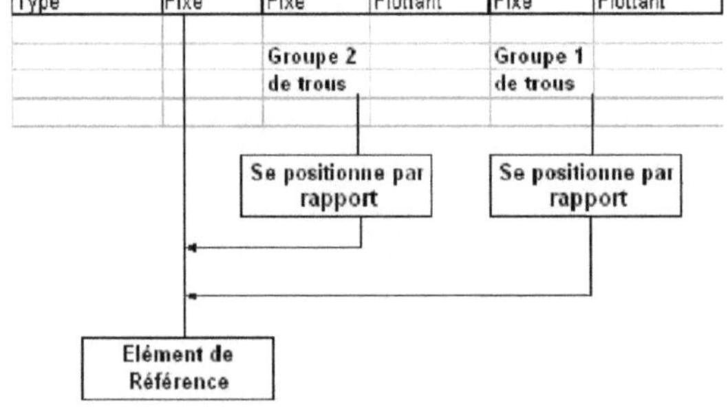

Figure 10; Tolerancing table

b. The cleat

The following relations are recorded for the cleat:

TTRS2 = S32 u S33, TTRS6= S31 u TTRS2, TTRS9= S34 u S35 and final_TTRS = TTRS6 u TTRS9, i.e. two principal set for this part. The links between the panel and the cleat (type L1j) and the cleat and beam (type L3j) show a descending direction and thus an orientation. This supposes that the bundle of surfaces stemming from links L3j will probably be the "Datum" bundle and the surfaces composing the second bundle the elements toleranced in relation to this datum. The surface relation structure is presented in the form given in Figure [11].

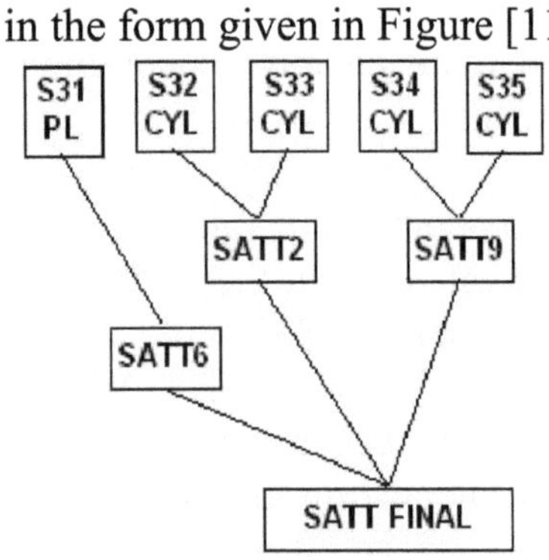

Figure 11; Cleat TTRS tree

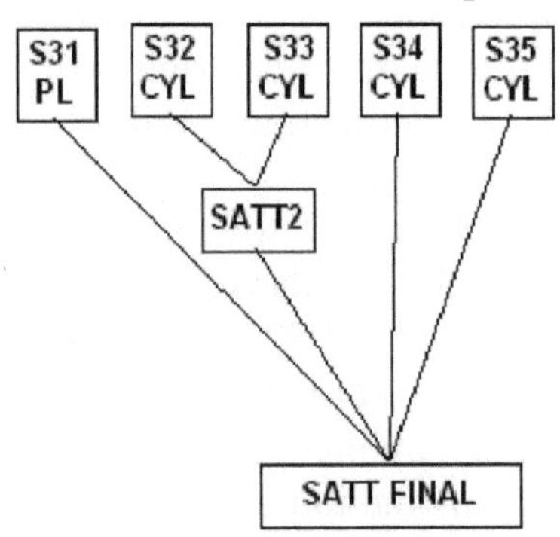

Figure 12; Modified cleat TTRS tree

This structuring leads to tolerancing that does not comply with the ISO standard. Thus the new structure proposed is presented in the more classic form of the case of several elementary surfaces referencing in relation to a reference plan. Figure [12] illustrates this new tolerancing.

c. The string

In the same manner as for the cleat, we record the following associations for the beam: TTRS4 = S72 u S73, TTRS8= S71 u TTRS4 , TTRS10= S74 u S75 and final_TTRS = TTRS10 u TTRS8, i.e. two principal set for this part. The links between the panel and the beam (type L3j) and the cleat and the beam (type L3j) show a descending direction and thus an identical orientation. This does not suppose a particular choice in favour of one or the other of the surface set.

In this case, we turn towards the type of referential constituted by the bundle in question and the case of tolerancing deduced in relation to the standard. The TTRS8-based bundle is more complete than the TTRS10. The orientation of the path will be given by choosing the TTRS8 as "Datum" and the TTRS10 as the toleranced element. The surface relation structure is presented in the form given by Figure[13]. In the same manner as for the cleat, the structuring here leads to tolerancing that does not comply with the ISO standard. Thus the new structure proposed is presented in the most typical form of the case where several elementary surfaces are referenced in relation to a plane. Figure [14] illustrates this new tolerancing.

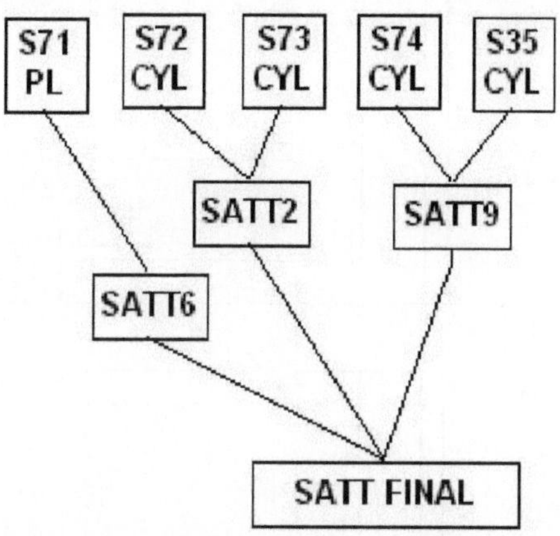

Figure 13; TTRS structure of the cleat beam

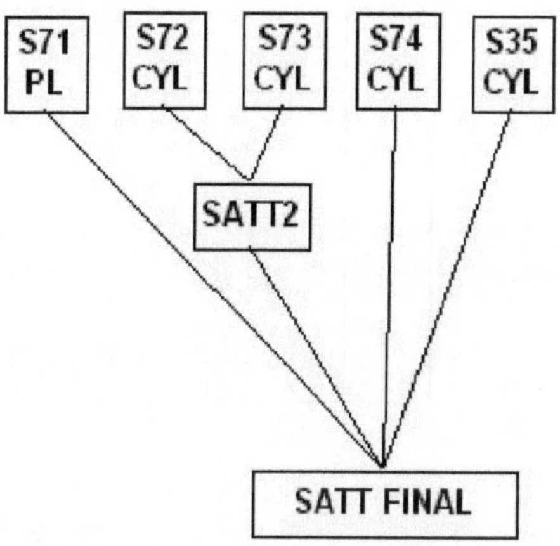

Figure 14; Modified TTRS structure

4.3. Implementation of specifications

The geometric specifications are obtained by using TTRS reclassification notions which thus enable the specifications on the linked surfaces and the intrinsic specifications of the surfaces (dimensional and shape) to be recorded. The latter are deliberately omitted.

a. The panel

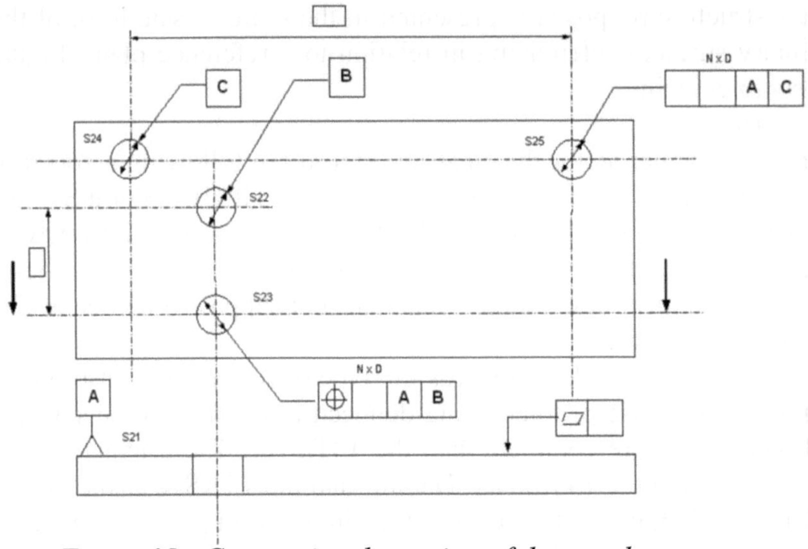

Figure 15; Geometric tolerancing of the panel

b. The cleat

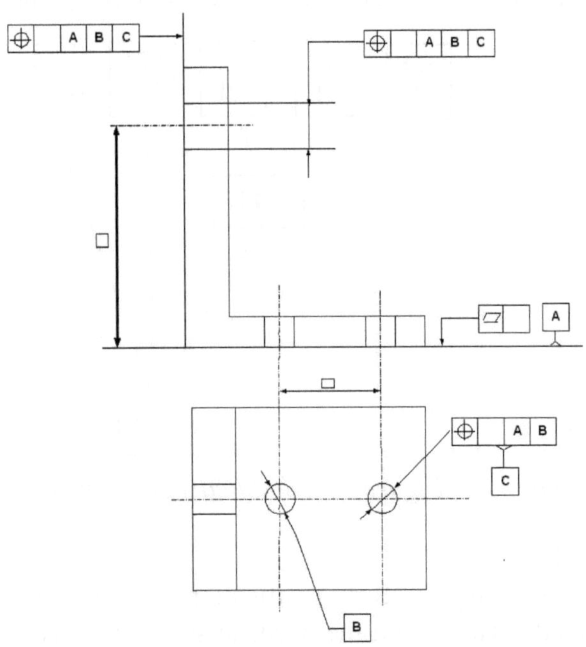

Figure 16; Geometric tolerancing tolerancing of the cleat

c. The string

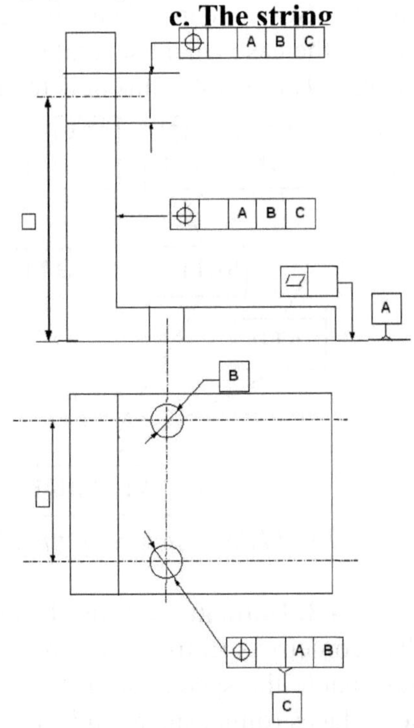

Figure 17; Geometric of the beam

5. CONCLUSION

The general objective of this research work was to develop and conduct experiments on a geometric tolerancing synthesis method using the specification of functional conditions and knowledge of the assembly plan.
The method presented is principally based on the TTRS model and on the path of oriented graphs.
It has been partially validated on examples for the aircraft industry. However, constraints linked to ISO language and the combinatorics resulting from the graph paths has led us to introduce "professional rules", the addition of which enable us to obtain a result more in line with user expectations.

REFERENCES

[Clément et al 94] Clement, A.; Rivière, A.; Temmerman, M.; "Cotation tridimensionnelle des systèmes mécaniques: Théorie et pratique" ("Three-dimensional tolerance specification of mechanical systems: Theory and practice"); PYC Edition; France ; ISBN 2-85330-132-X.

[Marguet 97] Marguet, B.; Mathieu, L.; " Tolerancing problems for aircraft industries" ; CIRP seminar on tolerancing ;Toronto ;1997.

[Marguet 99] Marguet, B.; Mathieu, L.;"Aircraft assembly analysis method taking into account part geometric variations"; CIRP seminar on tolerancing; Twente ;1999.

[Ballot 99] Ballot, E.; Bourdet, P.; Thiebaut, F.; " Contribution of mathematical model of specifications of a part to their coherence analysis"; CIRP seminar on tolerancing ;Twente;1999.

[Ballu 99] Ballu, A.; Mathieu, L.; " Choice of functional specifications using graphs within the framework of education" ; CIRP seminar on tolerancing;Twente ;1999.

[Ballu 99] Ballu, A.; Mathieu, L.; « Méthode de choix des spécifications fonctionnelles par des graphes » ; Technologie et Formations (Method for choosing functional specifications by graphs); N°84 ; pp12-19 ; France;May-June 1999.

[Sellakh 99] Sellakh, R.; « Contribution à l'intégration de la définition multiniveaux de la géométrie des systèmes mécaniques pour le tolérancement » ("Contribution to the integration of multilevel definition of the geometry of mechanical systems for tolerancing"); Thèse de Doctorat, INSA de Lyon, Juillet 1999, 255p. (Doctorate Thesis, Lyon INSA, July 1999, 255P).

Toward non-nominal virtual geometric verification by combining VR and CAT technologies

Casper Wickman M.Sc., PhD Student
Volvo Car Corporation
Exterior Design
Dept 93350 PV4B
SE-405 31 Göteborg, Sweden
and
Chalmers University of Technology
Department of Machine and Vehicle Design
SE-412 96 Göteborg, Sweden
E-mail: vcc2.casperw@memo.volvo.se

Rikard Söderberg, Ph.D., Associate Professor.
Chalmers University of Technology
Department of Machine and Vehicle Design
SE-412 96 Göteborg, Sweden
E-mail: riso@mvd.chalmers.se

Lars Lindkvist, Ph.D., Associate Professor
Chalmers University of Technology
Department of Machine and Vehicle Design
SE-412 96 Göteborg, Sweden
E-mail: lali@mvd.chalmers.se

Abstract: As the markets are becoming increasingly competitive and customer demands higher, virtual verification has become an important tool to cut cost and to shorten lead-time. So far, the main focus area for Virtual Reality (VR) tools has been to extend and support concurrent engineering between manufacturing and design in early stages of the design process. Today, when a styling concept of an artifact is evaluated in an esthetic manner, all models used are nominal. Variational aspects and design solutions, that greatly will influence the over-all quality appearance, will not be discovered until the first test series are made. By using non-nominal models during all stages of the design process, important geometric aspects can be issued and the needs for physical test series can be reduced. In the automotive industry, especially within body design, the relations between doors, hoods, fenders and other panels are critical. Today the quality appearance of vehicle is judged by these relations.

P. Bourdet and L. Mathieu (eds.),
Geometric Product Specification and Verification: Integration of Functionality, 301-310.

This paper presents how virtual reality techniques can be used to visualize simulation results gained from tolerance analysis. Combining traditional CAT tools with modern VR tools has the potential to enhance concurrency between styling and design and provide more powerful support for the geometry process in early phases. Traditional non-nominal verification can then be performed already in the concept phase using digital models instead of physical. An example of the rear end of a vehicle is used to illustrate how integrated CAT/VR tools can be used to support decision making and virtual verification throughout this process.

Keywords: virtual geometric verification, tolerance allocation, concurrent engineering, quality appearance, non-nominal visualization

1. INTRODUCTION

1.1. Background

In the automotive industry today, some of the most critical issues are to reduce cost, increase quality and decrease lead-time. To fulfill those issues, high demands has been set on the design process. Engineering tasks are performed more and more simultaneously between different departments in order to shorten design time and to be able to verify the design early in the developing process. Modern computer tools and flexible working methods have also played an important role in speeding up the design process and extending concurrent engineering.

One part of the process which has a great impact on the quality appearance is the interplay between styling and design within the geometry design process. This process involves many critical areas that can contribute greatly to decreased quality appearance. In this paper, quality appearance is referred to as all those aspects that make an impression of quality on the customer, just by observing the product. Those aspects can be relations between doors, hoods, fenders and other panels on the vehicle. The level of variation in those relations is affected by part variation, assembly variation but also with the robustness of the design concept. The robustness, i.e. the ability to suppress variation, is dependent on the design and style configuration. In the next section, a theoretical description of robustness is presented. Accurate information and tools for communication between styling and design will have a strong impact on the final quality appearance.

In [Maxfield *et al.*, 2000a and 2000b], a rather comprehensive outline of a computer system for visualization of cosmetic quality is presented. The system is based on a number of commercial tools combined with a visualization model, using Open GL 3D graphics. The paper reports on progress achieved during the first year of a project where the system is still in a very early prototype stage. However, evaluation of case studies recognizes that similar systems would greatly assist the setting of quality targets.

Physical prototypes are made today in order to perform nominal verification. This work is mostly conducted quite late in the design process, resulting in post-conceptual changes. If problems identified by the physical prototype can be highlighted earlier in the process, economic benefits are gained. Today, physical test series are made to verify the non-nominal geometry. This is highly expensive and is also conducted very late in the process, resulting in expensive post-conceptual changes.

1.2. Geometrical robustness and variation

Geometrical variation is introduced to the product as *component variation* and *assembly variation*. In most companies, component variation is controlled by the suppliers, whereas assembly variation is controlled in-house. An important contributor to final variation is also the *concept sensitivity*. In a sensitive concept, component and assembly variation is amplified, whereas in a robust concept variation is suppressed. Since the robustness of the product is governed by the shape of the parts and the placement of the locators, it is very important to try to achieve as high geometrical robustness as possible already in the concept phase. This may be achieved by the use of Computer Aided Tolerance (CAT) tools, see [Söderberg and Lindkvist, 1999a], and by early robustness analysis, see [Söderberg and Lindkvist, 1999b].

The geometrical aspects of an assembly can be judged by two characteristics, its *robustness* and its *variation* in critical dimensions, see [Söderberg and Lindkvist, 1999b]. The robustness reflects whether the assembly will suppress or amplify input variation whereas the variation in critical dimensions reflects what the customers ultimately see, which is governed both by the robustness of the concept and by the input variation.

The relation between robustness and variation can be illustrated with a simple beam support example as in figure 1 below. The robustness of the concept is defined by the relation between input and output variation, in this case totally controlled by the position of the support. Moving the support to the left will increase the robustness, whereas moving it to the right will decrease the robustness. The output variation is controlled by two parameters, the position of the support and the input variation. These relations can be described in matrix form as in figure 1. Since the position of the locator controls two important product characteristics, this should be treated first. Based on final requirements for the output variation and known sensitivity (relation between input and output), the tolerance for the input variation may then be determined

$$\begin{bmatrix} robustness \\ variation \end{bmatrix} = \begin{bmatrix} x & 0 \\ x & x \end{bmatrix} \begin{bmatrix} locators \\ tolerances \end{bmatrix}$$

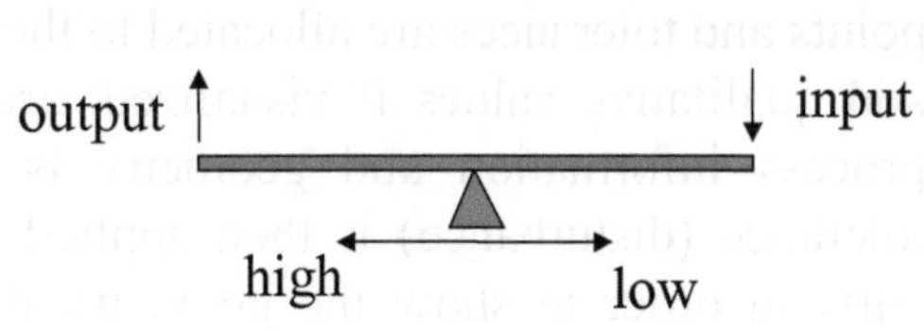

Figure 1; Robustness and variation

1.3. Scope of this paper

This paper focus on the how virtual reality technique can be used to verify and evaluate geometry robustness and variation aspects during *the geometry design process* in order to increase quality appearance. In this work, a tolerance analysis package is linked to a virtual reality package. By combining those programs, non-nominal models can be visualized in a realistic environment and manipulated in real-time. An example from the automotive industry will illustrate the use of the visualization of non-nominal models. The model structure and its impact on the cognitive aspects are also discussed.

2. VISUALIZING RESULTS FROM CAT SIMULATIONS WITH VIRTUAL REALITY

Computer Aided Tolerance software allows analysis of the geometric variation in assembled products. CAT tools allow the total variation in an assembled product to be predicted. One of the aims of tolerance analysis is to create a robust design, insensitive to geometrical variation. A number of commercial CAT systems are available on the market today, mainly based on Monte Carlo simulation or direct linearization. A review of commercial CAT systems was presented by [Salomonsen *et al.,* 1997].

Virtual reality is a progressing technology aimed at creating a illusion of reality using a computer environment. VR tools allow the user to interact with the virtual environment and objects submitted to the environment. Virtual reality has been defined by [Larijani, 1994] as "*a computer-synthesized, three-dimensional environment in which a plurality of human participants, appropriately interface, may engage and manipulate simulated physical elements and, in some forms, may engage and interact with representations if other humans, past, present or fictional or invented creatures.*" Virtual environments can be displayed using a variety of VR hardware, see [Vince, 1995]. For the applications in this paper power walls and desktop monitors has mainly been used. A review of commercial VR technologies and there applications is presented by [Kamble and Arora.,1998].

In order to visualize results from CAT simulations, two commercial systems have been integrated to create a virtual environment for non-nominal geometry models. The CAT program RD&T (Robust Design & Tolerancing) has been used to perform robustness and tolerance simulations. Geometry may be imported from any CAD system or created directly in RD&T. All parts are positioned with master locations points and tolerances are allocated to the geometry. Two kinds of simulation, which will add qualitative values if visualized, are performed. In the early design stages, when process information and geometry is restricted, stability analysis is used. A unit tolerance (disturbance) is then applied in the locating direction of all locators on all parts in order to show the geometrical sensitivity of the design. Later in the design process, when tolerances or process variation are known and allocated, variation analysis is performed. This simulation shows the final expected geometric variation in critical measures.

Opus Studio is the chosen VR tool, used for visualization of simulating results. It is a commercial program from Opticore AB. Opus Studio is a multi-platform toolkit for building applications, using Open GL Optimizer as 3D graphics technology. Opus Studio allows the user to interact with the virtual environment and to define behavior and test applications. Parameters may be changed during real-time simulations. The tolerance analysis, performed in RD&T, provides positioning data for all parts in the model. A specific identity is assigned for every part of the RD&T model and for the corresponding part in Opus Studio. The positioning and scaling data is then automatically transferred to Opus Studio.

2.1. Different scenarios

It is impossible to find a "worst case" for the whole vehicle, according to quality appearance. Different prominent areas must be studied separately. For body panels are two main relations are of interest, namely flush and gap. Both the gap relation and flush relation can vary along the *seam* see [Söderberg and Lindkvist, 2001] between two parts, causing a non-parallelism with respect to the flush or the gap. As a consequence of focusing on one area of the vehicle at a time, another relation is of interest. This is the derivative between two surfaces of two parts along a seam, here defined as the *seam derivative.* In figure 2 the seam derivative is illustrated as the angle between nominal position and the non-nominal position of the boot lid. The seam derivative could also be replaced by a flush relation along the opposite boot lid seam in the z-direction. But since the focus is on only one relation, it is useful to replace the flush relation with the seam derivative. One further aspect that occur du to flush and gap is see through. That is how details, seen through the seam, is affecting the quality appearance.

One of the most central questions is: W*hat kind of information should be*

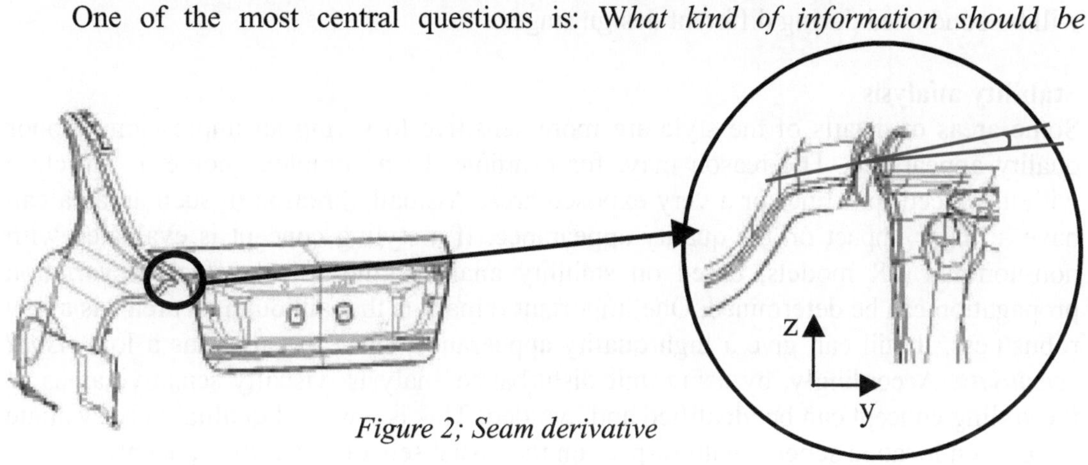

Figure 2; Seam derivative

visualized? For example, the relation between the rear tail and rear lamp of a vehicle has a substantial impact on the quality appearance. The relation can vary in flush and gap. The question is whether an immense gap or, for example, a small non-parallelism will have the worst impact on the quality appearance. Some areas of the vehicle might be more tolerant of variations than others, indicating that no general rules about how much variation will impact on the quality appearance can be established. Engineers often express the need to show the worst case situation for an area of the design. However, in many cases the worst case situation may not be described with one scenario. For instance, what is the worst case position of a vehicle front door? The front door must look nice in relation to the rear door, the fender, the roof and the sill and have a acceptable see through, which may not always be accomplished at the same time. Therefore, what a visual "worst case" is represented by is highly dependent on the area of the vehicle and the geometry of the model, i.e. the *visual sensitivity* of the model. Alternative scenarios, representing visualization of different variation cases, should be presented. All those scenarios can together give a final appreciation of the quality appearance.

2.2. Model structure

Since the result is supposed to be evaluated in a qualitative manner, the degree of realism of the model is of highest importance. Light, color, reflections, clear coats, textures and shadows are factors that will determine the realism and quality of the VR model. A great deal of effort must therefore be put into those factors. Moreover, all parts that cover up behind the seams have to be present since see through is one important aspect of the quality appearance. Even though the virtual model can be regarded as very accurate and realistic, the fact remains that cognitive aspects can have a huge impact on the interpretation of the model.

2.3. Benefits from visualization of CAT simulation

During different stages of the geometry design process, the level of details and information about the design and style varies. This indicates that different simulations will be conducted during different design stages.

Stability analysis

Some areas or details of the style are more sensitive to giving an impression of poor quality appearance. The reason may, for example, be a complex geometric structure with advanced split lines or a very exposed area. A small variation in such an area can have a major impact on the quality appearance. If a styling concept is evaluated with non-nominal VR models, based on stability analysis, the visual effect of variation propagation can be determined. One important remark is that although an area has a low robustness, it still can give a high quality appearance, since the area has a low *visual sensitivity*. Accordingly, by using unit disturbance analysis, visually sensitive areas of the styling concept can be identified and avoided. This is a way of qualitatively evaluate the chosen location scheme with respect on the visual sensitivity of the concept.

Variation analysis

When the locating scheme has been chosen and tolerances allocated, worst case scenarios can be created with simulation results from the variation analysis. Visualization of worst case scenarios, based on variation analysis, is a way to predict results that generally is conducted when the first physical test series are made. It is a verification of the geometry over a number of manufactured pieces/units. Aspects like flush, gap and see through can be evaluated. Moreover, the reflection quality, which is evaluated for every single part or surface when the styling surface is created, can also be evaluated. Depending on the styling concept, in some critical areas of the vehicle, small variations can contribute to a reduction in reflection quality.

2.4. Cognitive aspects

Although a realistic non-nominal model, representing the real part and assembly variation conducted during manufacturing is created, the interpretation, i.e. how the model is perceived, i.e. must be identified. That would represent a validation of the model from a perceptional point of view. Although cognitive studies of the human perception of virtual reality have already been conducted, the effect of geometric variation is such a specific application with enormously high demands on realistic and detailed reproduction that further studies are needed to ensure that VR can be used for this kind of geometric application. What is the influence on the appearance of geometric variations of parameters such as surface structure (degree of tessellation) color, light, number of light sources, glossiness, perspective, level of details, radius along seams and shadows? A cognitive study that identifies the effect of those parameters will have to be performed to verify the model for applications of this kind.

3. EXAMPLE OF VISUALIZATION

In this section, an example of visualization of a non-nominal model is presented. The rear end of a Volvo S60 has been chosen to be visualized, In this example the focus is on the split line between rear lamp, rear hood and the outer rear body side; see figure 3.

3.1. CAT simulation

Design models are imported into RD&T from CATIA. The locators for all parts are automatically transferred with the geometry. The first step in RD&T is to define the locating systems for all parts. Here, the 3-2-1 system or similar locating systems, used in the automotive industry, are available. Once the locating systems have been defined, stability analysis may be performed to evaluate the geometrical robustness and degree of coupling. The most important locating schemes and individual locators, with respect to overall robustness or specific critical dimensions, may then be found. When tolerances have been allocated for all parts and fixtures, variation analysis may

Figure 3; Evaluated area

be performed with respect to defined critical dimensions. In RD&T, variation analysis is based on Monte Carlo simulation.

A critical dimension of an automotive body could, for example, be a flush relation between two parts. The result of the variation analysis can be presented in RD&T as distributions showing the variation for all the defined measures, see figure 4. The result can also be presented in color coding where different colors represent the grade of variation, see [Lindkvist and Söderberg, 2000].

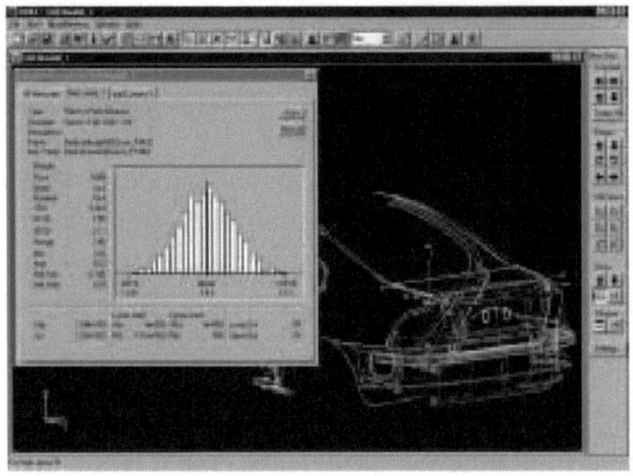

Figure 4; Variation analysis

This quantitative way of representing the result is not suitable for all receivers of the information. In order to reach a wider range of receiver and to present the result in a qualitative way, the result from the simulation is exported from RD&T to be viewed in Opus Studio. Position and part variation data for different "worst case" scenarios, defined as the worst case situation for a number of defined critical dimensions in RD&T, is imported to Opus. Fixture and assembly variation is visualized in Opus by changing the position of the part, corresponding to the defined and simulated "worst case" scenarios. Part surface variation is visualized by changing the scale of the part in gap or flush directions.

3.2. Visualization of results

The rear end parts are imported from Alias into Opus Studio. The number of parts (level of details) is of great interest for the quality of the VR model. As well as exterior parts, it is important to also include parts that cover up behind exterior parts to prevent see-

Figure 5; Non-nominal model of rear end

through and to give a realistic impression. Positioning and scalar information is transfers, through the local network by predefined virtual ports, from RD&T to Opus Studio in real-time during simulation. Figure 5 shows a scenario where the worst non-parallelism and flush between the boot lid and body is visualized.

This example clearly demonstrates that the same information or result can be shown with a totally different appearance in order to reach a wider range of receivers and to analyze the result in a qualitative manner. In Opus, the effect of variation can be evaluated in real time, from different angles and with high degree of realism, by engineers and industrial designers at the same time in order to support the geometry process and to increase the concurrency between styling and design.

4. CONCLUSIONS

In this paper we have proposed a tool that combines CAT simulation and virtual reality. The tool combines the visual advantages that VR offers and the simulation abilities that CAT technology provides in order to support the geometry design process and to extend the concurrency between design and styling. Benefits from visualization of non-nominal models are presented. What kinds of "worst case" scenarios, due to visual sensitivity and geometry, that can give a final appreciation of the total quality appearance, are presented. In addition, cognitive aspects that can have an immense impact on the interpretation of the non-nominal VR model, for virtual reality applications of this kind, have been discussed. With the proposed tool, aspects that are evaluated with physical prototypes today can be evaluated earlier in the design process with non-nominal models in virtual environments. The tool is illustrated with a example of a rear end of a Volvo S60.

5. ACKNOWLEDGMENTS

The authors would like to thank Volvo Car Corporation for the generous support and information access that has been sustaining this paper. The authors also like to acknowledge the Swedish foundation for Strategic Research through the ENDREA Research Program (The Swedish Engineering Design and Research Education Agenda) and the Swedish National Board for Industrial and Technical Development (NUTEK) for there financial support.

REFERENCES

[Kamble and Arora.,1998] Kamble, A.S.; Arora, R.K.; "Virtual Reality–A Technology and Application Review"; *Electronics Information and Planning*. vol.25, no.7; April 1998; p.360-77

[Larijani, 1994] Larijani, L. Casey "The Virtual Reality Primer."; New York: McGraw-Hill; ISBN 0-07-036417-6

[Lindkvist and Söderberg, 2000] Lindkvist, L.; Söderberg, R.; "Tool for Assembly Locating Scheme Definition"; *Proceedings of the 26th ASME Design Automation Conference*; September 10-13; Baltimore; Maryland; USA 2000

[Maxfield *et al.*, 2000a] Maxfield, J.; Dew, P.M. ; Zhao, J. ; Juster, N. P. ; Taylor, S. ; Fitchie, M.; Ion, W. J.; Thompson, M.; "Predicting Product Cosmetic Quality using Virtual Environment"; *Proc. Of 2000 ASME Design Engineering Technical Conferences and Computers and Information in Engineering Conference;* September 10-13 2000; Baltimore; Maryland; USA 2000

[Maxfield *et al.*, 2000b] Maxfield, J.; Dew, P. M.; Zhao, J.; Juster, N. P.; Taylor, S.; Fitchie, M.; Ion, W. J.; "Predicting Product Cosmetic Quality in the Automobile Industry"; *Proceedings of the 33rd International Symposium on Automotive Technology and Automation (ISATA 2000);* Dublin; Ireland September 25-29 2000

[Salomonsen *et al.*, 1997] Salomonsen, O. W.; van Houten, F.; Kals, H.; "Current Status of CAT Systems"; *5th CIRP Conference on Computer Aided Tolerancing*; pp.345-359; Toronto April 28-29 1997

[Söderberg and Carlson, 1999] Söderberg, R.; Carlson, J.; "Locating Scheme Analysis for Robust Assembly and Fixture Design"; *Proceedings of the ASME Design Automation Conference*; Las Vegas; Nevada; USA September12-15 1999

[Söderberg and Lindkvist, 1999a] Söderberg, R.; Lindkvist, L.; "Two-Step Procedure for Robust Design Using CAT Technology"; *Proceedings of the 6th CIRP International Seminar on Computer Aided Tolerancing*; Enschede; the Netherlands March 22-24 1999

[Söderberg and Lindkvist, 1999b] Söderberg, R.; Lindkvist, L.; "Computer Aided Assembly Robustness Evaluation", *Journal of Engineering Design*, Vol. 10, No. 2, pp. 165-181.

[Söderberg, R. and Lindkvist, L., 2001] Söderberg, R.; Lindkvist, L.; "Automated Seam Variation and Stability Analysis for Automobile Body Design*"; Submitted to The 7th CIRP International Seminar on Computer Aided Tolerancing*; ENS de Cachan; France April 24-25 2001

[Vince, 1995] Vince J.; "Virtual Reality Systems"; Addison-Wesley Publishing Company; ISBN 0-201-87687-6

A STEP Translator toward Computer Integrated Tolerancing

Jhy-Cherng Tsai
Department of Mechanical Engineering
National Chung-Hsing University
250 Koukwang Road
Taichung, Taiwan 40227
Republic of China
jctsai@mail.nchu.edu.tw

Abstract: Tolerancing plays an important role in product life cycle. Geometric dimensioning and tolerancing (GD&T) information, however, is needed to support different activities involved in product development such as tolerance analysis, tolerancing-based process planning, tolerancing control in manufacturing and tolerance evaluation in quality control. This paper investigates the feasibility to provide a uniform tolerancing information model, based on the ISO 10303 STEP product model, in order to support computer-integrated tolerancing. Mapping between current tolerancing standard in engineering practice, such as ASME Y14.5M, and tolerancing data models in STEP is investigated. This leads to the development of a software prototype for translating tolerancing specifications into STEP-based tolerancing data models. The translator takes GD&T specifications from a CAD system and maps them into STEP data models by creating instances of these data models. System architecture and information flow are also illustrated and discussed.
Keywords: tolerancing information, GD&T, STEP, data model, translator

1. BACKGROUND

In the age of fast pace of technologies, competition of a product nowadays is based on its performance/price ratio and service rather than based on its cost alone. Among many factors, quality is one of the best indices for product performance. Tolerancing, however, has been used for a long period as the index of the quality of a product. Tolerancing technologies, including tolerance design, tolerance analysis, tolerance control, and tolerance inspection, have been emphasized in product life cycle and therefore are part of the most fundamental technologies in product development.

While product development is an inter-disciplinary task, it involves different departments at each stage in the process. A unified and consistent product data is essential to support activities involved in the process, from product concept to design, analysis, process planning, production, and to customer service. Tolerancing information, nevertheless, plays an important and critical role in product life cycle as

P. Bourdet and L. Mathieu (eds.),
Geometric Product Specification and Verification: Integration of Functionality, 311-320.

this information is associated with product geometry and dimensioning as well as with manufacturing process planning and quality control.

As computer-aided design/manufacturing/engineering (CAD/CAM/CAE) systems are widely employed in product development in the past decades, exchange of product data among different computer-aided systems becomes a critical issue. Noticed the requirements, the International Standard Organization (ISO) formed the technical committee 184 (TC184: Industrial Automation Systems) in the 1980s in order to develop an international standard for the exchange of product data, called STEP (STandard for the Exchange of Product model data) and later assigned as ISO 10303. STEP is the standard for a computer-interpretable representation and exchange of product data to provide a mechanism capable of describing product data throughout its life cycle while independent from any specific system. STEP consists of different parts including describing languages, core data models, application protocols as well as implementation methods and abstract test suites. Each part further consists of one or more schemes. A schema is a set of data models related to an information field. Several parts of STEP have been published as international standards though many other parts are in preparation or in early stage for development. It can be expected that STEP will become the industrial standard for product data exchange. Seeing this trend, many computer-aided systems now provide interfaces to STEP data models in order to keeping their systems in the market. However, none of current commercial systems supports data exchange for STEP tolerancing standards though many of them support geometry and dimensioning.

While previous work mostly emphasized on dealing with tolerancing information in CAD/CAM systems, such as Salomons et al. [1996], Mostefai & Bonney [1997], and Tsai & Wang [1999], not much work was stressed on the exchange of tolerancing information. Gu & Chen [1995] noticed STEP data models for tolerancing but did not construct tolerancing data compliant to STEP. Feng & Yang [1995] described detail data models proposed for STEP, though these data models were modified before published [ISO 10303-47, 1997]. Tsai et al. [1998], based on data models described in Feng & Yang [1995], developed a geometric dimensioning and tolerancing (GD&T) information model for constructing product GD&T data and applications. Although previous work emphasized on data sharing for product design, work on transferring GD&T information to standard data models is lacking. This paper is aimed to investigate the requirements to develop a prototype for translating tolerancing information in product development to STEP-based form such that the information can be used to support different activities in product design, analysis, and manufacturing.

In the following sections, mapping between current tolerancing practice, the ASME Y14.5M [1995], and STEP tolerancing data models will be investigated. Construction of the GD&T data models and implementation of a prototype system for STEP translator are then discussed and followed by conclusion and discussions.

2. MAPPING BETWEEN ASME Y14.5M AND STEP GD&T DATA MODELS

As STEP is intended to support information exchange, it must be compliant with current GD&T standards or current engineering practice. Among many different standards, GD&T standards defined by ISO and ANSI (American National Standard Institute) are the most commonly used ones. GD&T standards defined by the two units are about the same. Compared with GD&T standards defined by ISO being in different parts, the ANSI standard collected related issues and defined its GD&T standard in ASME Y14.5M [1995]. It is therefore a popular standard commonly used in engineering practice and thus is used as the reference for establishing GD&T data models in this paper.

The generic GD&T data models in STEP are described in part 47 -- Shape Variation Tolerance which, afetr many years' efforts, was published as an international standard in 1997 [ISO 10303-47, 1997]. An applications protocol (AP), such as part 214, can use data models defined in part 47 for it application without redefining these models. In order to develop a STEP-based translator, the relationship between STEP GD&T data models and engineering GD&T standards and/or practice must be investigated. In this section, mapping of a product GD&T information into data models defined in STEP part 47 and other related application protocols are investgated and discussed.

2.1. GD&T data models in STEP

GD&T data models of part 47 in STEP are defined in three schemes – the shape_dimension_schema, the shape_aspect_definition_schema, and the shape_tolerance_schema. The first schema describes the representation structure of dimensioning scheme and dimensioning locations for engineering design. The shape_aspect_definition_schema describes data structures for geometric elelments or features used in tolerancing. In additional to data models for geometric elements defined in part 42 – geometry and topology, it also defines aggregated and complicated geometric features. Data models for datum and datum feature are also defined in this schema. The shape_tolerance_schema defines data models that represent basic shape tolerances and allowances for production, *i.e.*, dimensional and geometric tolerances. The location of a dimensioninal tolerance and the shape_aspect (geometric element) it applied in this schema, however, refer to data models defined in the shape_dimension_schema.

The GD&T data models in part 47 of STEP defined only the basic data structures of tolerancing, including both dimensional and geometric tolerances with and without datum. Extended geometric tolerances, such as positioning and form tolerances, are further defined in APs that refer to the information. Among current published STEP standards, part 214 is the only AP that uses GD&T data models. As part 47 defines only basic data models for GD&T, data models for extended tolerances, including positioing, form, orientational, and runout tolerances, are defined in part 214. These extended tolerances inherit properties of basic data models defined in part 47. For instance, the positioning tolerance defined in AP214 is a subtype of geometric_tolerance_with_datum_reference, which is further a subtype of geometric_tolerance that is defined in the shape_tolerance_schema in part 47. Therefore, mapping of GD&T data models in this paper consists of those defined in both part 47 and part 214.

2.2. Mapping between ASME Y14.5M and STEP GD&T data models

The two major parts of GD&T defined in ISO and ANSI standards are dimensioning and tolerancing. We investigated the mapping of the two parts between the ANSI standard [ASME Y14.5M, 1995] and data models defined in STEP as the foundation for implementing a translator.

Dimensioning is an important information of a geometric element, or shape_aspect defined in STEP. It specifies the size, location, and geometric properties associated with a geometric element. A data model defined in STEP is called an entity that is a data structure similar to a class in object-oriented programming. Table 1 shows the mapping of the information for 'dimension', defined in ASME Y14.5M, into correspondant entities that represent the 'dimension' information. In table 1, a data model with a * symbol means that it is defined in part 47. Otherwise it is defined in STEP part 214. Dimensional specifications of a product can be translated into entities of STEP based on this mapping table.

Table 1; Mapping of dimensional information to entities of STEP.

ASME Y14.5M	entities of STEP	
dimension	dimensional_characteristic_representation* dimensional_characteristic*	
	dimension_size* shape_aspect dimension_size_with_path* (angular_size, angular_relator)*	dimension_location* dimensional_location_with_path* (angular_location, angular_relator)* shape_aspect
	shape_dimension_representation* measure_representation_item measure_with_unit unit named_unit SI_unit measure_value positive_length_measure	

* entity defined in part 47.

Tolerancing information associated with a dimension can be mapped into data models defined in STEP as shown in Table 2. A dimensional tolerance specifies the size or the location of a shape_aspect, as shown in the right side in table 2. The value and unit of the tolerancing specification are mapped into properties 'tolerance_value', 'measure_with_unit', 'unit', and 'SI_unit'. Two properties, 'measure_value' and 'length_measure', in STEP are used for representing measured values.

Unlike dimensional tolerancing, a geometric tolerance specifies either the allowed shape variation of a shape_aspect or the allowable variation of relative location of a shape_aspect with respect to another one. Tolerance of the first type is called a self-referenced tolerance as it specifies the allowable variation of the shape_aspect itself. The latter, on the other hand, is called cross-referenced tolerance as it specifies the relative location of a shape_aspect with respect to another one, called datum(s). A cross-referenced tolerance is often associated with a positioning or an orientation constraint, sometimes in implicit form. Mapping of self-referenced and cross-referenced geometric tolerances to STEP GD&T data models are shown in tables 3 where '*_tolerance' represents certain tolerance item defined in part 214. The major difference of data models for the two types of tolerances is the 'datum'. Data model for a cross-referenced tolerance uses the data model 'geometric_tolerance_with_datum_reference' rather than 'geometric_tolerance', which is used for a self-referenced tolerance.

Table 2; Mapping of dimensional tolerancing information to entities of STEP.

ASME Y14.5M	entities of STEP		
dimensional tolerance	plus_minus_tolerance tolerance_method_definition		
	tolerance_value measure_with_unit measure_value length_measure	limit_and_fit	
		form_variance	zone_variance

Table 3; Mapping of geometric tolerancing information to entities of STEP.

ASME Y14.5M	entities of STEP				
geometric tolerance (self-referenced)	*_tolerance geometric_tolerance				
	label	shape_aspect	measure_with_unit measure_value length_measure		
geometric tolerance (cross –referenced)	*_tolerance geometric_tolerance_with_datum_reference				
	label	shape_aspect	measure_with_unit measure_value length_measure	datum_reference datum	
				identifier	shape_aspect

*represents geometric tolerances defined in Part 214.

3. CONSTRUCTION OF GD&T DATA MODELS

To translate GD&T specifications to STEP-based data models, one has to construct the corresponding GD&T information of the product based on STEP GD&T data models discussed in previous section. Figure 1 shows an approach to construct such information for dimensional data. The construction starts with the data entity 'dimensional_characteristc_representation' that has two properties, 'dimension' and 'representation'. The *dimension* is a type of 'dimensional_characteristic' for representing the name and the location of the shape_aspect it applies. It links to two data entities 'dimensional_size' and 'dimensional_location' for the size and location of this dimensioning data. The other property *representation* is a type of 'shape_dimension_representation' which further links to an instance of 'measure_representation_item' used to describe the value and unit of this dimensioning data.

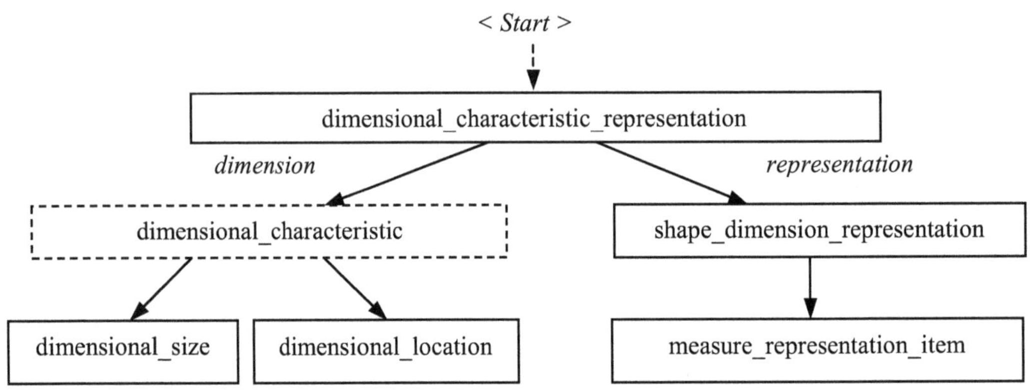

Figure 1; Flow chart for establishing dimensional data models.

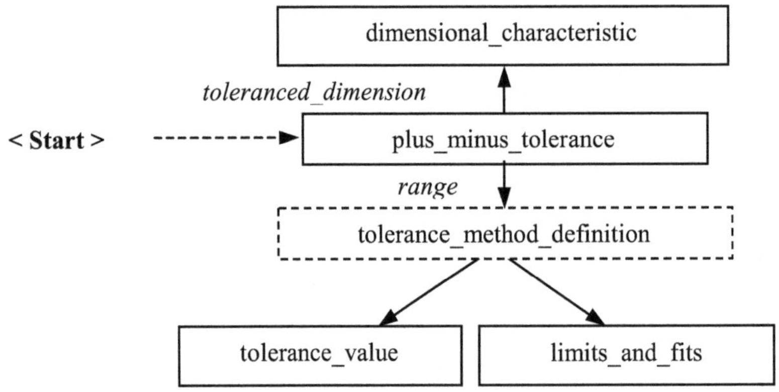

Figure 2; Flow chart for establishing data models of dimensional tolerancing.

Similary, to construct a dimensional tolerance data, it begins from the entity 'plus_minus_tolerance' as shown in figure 2. This entity has two properties, the *toleranced_dimension* that points to an instance of 'dimensional_characteristic' and the *range* that points to an instance of 'tolerance_method_definition'. The instance of 'tolerance_method_definition' is the method that defines the tolerancing range by either giving a 'tolerance_value' or defining a 'limits_and_fits' grade.

Data model for geometric tolerance is defined in another way. The type, self-referenced or cross-referenced, of the tolerance must be verified first, as shown in figure 3. A self-referenced tolerance starts from the entity 'geometric_tolerance' while a cross-referenced tolerance starts from the entity 'geometric_tolerance_with_datum_reference'.

As shown in figure 3, a self-referenced tolerance describes the shape_aspect it applied to, shown as the property *toleranced_shape_aspect*, the value, shown as the property *magnitude*, and the name of this tolerance, shown as the property *name*. Cross-referenced tolerance starts from the entity 'geometric_tolerance_with_datum_reference', which is a subtype of, and therefore inherits properties from, the entity 'geometric_tolerance'. In addition to properties of 'geometric_tolerance' an instance of 'geometric_tolerance_with_datum_reference' also links to 'datum_reference' that further points to an instance of 'datum'. Following the flow shown in figure 1 to 3, data models for a geometric tolerance are generated as instances used for recording product GD&T data.

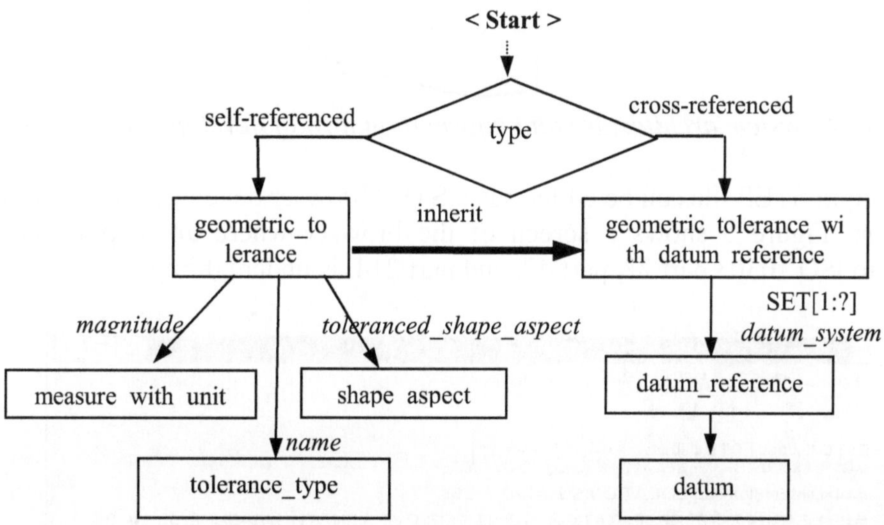

Figure 3; Flow chart for establishing data models of geoetric tolerancing.

4. SYSTEM IMPLEMENTATION

Based on the mapping discussed in section 2 and creation of data models for GD&T information discussed in section 3, a prototype of STEP translator is implemented to translate GD&T speficications into STEP-defined data. Figure 4 is the system architecture and information flow of the translator. The translator, programmed in the Java language compliant to STEP Java SDAI (Standard Data Access Interface) [ISO 10303-27, 1999], reads tolerancing specifications from a GD&T database and translates these specifications into STEP-formatted file [ISO 10303-22, 1995]. The GD&T database stores GD&T information extracted from a CAD/CAM system [Tsai & Wang, 1999]. The translation is based on a mapping scheme discussed in section 2 and on instances of data models compiled from STEP part 47 and part 214. The output is a STEP-formatted neutral file that can be further used for other tolerancing-related applications.

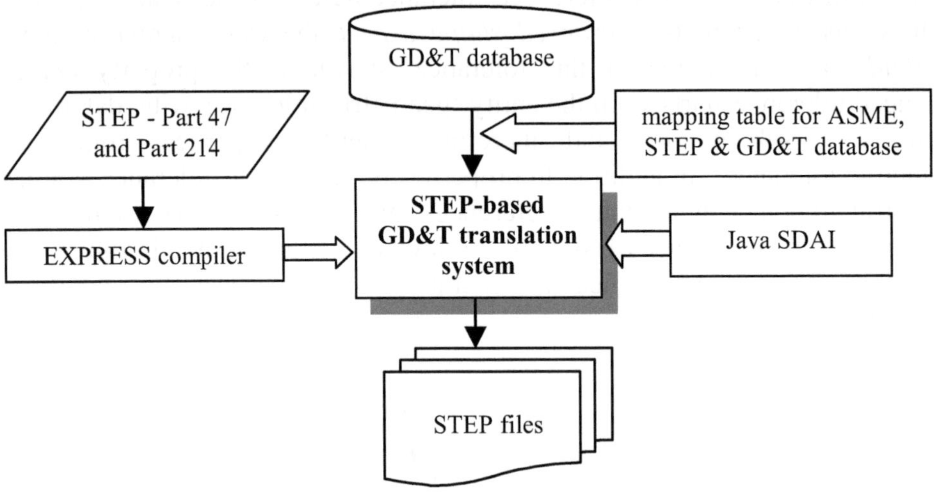

Figure 4; System architecture and information flow of developed translator.

The output STEP file can be edited via a STEP browser developed in MS Windows environment. Figure 5 shows a screen of the browser where an output STEP file, compliant to ISO 10303 part 22, part 47, and part 214, is under editing.

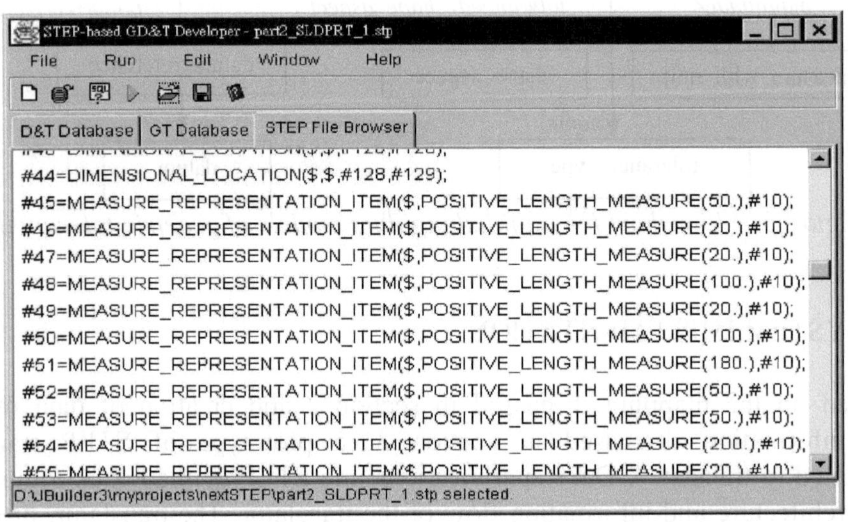

Figure 5; User Interface of the STEP file browser.

5. CONCLUSION AND DISCUSSIONS

Tolerancing data of a product is an important information that involved in the product life cycle. This paper presents a prototype of STEP translator for translating GD&T specifications into a STEP-formatted neutral file as a unified product data supporting activities in the product life cycle. Mapping between current engineering

GD&T standard and engineering practice into data models defined in STEP GD&T data models are first investigated as the foundation for the translator. System architecture and information flow of the developed translator are also discussed. Although completeness and ambiguity of the developed translator need further investigation, this work combined GD&T information from CAD/CAM system, STEP data models, and Java SDAI technology to establish a foundation supporting computer integrated tolerancing. Interfaces to the translated STEP file for other applications, including tolerance analysis and tolerancing-based process planning, are under investigation to take the advantage of the unified product information.

6. ACKNOWLEDGEMENTS

The author wants to express his appreciations to Mr. Tung-Hung Lu of the Mechanical Industrial Research Laboratory, Industrial Technologies Research Institute for his thoughtful discussions. Mr. Shao-Yuan Chang helped the implementation in the past two years. Mr. Yih-Ching Chang and Mr. Jia-Wei You also provided valuable suggestions. Funding for this research has been supported by the National Science Council, Taiwan, ROC under contracts NSC89-2212-E-005-012 and NSC89-2218-E-005-004。

REFERENCES

[ASME Y14.5M, 1995] ASME Y14.5M-1994, *Dimensioning and Tolerancing*, The American Society of Mechanical Engineers, New York, U.S.A., 1995.

[Feng & Yang, 1995] S.-C. Feng and Y. Yang, "A Dimension and Tolerance Data Model for Concurrent Design and Systems Integration," *Journal of Manufacturing Systems,* Vol. 14, No.6, pp.406-426, 1995.

[Gu & Chan, 1995] P. Gu and K. Chan, "Product Modelling Using STEP, " *Computer Aided Design*, Vol. 27, No. 3, pp.163-179, March 1995.

[ISO 10303-22, 1995] ISO 10303-22, *STEP Part22：Standard Data Access Interface*, 1995.

[ISO 10303-27, 1999] ISO 10303-27, *STEP Part27：Java Language Binding*, 1999.

[ISO 10303-47, 1997] ISO 10303-47, *STEP Part47：Shape Variation Tolerances*, 1997.

[ISO 10303-214, 1997] ISO 10303-214, *STEP Part214：Core Data for Automotive Design Processes*, 1997.

[Mostefai & Bonney, 2000] Y. Mostefai and M. C. Bonney, "Computer aided system for tolerance allocation and verification- CASTAV, " *Proceedings of the 5th International Conference on FACTORY 2000*, Conference Publication No. 435, pp346-355, 1997.

[Salomons et al., 1996] O. W. Salomons, H. J. Jonge Poerink, F. J. Haalboom, F. van Slooten, F. J. A. M. van Houten and H. J. J. Kals, "A Computer Aided Tolerancing Tool I: Tolerance Specification," *Computers in Industry*, Vol. 31, pp.161-174, 1996.

[Tsai et al., 1998] J.-C. Tsai, T.-C. Chuang and D.-N. Guo, "Development of A STEP-based Dimensioning and Tolerancing Data Model," *Proceedings of the National Science Council Part A: Physical Science and Engineering,* Vol. 22, No. 6, pp.831-840, Nov. 1998.

[Tsai & Wang, 1999] J. -C. Tsai and W. -W. Wang, "Development of a Computer-Aided Tolerancing System in CAD Environment, " in F. van Houten & H. Kals Eds. *Global Consistency of Tolerances: Proceedings of the 6th CIRP International Seminar on Computer-Aided Tolerancing*, pp.47-54, Kluwer Academic Publishers, Dordrecht, The Netherlands, ISBN 0-7923-5654-3.

Author index

Keyword index

MIX
Papier aus verantwortungsvollen Quellen
Paper from responsible sources
FSC® C105338

If you have any concerns about our products,
you can contact us on
ProductSafety@springernature.com

In case Publisher is established outside the EU,
the EU authorized representative is:
Springer Nature Customer Service Center GmbH
Europaplatz 3, 69115 Heidelberg, Germany

Printed by Libri Plureos GmbH
in Hamburg, Germany